Epigenetics for Drug Discovery

RSC Drug Discovery Series

Titles in the Series:

1: Metabolism, Pharmacokinetics and Toxicity of Functional Groups
2: Emerging Drugs and Targets for Alzheimer's Disease; Volume 1
3: Emerging Drugs and Targets for Alzheimer's Disease; Volume 2
4: Accounts in Drug Discovery
5: New Frontiers in Chemical Biology
6: Animal Models for Neurodegenerative Disease
7: Neurodegeneration
8: G Protein-Coupled Receptors
9: Pharmaceutical Process Development
10: Extracellular and Intracellular Signaling
11: New Synthetic Technologies in Medicinal Chemistry
12: New Horizons in Predictive Toxicology
13: Drug Design Strategies: Quantitative Approaches
14: Neglected Diseases and Drug Discovery
15: Biomedical Imaging
16: Pharmaceutical Salts and Cocrystals
17: Polyamine Drug Discovery
18: Proteinases as Drug Targets
19: Kinase Drug Discovery
20: Drug Design Strategies: Computational Techniques and Applications
21: Designing Multi-Target Drugs
22: Nanostructured Biomaterials for Overcoming Biological Barriers
23: Physico-Chemical and Computational Approaches to Drug Discovery
24: Biomarkers for Traumatic Brain Injury
25: Drug Discovery from Natural Products
26: Anti-Inflammatory Drug Discovery
27: New Therapeutic Strategies for Type 2 Diabetes: Small Molecules
28: Drug Discovery for Psychiatric Disorders
29: Organic Chemistry of Drug Degradation
30: Computational Approaches to Nuclear Receptors
31: Traditional Chinese Medicine
32: Successful Strategies for the Discovery of Antiviral Drugs

33: Comprehensive Biomarker Discovery and Validation for Clinical Application
34: Emerging Drugs and Targets for Parkinson's Disease
35: Pain Therapeutics; Current and Future Treatment Paradigms
36: Biotherapeutics: Recent Developments using Chemical and Molecular Biology
37: Inhibitors of Molecular Chaperones as Therapeutic Agents
38: Orphan Drugs and Rare Diseases
39: Ion Channel Drug Discovery
40: Macrocycles in Drug Discovery
41: Human-based Systems for Translational Research
42: Venoms to Drugs: Venom as a Source for the Development of Human Therapeutics
43: Carbohydrates in Drug Design and Discovery
44: Drug Discovery for Schizophrenia
45: Cardiovascular and Metabolic Disease: Scientific Discoveries and New Therapies
46: Green Chemistry Strategies for Drug Discovery
47: Fragment-Based Drug Discovery
48: Epigenetics for Drug Discovery

How to obtain future titles on publication:
A standing order plan is available for this series. A standing order will bring delivery of each new volume immediately on publication.

For further information please contact:
Book Sales Department, Royal Society of Chemistry, Thomas Graham House, Science Park, Milton Road, Cambridge, CB4 0WF, UK
Telephone: +44 (0)1223 420066, Fax: +44 (0)1223 420247,
Email: booksales@rsc.org
Visit our website at www.rsc.org/books

Epigenetics for Drug Discovery

Edited by

Nessa Carey
PraxisUnico, Cambridge, UK
Email: nessacarey5000@hotmail.com

ROYAL SOCIETY OF CHEMISTRY

RSC Drug Discovery Series No. 48

Print ISBN: 978-1-84973-882-8
PDF eISBN: 978-1-78262-848-4
ISSN: 2041-3203

A catalogue record for this book is available from the British Library

Published by The Royal Society of Chemistry,
Thomas Graham House, Science Park, Milton Road,
Cambridge CB4 0WF, UK

Registered Charity Number 207890

For further information see our web site at www.rsc.org

Printed in the United Kingdom by CPI Group (UK) Ltd, Croydon, CR0 4YY, UK

Preface

Much of the field of epigenetics arose from developmental biology and the attempt to answer some conceptually simple questions. The main ones were probably:

- Do all the cells in the body contain the same genomic information, and if so, how do so many different stable cell types develop?
- Why do you need an egg and a sperm to create a new mammal, why won't two sperm or two eggs do the same job?

Conceptually simple questions, but ones which have led to a field with impact on multiple areas of biology, human health and new treatments for disease.

Epigenetics refers broadly, at a molecular level, to heritable modifications to DNA and its associated proteins which do not alter gene sequence but do alter how genes are expressed. The modifications may influence gene expression transiently, mediating responses to passing environmental stimuli. Or they may control gene expression for the entire length of a human life, maintaining multiple cellular types from one unchanging genome.

Epigenetics has shifted in the last 5 to 10 years, from a somewhat niche research area to centre stage, both in fundamental research and in drug discovery. This volume aims to cover both the pure and applied areas of this discipline, because in such a rapidly evolving science integration is vital. So this publication highlights not just the successes in epigenetic drug discovery, but also the gaps in our understanding. These range from applying personalised medicine strategies to designing reliable assays; from exploring new chemical space to changing the dosing paradigm in the clinic; from identifying relevant biomarkers to rethinking how we identify and de-risk new theoretical toxicological issues.

RSC Drug Discovery Series No. 48
Epigenetics for Drug Discovery
Edited by Nessa Carey

Published by the Royal Society of Chemistry, www.rsc.org

The cross-section of scientists from academia and industry who have contributed the chapters for this volume represent the multidisciplinary approach that is driving this field.

I am hugely grateful to them all, and to the editors at the Royal Society of Chemistry.

Nessa Carey

Contents

RSC Drug Discovery Series No. 48
Epigenetics for Drug Discovery
Edited by Nessa Carey

Published by the Royal Society of Chemistry, www.rsc.org

CHAPTER 1

Epigenetics – What it is and Why it Matters

KARL P. NIGHTINGALE*[a]

[a]Institute of Biomedical Research, Birmingham Medical School, University of Birmingham, Birmingham B15 2TT, UK
*E-mail: k.p.nightingale@bham.ac.uk

1.1 Introduction

The last few years have witnessed explosive growth in the broad field of epigenetics, with the emergence of new paradigms, and the discovery of novel regulatory processes and molecules. This offers new approaches to understanding fundamental biology, but also promises new insight into disease processes, and the potential to modulate the genes involved.

What then are epigenetic phenomena? In the broadest sense 'epigenetics' describes a layer of information and processes, which act in combination with the DNA sequence to determine an organism's characteristics (*i.e.* the 'phenotype' – hair colour, height, *etc.*). This concept is encompassed in the word epigenetics itself, where the Greek prefix 'epi-' suggests these are processes that are 'on top of' or 'in addition to' genetic effects. As such, genetic effects describe a change in gene activity and phenotype, due to a change in the DNA sequence (*i.e.* when genes are rearranged or mutated). In contrast, an epigenetic change is where gene expression and the phenotype are altered, but without a change in the DNA sequence – although there may be a change in the DNA's chemical modification or its packaging within the nucleus.

RSC Drug Discovery Series No. 48
Epigenetics for Drug Discovery
Edited by Nessa Carey

Published by the Royal Society of Chemistry, www.rsc.org

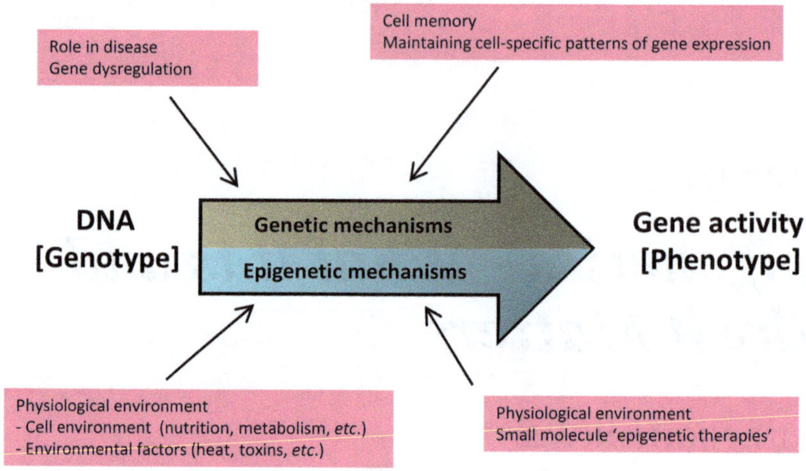

Figure 1.1 Genetic and epigenetic processes link the genotype and phenotype. Epigenetic mechanisms are an integral part of gene regulation, and play a role in initiating and maintaining gene activation or silencing, the misregulation seen in many disease processes, and determining the responses of genes to changes in the environment. Epigenetic therapies are designed to impact on gene activity by targeting these processes.

There are three reasons for the intense interest in this field: (i) epigenetic mechanisms are involved in many fundamental areas of biology, and are often underpinned by novel, interesting mechanisms. (ii) Epigenetic regulation plays a central role in gene expression, so is involved in (and occasionally responsible for) initiating disease processes. (iii) A large number of novel enzymes and protein complexes are involved in epigenetic processes, and are potential targets for small-molecule inhibition. To date, two classes of drugs with epigenetic targets have been approved for cancer chemotherapy, but similar 'epigenetic therapies' are likely to be developed for a broad range of diseases in the medium term (Figure 1.1).

1.2 Beyond DNA: A Broad Picture of Epigenetic Mechanisms

The term 'epigenetics' was originally developed by Conrad Waddington in the 1940s to describe the processes taking place during cell differentiation to develop functionally and morphologically distinct cell types (*e.g.* neurones, red blood cells). In this concept, the 'epigenetic landscape' described the network of gene regulatory decisions that a cell takes when progressing from a pluripotent stem cell to a terminally differentiated cell (Figure 1.2). This concept was developed before the recognition that DNA was the genetic material, but its insight remains valid, as mechanisms must exist to allow the hundreds of distinct cell types present in typical multicellular organisms

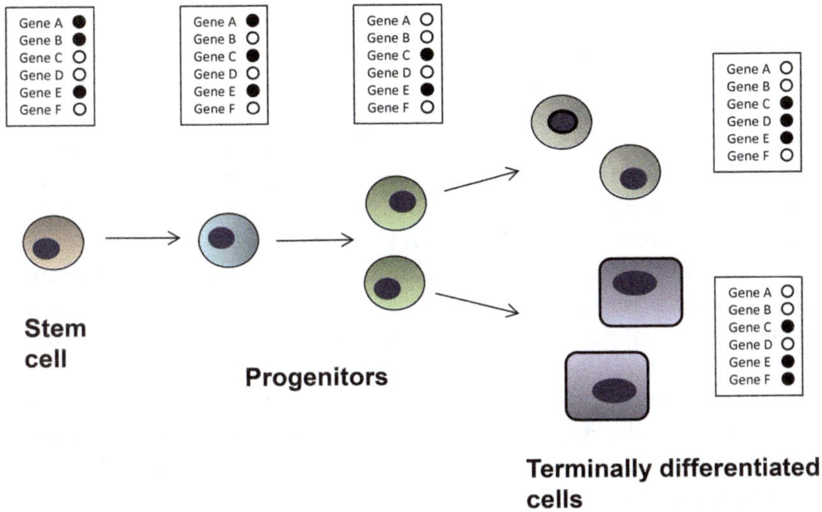

Figure 1.2 Patterns of gene activity change during cell differentiation. Differentiation from a pluripotent stem cell to terminally differentiated cells is associated with changes in the transcriptional activity of multiple genes. For example, differentiation is associated with the silencing of genes associated with the pluripotent state (*i.e.* Gene A, B), and the up-regulation of genes required in specialised cells (*i.e.* Genes C, D and F). In contrast, some gene products are required in all cell types (*i.e.* Gene E, termed 'housekeeping genes'), and these are expressed at a constant level.

to be expressed from a single genotype. This requires that the common DNA sequence is interpreted differently in different types of cells – to establish a single cell-type specific pattern of gene activity, and maintain this throughout the cell's lifespan. Similarly, processes like ongoing tissue turnover require the replacement of differentiated cells, so cell-type specific patterns of gene activity must be duplicated and passed onto daughter cells. It is now clear that epigenetic processes underpin the establishment and maintenance of these patterns of gene activity, and the 'cell memory' that allows them to be passed through cell division.

Epigenetic mechanisms have a role in determining the interactions that link genotype and phenotype at the cellular level, but this influence is also apparent on the human scale. Our inherited parental DNA is the basis of our genetic identity and defines many of our characteristics, but it is also clear that it does not predetermine all the elements of who we are. This is most obvious with identical (*i.e.* monozygotic) twins – people who were initially a single fertilised egg, but by an accident of nature ultimately became two separate but genetically identical individuals. If the DNA sequence is wholly responsible for determining our physical appearance and personalities, then identical twins would indeed be indistinguishable, but in fact they very rarely are. This suggests there is a role for something other than the genome in determining phenotype. This is typically attributed to the environment an organism is exposed to, and is assumed to be the result of a range of factors in the environment (*e.g.* nutrition in the womb, environmental toxins, *etc.*).

One aspect of the phenotype where twin studies show a clear role for the environment is in the susceptibility to multifactorial disease. For example, identical twins have an equal chance of developing diabetes mellitus, indicating its strong genetic basis, whereas the risk of both twins experiencing epilepsy or schizophrenia has a much lower concordance, showing a greater role for environmental factors in the development of these diseases (Figure 1.3).

This, and the finding that some forms of epigenetic information associated with individual genes diverge between twins as they age (DNA methylation, discussed later[1]), suggests that epigenetic processes play a central role in how genes respond to the distinct environmental conditions each twin experiences. As such, epigenetics is believed to play a role in translating the effect(s) of the environment to determine many aspects of cellular and, ultimately, organismal phenotype.

This suggests that epigenetics plays a role in both the minute-by-minute regulation of gene activity in response to changing environmental cues, and the long-term maintenance of patterns of gene expression throughout a cell's lifespan. In both these cases the pattern of gene expression is determined in response to extra- or intracellular conditions. However, this is not the only way that genes can be regulated – developmental or cellular processes that are determined by stochastic (or random) switches are also maintained by epigenetic mechanisms. This is clearest in the case of X chromosome inactivation, a process driven by the need to balance transcriptional activity at genes located on the X chromosome in males (XY, *i.e.* 1 X chromosome) and females (XX, *i.e.* 2 X chromosomes). In most mammals this is achieved by the random silencing of all of the genes on one of the X chromosomes in

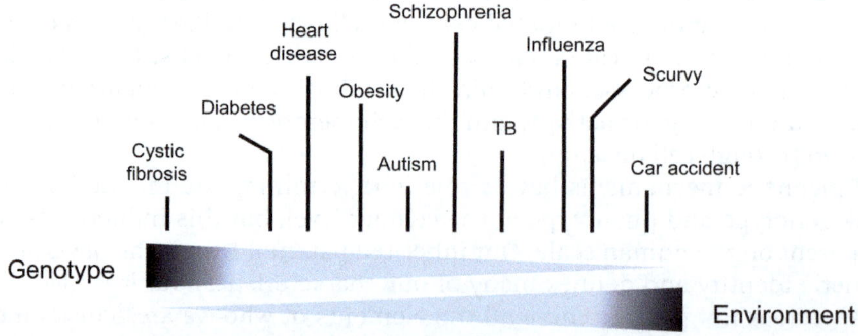

Figure 1.3 Genetic and environmental factors and their impact on disease. Diseases can be wholly genetic or wholly environmental in their origin, but more typically arise due to the combined effect of multiple genes, and the impact of environmental factors on their activity. Epigenetic mechanisms play a role in transmitting environmental signals to determine gene output.

females, in a process driven by epigenetic changes in the chromatin and DNA.[2] Likewise, position effect variation, the switching of a gene's activity, determined by its proximity to transcriptionally silent regions of the chromosomes,[3] is a stochastic 'on' or 'off' process that is initiated and maintained by epigenetic processes. The next section discusses the mechanistic basis of these processes.

1.3 Gene Regulation and the Impact of Chromatin Assembly

Gene regulation is an inherently complicated process given the large number of proteins involved in the mechanics of transcription. However, the fundamental regulatory decision(s) that drive gene activity: '*How much of gene product X does the cell need at this point in time?*' and thus, '*Should gene X be active or silent?*' seem straightforward. The complexity arises due to (i) the size of eukaryotic genomes (*i.e.* ~23 000 genes in the human genome), (ii) the distinct, coordinated patterns of transcriptional activity required to generate specialised cell types, and (iii) the range of factors in the cellular environment to which cells need to respond. Higher eukaryotes manage this by multiple regulatory steps throughout the transcription–translation process, but for many genes, control is primarily applied at the initial step – the recruitment of RNA polymerase and its subsequent release from the transcriptional start site.

Genes consist of two functionally distinct types of DNA: (i) the coding sequence that defines the amino acid sequence of its protein product, and (ii) the regulatory sequences that ensure it is expressed appropriately. It is this second class of sequences that allow genes to respond to a range of intra- and extracellular factors, and ensure they are transcriptionally active or silent at the right place and time. This control is performed by DNA-binding proteins that recognise specific sequence motifs within the gene-regulatory regions, which are typically located adjacent to the transcription start site (termed 'promoters') or at considerable distances from the gene of interest ('enhancers'). Thus, transcription at individual genes is activated or silenced by the binding of regulatory 'transcription factors' (Figure 1.4A), where the act of DNA binding is regulated by signals from the cellular environment (*e.g.* following steroid hormone binding). In this way the activity of a gene is regulated by a 'committee' of transcription factors that interact with RNA polymerase and the general transcription machinery to ensure initiation takes place. As such, gene activity is set at an appropriate level as it reflects the coordinated response to multiple transcription factors, allowing the integration of information from a variety of intra- and extracellular signals (Figure 1.4A).

Undergraduate biologists are introduced to gene regulation using a simplified model – that transcription takes place on protein-free DNA, suggesting that DNA sequences throughout the genome are equally available for protein binding. This is useful for understanding the protein–DNA interactions that underpin transcription, but ignores a fundamental aspect of gene regulation in eukaryotes – that the DNA template is assembled into chromatin. *In vitro* experiments show this plays a key regulatory role, as many transcription factors will not bind their binding sites when assembled into chromatin (Figure 1.4B), suggesting this is a highly repressive environment for transcription.

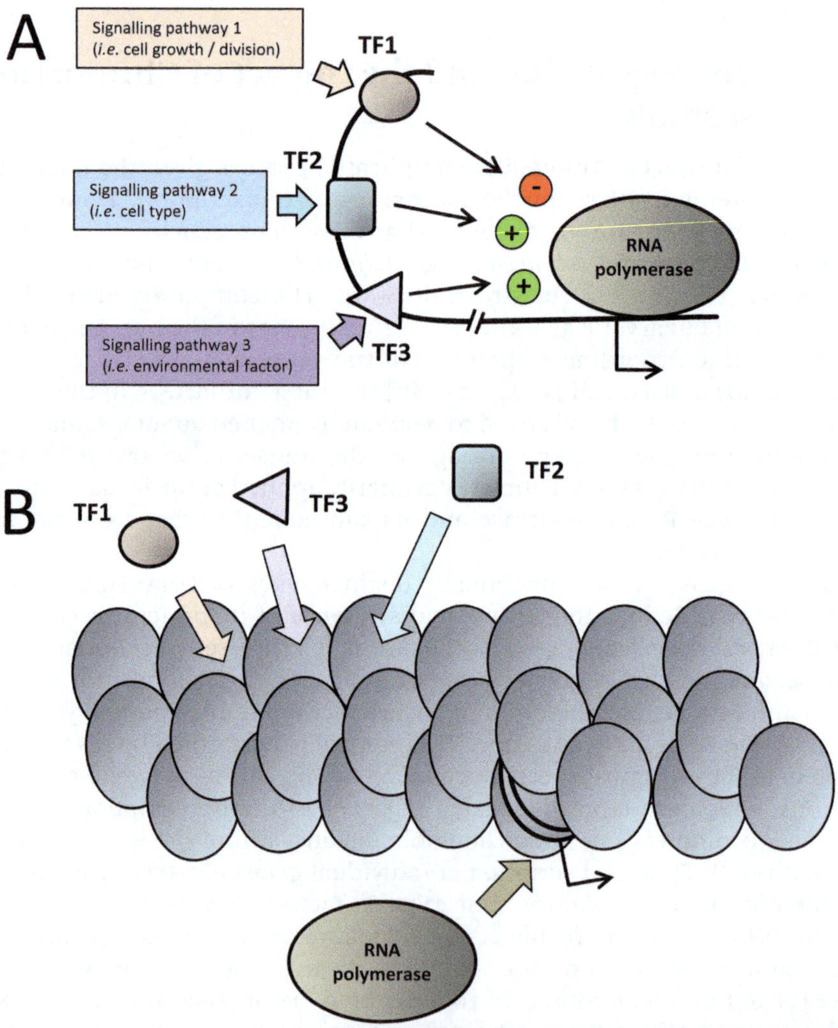

Figure 1.4 Gene regulation and the impact of chromatin assembly. (A) Gene regulation modelled on a protein-free DNA template. Gene regulation is driven by a number of gene-specific transcription factors (TF1–3), which bind to sequence binding sites and together ensure the efficient recruitment and release of RNA polymerase. (B) The impact of chromatin on this process. The assembly of the DNA template into chromatin presents a barrier to the binding of regulatory factors and RNA polymerase to their binding sites. This is overcome by epigenetic processes.

This has advantages for gene regulation as: (i) it suppresses 'adventitious' transcription (*i.e.* from spurious RNA polymerase binding at non-genic regions), but also (ii) introduces an unavoidable and powerful regulatory step – that chromatin-mediated repression has to be overcome for transcription to occur. The next section gives an overview of chromatin structure, before describing how epigenetic processes exploit this to generate regulatory steps.

1.4 Chromatin Structure and the Basis of Epigenetic Mechanisms

Cells in higher eukaryotes are characterised by the presence of a nucleus, the division of the genome into chromosomes, and the packaging of the DNA into an organised protein superstructure termed chromatin. This consists of an enormous range of proteins, which either contribute to chromosome structure or the functional processes that use the DNA template. At its core, chromatin has a straightforward modular structure, based on a single repeating building block – the nucleosome (Figure 1.5). This large (~100 kD) globular protein complex consists of eight arginine and lysine-rich histone proteins (two each of histones H2A, H2B, H3 and H4), around which 147 base pairs of DNA wrap in almost two complete turns of DNA.[4] Nucleosomes have a degree of preference for specific DNA sequences, but this is relatively weak,

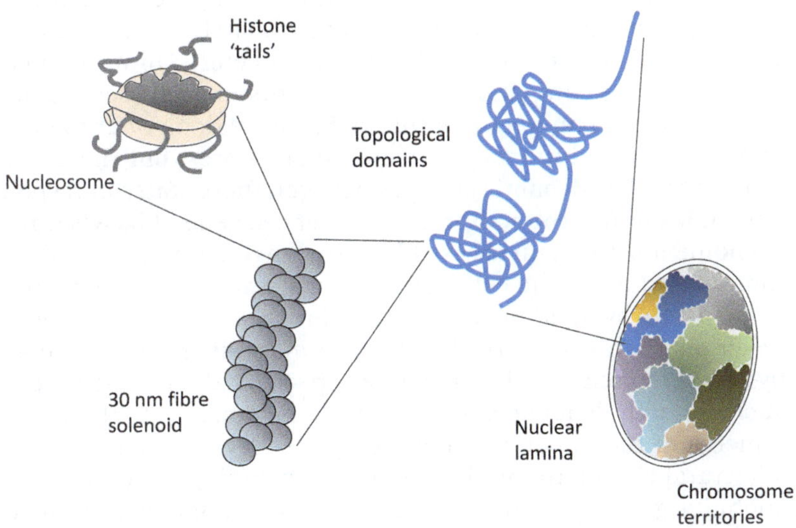

Figure 1.5 The hierarchy of chromatin structures in the interphase nucleus. This cartoon indicates the nucleosome, the unit of chromatin, and how this is assembled into a higher-order solenoid termed the 30 nm fibre. The molecular detail of the relationship between this structure and the stable topological domains that define the interactions of large regions of chromatin and/or how chromosomal DNA is constrained into 'territories' remains unclear.

allowing their assembly at ~200 base pair intervals throughout the genome.[5] This regular array is disrupted at regulatory sites (*i.e.* promoters, enhancers), which often adopt starkly different patterns of transcription factor and nucleosome binding, depending on whether a gene is active or silenced.[6] These differences are driven by factor binding, with the resultant assembly of positioned nucleosomes at defined sequences and chromatin looping to allow long-range interaction between distant regulatory sites.

Much of our current knowledge of the nucleosome comes from structural studies,[4] which revealed the histone–histone and histone–DNA interactions that build the central globular core and scaffold the assembled DNA. In contrast, the structure of the histone N- and C-terminal 'tails' are unclear – suggesting that these highly conserved domains are flexible, and extend beyond the bound DNA to interact with other components of chromatin. Residues on these domains stabilise the assembly of higher-order structures in chromatin,[7,8] initially forming a coiled solenoid termed the 30 nm fibre (Figure 5). The structure of this solenoid, and its subsequent refolding to generate as yet poorly defined higher-order chromatin structures, remains an area of active research, as does the nature of the chromosomal loops, which appear to contribute to the condensation of chromatin in metaphase chromosomes.

This focus on the role of the nucleosome in assembling DNA and stabilising higher-order chromatin structures emphasises one of the main roles of chromatin – in packaging the DNA, and enabling the condensation of the human genome (*i.e.* 3×10^9 bases, approximately 2 metres in length) into the eukaryotic nucleus (typically 10 μm diameter). However, this compaction must also be compatible with the functional processes that use the DNA as a template. This requires that different regions of chromatin are maintained in transcriptionally active or silent states, but there must be mechanisms to allow dynamic changes between these states in response to cellular conditions (*i.e.* the activation of a silenced gene). Likewise, all areas of the genome must remain accessible to the ongoing processes of replication and repair. Eukaryotic cells have evolved several ways to resolve this apparent tension, by building on and adapting the framework of the repeating nucleosome array by recruiting chromatin-binding proteins in a functionally-sensitive context. This generates specialised chromatin domains, but which are both dynamic and reversible – responding to the functional requirements of the underlying DNA template. Several abundant families of architectural chromatin-binding proteins, including linker histones and high mobility group proteins, are thought to stabilise and regulate the accessibility of chromatin and the formation of higher-order chromatin structures, whereas other protein complexes, notably the Polycomb group (PcG) or Trithorax group (TxG) proteins are known to stabilise transcriptionally silent or active chromatin.

Finally, an emerging area of research interest focuses on the role that nuclear structure and 3D chromatin interactions play in regulating gene activity. Early light microscopy studies recognised that chromatin is not uniformly distributed in the nucleus, but forms morphologically distinct regions, which were later shown to reflect the function of the underlying DNA (*i.e.* transcriptionally active euchromatin, silent heterochromatin). More recently, immunofluorescence microscopy has shown that nuclei have a small number of punctate 'transcription factories' containing high concentrations of RNA polymerase II, suggesting that the DNA template is mobile, and genes move to fixed structures when transcribed.[9] Similar approaches show that individual chromosomes are restricted to defined regions of the nucleus ('chromosomal territories'), and that transcriptional activation is associated with genes 'looping out' from their corresponding territory, confirming that gene regulation is associated with intra-nuclear movement.[10] This, and findings that adjacent regions of chromosomes interact to form large (~880 kb), apparently stable 3D 'topological domains',[11] stresses that chromatin adopts large, functionally related structures in the interphase nucleus, yet regions of chromatin remain mobile, even within short timescales.

1.5 Gene Regulation in Chromatin: The Role of Epigenetic Mechanisms

The complexity and nature of regulation at individual genes is driven by the cellular requirement for the gene product and the range of environmental factors that influence this demand. This varies from gene to gene, such that the transcription factors acting at specific promoters and enhancers will have evolved to match a gene's regulatory needs. These factor-binding sites, and the propensity of the DNA sequence to assemble positioned nucleosomes, will ultimately define the promoter architecture and function, but the DNA is also overlaid by several forms of epigenetic information which contributes to gene regulation. As such, genetic and epigenetic information routinely integrate and complement each other at all genes – acting at the same regulatory sequences, often *via* common protein regulators in coordinated patterns of activity. In this context, distinguishing 'epigenetic' from 'genetic' processes is misleading, as it does not capture the seamless nature of gene regulation. The next four sections introduce the key epigenetic processes that act at most genes, and discuss how they interact with transcription factors and each other to generate an integrated process of gene control.

1.5.1 Chromatin Remodelling

In vitro experiments show that nucleosomes are sensitive to aspects of the DNA sequence, and adopt tightly defined positions on some sequences.[5] This is also seen *in vivo*, but the promoter architecture at gene-regulatory regions frequently changes upon gene activation, corresponding with the binding of transcription factors.[6,12] This initial recruitment phase represents

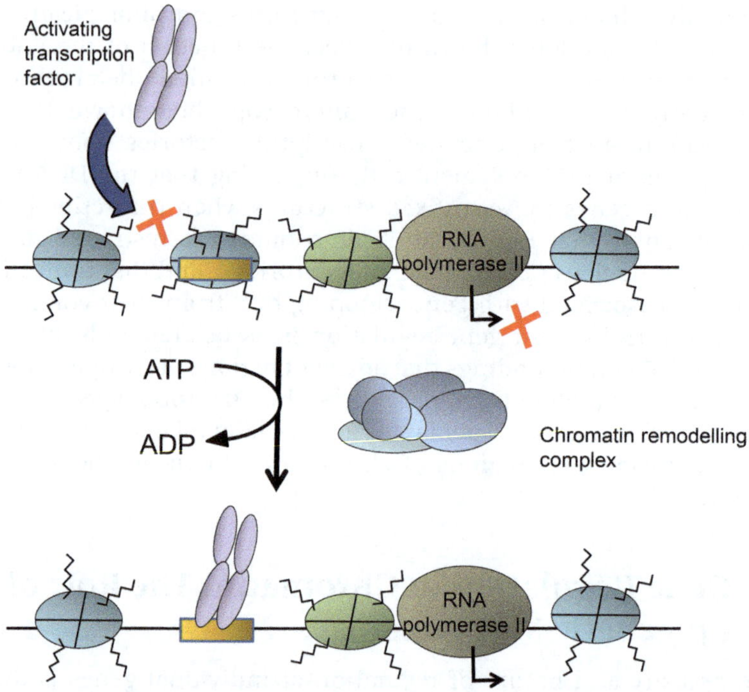

Figure 1.6 Chromatin remodelling facilitates transcription factor access. Steric factors, and the distortion of DNA associated with nucleosome assembly, prevents many transcription factors gaining access to their binding sites [yellow box]. Chromatin remodelling complexes are large, multi-subunit complexes that use ATP hydrolysis to slide or evict nucleosomes from these sites, and thereby allow factor binding.

a regulatory step, as many transcription factors cannot bind their recognition sites when these are assembled on nucleosomes. Thus a class of enzymes, chromatin remodelling complexes, are required to rearrange (or 'remodel') the positions of occluding nucleosomes to allow factors to bind.[13] Several families of remodelling complexes have been identified, and use ATP hydrolysis to reposition or evict nucleosomes, allowing transcription factors to bind their sites in the newly accessible DNA (Figure 1.6). These large, multi-subunit complexes (*e.g.* NURF: 520 kD complex;[14] ACF: 450 kD[15]), contain a core ATPase subunit that performs the nucleosome remodelling, and additional components involved in targeting and/or regulating this activity, and which recruit the complexes to target promoters by interacting with transcription factors. Other classes of remodellers (human SWI/SNF, dCHD1) are found to co-localise with RNA polymerase, and play a role in facilitating RNA polymerase II elongation through chromatin.[16,17] Interestingly, chromatin remodelling is reversible, as in *S. cerevisiae* the transcriptional repressor SSn5-Tup1 recruits the remodeller yISW2 to reassemble nucleosome arrays and evict transcription factors, thereby generating transcriptionally repressive chromatin.[18]

1.5.2 Histone Variants

Eukaryotic cells contain a diverse range of highly conserved histone variants, which contain small changes in primary amino acid sequence (*e.g.* the variant H3.3 differs from H3.1 by 4–5 amino acids), or substantial differences (*e.g.* macro-H2A contains an additional 25 kD C-terminal domain), from the standard 'canonical' histone sequence.[19] Variants represent a small percentage of the total histones in the cell, but are concentrated in chromatin regions with specific functional roles,[20] including marking the inactive X chromosome (macroH2A), or centromeric heterochromatin (CENP-A), or are positioned adjacent to transcriptional start sites (H2A–Z) and sites associated with transcriptionally active or paused RNA polymerase II (H3.3).

The standard 'canonical' histone isoforms are assembled into chromatin in a replication-dependent manner, and are deposited onto the DNA by components of the replication machinery.[21] In contrast, histone variants are deposited by variant-specific chaperone complexes in a replication-independent manner,[22,23] generating localised regions of chromatin where nucleosomes have distinct physicochemical properties or novel sites for post-translational modification. For example, H2A–Z differs from H2A by an 'acidic patch', which appears to make variant-containing nucleosomes more unstable, and thus easily displaced by transcription factors.[24] Likewise, H3.3 contains a novel site in the N-terminal for post-translational phosphorylation, although the key differences from H3.1/2 appear to be those that permit replication-independent assembly following RNA polymerase II-mediated eviction of nucleosomes.[19] In contrast, the extended N-terminal domain of macro-H2A is thought to stabilise chromatin and prevent transcription factors gaining access to their binding sites,[25] consistent with the transcriptional silencing of the inactive X chromosome. In summary, histone variants are used to pre-set the structural and functional properties of chromatin, to facilitate or preclude subsequent factor binding, and their associated regulatory processes.

As such, chromatin remodelling and histone variants act to modulate local chromatin structure (*i.e.* altering nucleosome positions, establishing regions of specialised chromatin), and exert functional effects indirectly in a non-specific manner to permit or occlude the binding of a broad range of regulators. There is evidence that some classes of histone modification (*i.e.* acetylation, phosphorylation) also work in this way, but these seem to be exceptions – typically these modifications act in a specific and directed manner, by recruiting a single protein, or narrow class of partner proteins, to chromatin.

1.5.3 Histone Modification

Histones are post-translationally modified at sites within the central nucleosome core, and on the N- and C-terminal 'tails' which are subject to unprecedented levels of modification, with a large proportion of residues subject to modification by a diverse range of chemical moieties. These domains are highly conserved, yet do not appear to adopt a stable structure[4] (although this remains unclear[26]), suggesting they are dynamic, and evolved to interact with a broad range of proteins that contribute to the structural and functional properties of chromatin. This is consistent with our current understanding of histone modifications – that these act as regulatory 'marks' that modulate the ability of the histone tails to interact with other proteins, and regulate their recruitment to chromatin.

How do histone modifications exert effects in chromatin? We can take histone acetylation as an example. Histone acetylation is a highly abundant, extensively characterised modification, with a large number of lysine residues subject to acetylation on all four histone tails, but also within the nucleosome core. Acetylation neutralises the positive charge associated with lysine, and is assumed to reduce histone tail interactions with the DNA and impact on chromatin structure. This has functional consequences as *in vitro* experiments show histone acetylation facilitates transcription factor binding to nucleosome-bound DNA,[27] and reduces the ability of chromatin to form the 30 nm fibre, a transition associated with transcriptional repression.[28] Likewise, chromatin immunoprecipitation (ChIP) indicates many acetylated lysine isoforms are associated with transcriptionally active loci (*e.g.* at promoters and enhancers) and are absent in silenced regions of chromatin. This suggests histone acetylation exerts functional effects by influencing local chromatin structure, but other histone modifications do not seem to work this way, as even small chemical differences with no apparent impact on chromatin structure have functional outcomes.

For example, lysine residues in histones are subject to different degrees of methylation (*i.e.* me1, me2 or me3), an apparently minimal difference, but the H3 K 4 mono- and trimethylated isoforms show distinct genomic distributions in cells.[29,30] Likewise, the methylation of lysine residues on the histone H3 tail can have functionally opposing roles – methylation of H3K4 locating to active genes (termed an 'activating mark'), whereas methylation of H3 K 9 is found in transcriptionally silent chromatin ('repressive mark'). This suggests that the genomic distribution of a histone modification (and by inference, its function) is determined by both the precise chemical moiety, and residue modified. This is consistent with the 'epigenetic code'[31] and 'histone code' hypotheses[32] that propose these modifications represent a layer of epigenetic information (or 'marks') that determine the functional status of the underlying DNA. As such, they act by playing a 'signalling' role – acting as precise marks that are recognised by a class of specialised 'reader' protein domains which bind histone modifications with a high degree of modification and residue specificity, and thereby co-recruit functional and/or structural 'effector' proteins to the marked region of chromatin[33] (Figure 1.7). The last decade has seen the identification and characterisation of many classes of modification-specific reader domains (*e.g.* acetylation-binding 'bromodomains'), and the protein complexes that these recruit (*e.g.* chromatin remodelling complexes, components of the general transcriptional machinery).

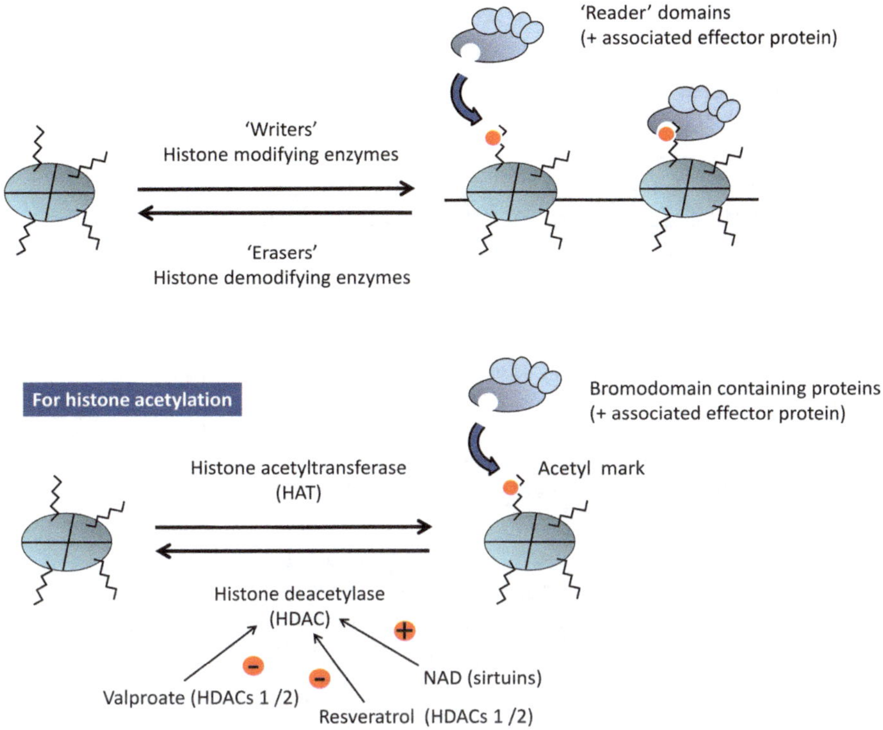

Figure 1.7 Regulatory components of the epigenetic/histone code. [Upper panel] The abundance of a specific form of histone modification at a site is determined by the recruitment and activity of opposing classes of modification enzymes: 'writers' that deposit the mark(s), and 'erasers' that subsequently remove them. Modification(s) are recognised by specific 'reader' domains that bind them and co-recruit the associated complex to chromatin, thereby targeting a functional or structural activity. [Lower panel] Enzymatic regulation of histone acetylation. Histone acetyl isoforms are regulated by large families of histone acetyltransferases (HATS) and histone deacetylases (HDACs), with bromodomain-containing effector complexes binding these sites. Enzyme activity and modification abundance is modulated by factors in the intra- and extracellular environment.

1.5.3.1 Drivers of Dynamic Gene Regulation: Histone-Modifying Enzymes

Histone modification(s) are a diverse range of chemical tags, but those that have been characterised to date act as a short-term layer of epigenetic information to regulate local protein recruitment to chromatin. As such, the abundance and distribution of these marks at a gene reflects the ongoing processes that are taking place, and emphasise the regulatory importance of the recruitment and activity of two or more opposing classes of histone-modifying enzymes that maintain this. This common theme suggests

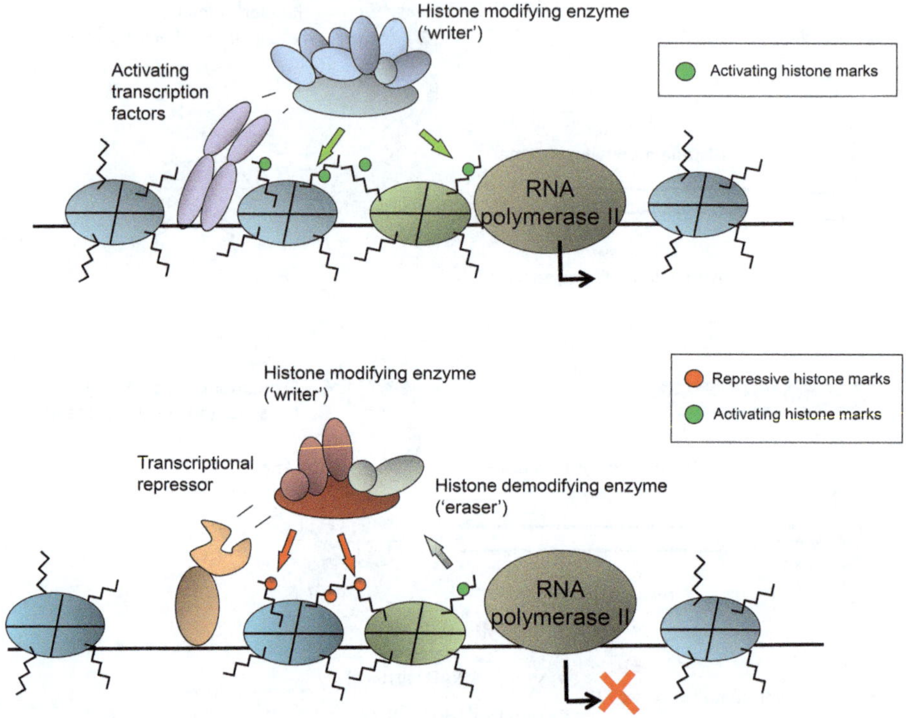

Figure 1.8 Gene regulation in chromatin. (Upper panel) Activating transcription factors bind in chromatin and recruit transcriptional co-activator complexes containing chromatin remodellers and/or histone modifying enzymes which deposit 'activating' marks on the adjacent chromatin. Note the nucleosome at the transcriptional start site may contain histone variants. (Lower panel) Transcriptional repressors act *via* recruiting regulatory complexes that deposit 'repressive' marks and/or erase activating marks.

histone-modifying enzymes can be divided into two classes: 'writers', which deposit a class of histone modifications (*e.g.* histone acetyl transferases) and 'erasers', the enzymes that remove these marks (*e.g.* histone deacetylases (HDACs): Figure 1.7). These activities are recruited to genes in the context of large co-transcriptional regulator complexes, often by protein–protein interactions with activating or repressive transcription factors, or other components of the transcriptional process (Figure 1.8). Analysis of the epigenetic regulators recruited to steroid hormone responsive genes shows the binding of these enzyme complexes is highly dynamic, changing on a minute-by-minute basis, consistent with the rapid turnover of histone modifications, and their associated reader proteins.[34,35] For example, histone acetyl isoforms have a half-life measured in minutes, suggesting individual genes are regulated on this timescale. The recent finding that the activity of several classes of HDACs are sensitive to intracellular metabolic intermediates (*e.g.* NAD^+), naturally occurring dietary components (*e.g.* resveratrol, a component of red wine), or by-products of the intestinal flora (*e.g.* the fatty acid, butyrate) suggests that these enzymes may have evolved to modulate their activity (and gene expression) in response to these agents (Figure 1.7, lower panel). This suggests that histone modification represents a second layer of mechanisms that integrate environmental signals and modulate gene activity – in addition to the transcription factors (Figure 1.4A). A detailed discussion of the regulators, reader domains and effector proteins involved in gene regulation, and the implications of this for modulation by small-molecule inhibitors, is contained elsewhere in this volume.

1.5.4 DNA Methylation

The other major epigenetic mechanism, DNA methylation, is restricted to metazoans, suggesting that it co-evolved with the increased complexity of higher eukaryotic genomes. Methylation takes place at the 5-carbon of cytosine residues in cytosine–guanine (CpG) sites, a symmetrical sequence found on both strands of the DNA. The human genome has a low incidence of CpG dinucleotides, but is interspersed with regions containing elevated levels of these sequences (termed 'CpG islands'), which associate with gene-regulatory regions. Typically, CpG sites throughout the genome are methylated, but sites within CpG islands remain non-methylated, unless the corresponding gene is silenced. DNA methylation is associated with long-term transcriptional silencing in processes which are not normally reversed including: (i) imprinting, where maternal or paternal alleles of genes are silenced, (ii) dosage compensation, the inactivation of one X chromosome in females, or (iii) the silencing of developmentally regulated genes with cell differentiation, and/or potentially damaging transposon and virally integrated sequences.

DNA methylation and its associated transcriptional repression is assumed to be an irreversible process, but there are biological exceptions when DNA demethylation must occur – notably the global wave of demethylation observed in early development,[36] and the experimental reversal of differentiation-associated gene silencing when terminally differentiated cells are reprogrammed to induced pluripotent stem cells.[37] The recent finding that DNA methylation is a substrate for the ten-eleven translocation (TET) proteins,[38] which act as methylcytosine oxidases to generate a number of oxidised intermediates including 5-hydroxymethylcytosine (5hmC), 5-formylcytosine (5fC) and 5-carboxylcytosine (5caC, Figure 1.9), suggest that this may be a route to DNA demethylation, potentially *via* the dilution of 5mC marks during replication, or removal by base-excision repair.[39,40] This remains an area of active research, but the finding that these intermediates are present in nuclei, that TET proteins are recruited to CpG islands, and that their activity correlates with transcriptional activation or repression, suggests that this plays a role in gene regulation. The function of these oxidised 5mC intermediates, and

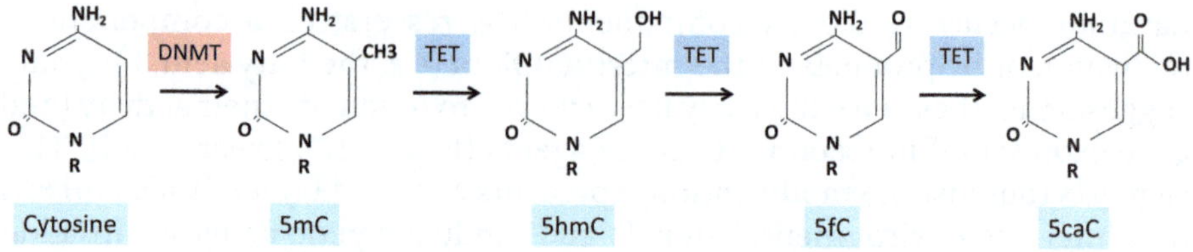

Figure 1.9 Cytosine methylation and TET-mediated oxidation. The methylation of deoxycytosine residues to 5-methylcytosine (5mC) is mediated by DNA methyltransferase (DNMT) enzymes, and sequentially oxidised by ten-eleven translocation (TET) enzymes *via* 5-hydroxymethylcytosine (5mC), 5-formylcytosine (5fC) and 5-carboxylcytosine (5caC).

whether they represent distinct epigenetic marks that are able to recruit a subset and/or specialised methyl-CpG binding domain (MBD) proteins remains an area of active research.[41]

1.6 Epigenetic Crosstalk – Integrating Histone Modification and DNA Methylation

A final topic is the level of integration between the different epigenetic processes. This occurs not just between mechanisms that regulate short-term gene expression (*e.g.* chromatin remodelling enzymes can be recruited by histone modifications), but also in long-term gene silencing *via* integration of DNA methylation with short-term histone modifications like acetylation and lysine methylation.

This crosstalk is apparent in the mechanism of mark deposition, but also the ways the various marks impact on transcription. DNA methylation is regulated by one of three classes of DNA methyltransferase (DNMT) enzymes, which maintain the pattern of methylation after replication or excision repair by acting at hemi-methylated sites (DNMT1), or deposit *de novo* methylation at unmethylated CpGs (DNMT3a,b). DNMT activity can be recruited to chromatin by heterochromatin protein 1 (HP1), a protein that binds the repressive histone modification H3K9me3. Likewise, the primary functional effect of DNA methylation is *via* the recruitment of specialised MBD proteins (*i.e. MBD1*, *MBD2*, *MeCP2* and *Kaiso*[42,43]) which co-recruit large repressor complexes. Many of these complexes contain the histone deacetylases HDAC1 and HDAC2, which act on adjacent chromatin and maintain transcriptional silencing (Figure 1.10), although the MBD2 complex also contains Mi2, a chromatin remodeller involved in transcriptional repression.

This coordination and integration of histone modification and DNA methylation processes is consistent with multiple examples of large epigenetic regulators containing remodelling and/or DNA or histone (de)modifying enzymes, and contributes to the regulation and maintenance of gene expression patterns over the short and long term.

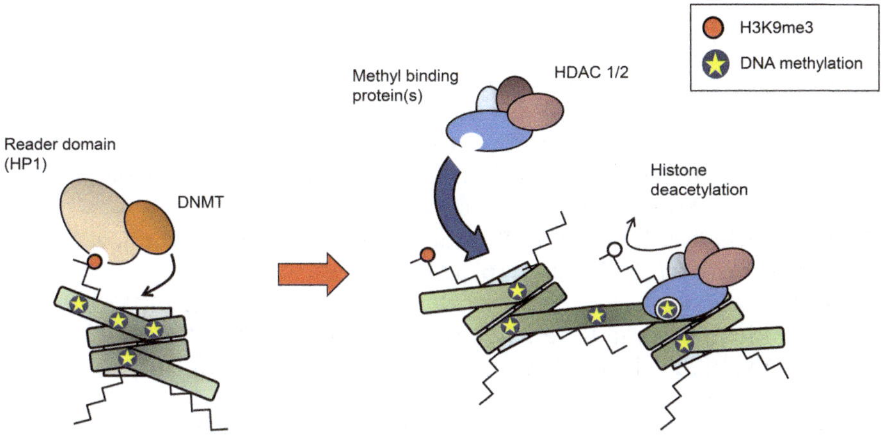

Figure 1.10 Integrating histone modification and DNA methylation. Heterochromatin protein 1 binds the repressive mark H3K9me3, and co-recruits a complex containing DNMT activity which methylates CpG dinucleotides. This contributes to establishing silencing at these loci *via* recruiting one of several classes of methyl CpG binding domain proteins, which co-recruit HDAC activity to the locus.

1.7 Summary

Epigenetic regulation represents an integrated layer of regulatory information – ranging from short-term and highly dynamic processes involved in the minute-to-minute activation or silencing of genes, to the long-term maintenance of these patterns. There are now many examples where mutated epigenetic regulators, for example following gene rearrangements of the MLL (mixed lineage leukaemia) histone methyltransferase,[44] or MOZ/MYST3 (Monocytic leukaemia zinc-finger protein) acetyltransferase, or mutation of components of chromatin remodellers (SNF2) are known to be causal for gene deregulation and initiating disease processes.

The realisation that epigenetic regulation is driven by multiple families of regulators, many acting *via* multi-subunit complexes, to allow tight integration with each other, represents a challenge for developing gene-specific agents. Nonetheless, the emergence of successful epigenetic therapies suggests that this is possible, particularly as only a small proportion of epigenetic regulators appear to have been targeted to date.

References

1. M. Ollikainen and J. M. Craig, *Epigenomics*, 2011, **3**, 295.
2. F. Ferrari, A. A. Alekseyenko, P. J. Park and M. I. Kuroda, *Nat. Struct. Mol. Biol.*, 2014, **21**, 118.
3. S. C. Elgin and G. Reuter, *Cold Spring Harbor Perspect. Biol.*, 2013, **5**, a017780.

4. K. Luger, A. W. Mäder, R. K. Richmond, D. F. Sargent and T. J. Richmond, *Nature*, 1997, **389**, 251.
5. E. Segal, Y. Fondufe-Mittendorf, L. Chen, A. Thåström, Y. Field, I. K. Moore, J. P. Wang and J. Widom, *Nature*, 2006, **442**, 772.
6. A. Almer and W. Hörz, *EMBO J.*, 1986, **5**, 2681.
7. B. Dorigo, T. Schalch, A. Kulangara, S. Duda, R. R. Schroeder and T. J. Richmond, *Science*, 2004, **306**, 1571.
8. M. Kruithof, F. T. Chien, A. Routh, C. Logie, D. Rhodes and J. van Noort, *Nat. Struct. Mol. Biol.*, 2009, **16**, 534.
9. M. S. Buckley and J. T. Lis, *Curr. Opin. Genet. Dev.*, 2014, **25**, 126.
10. W. A. Bickmore and B. van Steensel, *Cell*, 2013, **152**, 1270.
11. J. R. Dixon, S. Selvaraj, F. Yue, A. Kim, Y. Li, Y. Shen, M. Hu, J. S. Liu and B. Ren, *Nature*, 2012, **485**, 376.
12. H. Richard-Foy and G. L. Hager, *EMBO J.*, 1987, **6**, 2321.
13. C. R. Clapier and B. R. Cairns, *Annu. Rev. Biochem.*, 2009, **78**, 273.
14. T. Tsukiyama, C. Daniel, J. Tamkun and C. Wu, *Cell*, 1985, **83**, 1021.
15. T. Ito, M. Bulger, M. J. Pazin, R. Kobayashi and J. T. Kadonaga, *Cell*, 1997, **90**, 145.
16. L. L. Corey, C. S. Weirich, I. J. Benjamin and R. E. Kingston, *Genes Dev.*, 2003, **17**, 1392.
17. E. Batsché, M. Yaniv and C. Muchardt, *Nat. Struct. Mol. Biol.*, 2006, **13**, 22.
18. J. P. Goldmark, T. G. Fazzio, P. W. Estep, G. M. Church and T. Tsukiyama, *Cell*, 2000, **103**, 423.
19. P. J. Skene and S. Henikoff, *Development*, 2013, **140**, 2513.
20. C. M. Weber and S. Henikoff, *Genes Dev.*, 2014, **28**, 672.
21. A. Corpet and G. Almouzni, *Trends Cell Biol.*, 2009, **19**, 29.
22. S. J. Elsässer, H. Huang, P. W. Lewis, J. W. Chin, C. D. Allis and D. J. Patel, *Nature*, 2011, **491**, 560.
23. V. Q. Nguyen, A. Ranjan, F. Stengel, D. Wei, R. Aebersold, C. Wu and A. E. Leschziner, *Cell*, 2013, **154**, 1220.
24. C. Bönisch and S. B. Hake, *Nucleic Acids Res.*, 2012, **40**, 10719.
25. D. Angelov, A. Molla, P. Y. Perche, F. Hans, J. Côté, S. Khochbin, P. Bouvet and S. Dimitrov, *Mol. Cell*, 2003, **11**, 1033.
26. L. L. du Preez and H. G. Patterton, *Subcell. Biochem.*, 2013, **61**, 37.
27. M. Vettese-Dadey, P. A. Grant, T. R. Hebbes, C. Crane- Robinson, C. D. Allis and J. L. Workman, *EMBO J.*, 1996, **15**, 2508.
28. G. Li and D. Reinberg, *Curr. Opin. Genet. Dev.*, 2011, **21**, 175.
29. H. Santos-Rosa, R. Schneider, A. J. Bannister, J. Sherriff, B. E. Bernstein, N. C. Emre, S. L. Schreiber, J. Mellor and T. Kouzarides, *Nature*, 2002, **419**, 407.
30. D. K. Pokholok, C. T. Harbison, S. Levine, M. Cole, N. M. Hannett, T. I. Lee, G. W. Bell, K. Walker, P. A. Rolfe, E. Herbolsheimer, J. Zeitlinger, F. Lewitter, D. K. Gifford and R. A. Young, *Cell*, 2005, **122**, 517.
31. B. M. Turner, *Cell*, 1993, **75**, 5.
32. B. D. Strahl and C. D. Allis, *Nature*, 2000, **403**, 41.
33. S. Glatt, C. Alfieri and C. W. Müller, *Curr. Opin. Struct. Biol.*, 2011, **21**, 335.

34. S. Kangaspeska, B. Stride, R. Métivier, M. Polycarpou-Schwarz, D. Ibberson, R. P. Carmouche, V. Benes, F. Gannon and G. Reid, *Nature*, 2008, **452**, 112.
35. R. Métivier, R. Gallais, C. Tiffoche, C. Le Péron, R. Z. Jurkowska, R. P. Carmouche, D. Ibberson, P. Barath, F. Demay, G. Reid, V. Benes, A. Jeltsch, F. Gannon and G. Salbert, *Nature*, 2008, **452**, 45.
36. M. S. Bartolomei and A. C. Ferguson-Smith, *Cold Spring Harbor Perspect. Biol.*, 2003, **3**, a002592.
37. K. Takahashi and S. Yamanaka, *Development*, 2013, **140**, 2457.
38. G. Ficz, M. R. Branco, S. Seisenberger, F. Santos, F. Krueger, T. A. Hore, C. J. Marques, S. Andrews and W. Reik, *Nature*, 2011, **473**, 398.
39. W. A. Pastor, L. Aravind and A. Rao, *Nat. Rev. Mol. Cell Biol.*, 2013, **14**, 341.
40. R. M. Kohli and Y. Zhang, *Nature*, 2013, **502**, 472.
41. M. Iurlaro, G. Ficz, D. Oxley, E. A. Raiber, M. Bachman, M. J. Booth, S. Andrews, S. Balasubramanian and W. Reik, *Genome Biol.*, 2013, **14**, R119.
42. J. Ausió, A. M. Paz and M. Esteller, *Trends Mol. Med.*, 2014, **1471**, 00056.
43. O. J. Sansom, K. Maddison and A. R. Clarke, *Nat. Clin. Pract. Oncol.*, 2007, **4**, 305.
44. R. K. Slany, *Haematologica*, 2009, **94**, 984.

CHAPTER 2

Enzyme and Protein Families that Regulate Histone Modifications and Crosstalk

CHARLOTTE E. RUTLEDGE[a] AND BRYAN M. TURNER*[a]

[a]School of Cancer Sciences, College of Medical and Dental Sciences, University of Birmingham, Edgbaston, Birmingham B15 2TT, UK
*E-mail: b.m.turner@bham.ac.uk

2.1 Histones: Endless Variations on a Highly Conserved Theme

2.1.1 Organising DNA

The DNA of all eukaryotic organisms is contained within a membrane-delimited organelle, the cell nucleus. In mammals this means packing around 2 metres of DNA into a space only 10 µm in diameter. These figures suggest that a huge degree of compaction of the genetic material is required but in fact there is plenty of space in the nucleus to accommodate the DNA. (Some simple calculations reveal that it actually takes up less than 10% of the nuclear volume.) What is amazing is that DNA can be packaged so as to reduce its length up to 10 000-fold, yet still retain sufficient organisation to allow access by protein complexes required for transcription, replication and repair exactly when and where they are required.

RSC Drug Discovery Series No. 48
Epigenetics for Drug Discovery
Edited by Nessa Carey

Published by the Royal Society of Chemistry, www.rsc.org

To achieve these demanding functional objectives, DNA is packaged around histone proteins. These proteins interact to form an octamer composed of the histone core proteins (two H2A/H2B heterodimers and an H3/H4 heterotetramer), around which the DNA is wrapped in 1¾ superhelical turns. Each histone octamer with its enwrapping DNA is called a nucleosome, and this constitutes the basic unit of chromatin structure in almost all eukaryotes. The structure of the nucleosome core particle was defined at low (25 Å) resolution by X-ray crystallography in 1980[1] and at 2.8 Å resolution in 1997.[2] The linker histones (H1.1 and variants) associate with the nucleosome at the points of DNA exit and entry.

The core histone proteins show a very high degree of evolutionary conservation from mammals to insects and even plants, suggesting a strong selective pressure to conserve their function. All core histones have a globular central region, containing the characteristic histone fold domain, which functions in heterodimerisation and interacts with the phosphodiester backbone of the DNA in the nucleosome.[2–4] Each core histone also contains an unstructured N-terminal tail, which protrudes from the nucleosome and is only partially resolved by X-ray crystallography.[2] The structure of histone H1 is significantly different, comprising a long C-terminal tail and a short N-terminal tail, without a histone fold domain.[5]

2.1.2 Histone Genes

Mammalian genomes contain multiple copies of the genes encoding histone proteins, organised into co-regulated clusters. These gene clusters show a cell cycle-dependent pattern of expression, being highly expressed during S-phase, when more histones are required to package the newly replicated DNA.[6] Histone genes lack introns and their transcripts are unique in being the only eukaryotic mRNAs that are not polyadenylated. Instead, the 3′UTR contains a histone-specific RNA hairpin element,[7] which is required for normal processing of the mRNA in the nucleus[8] and which also controls histone translation.[9] In addition to the canonical linker and core histones, there are also several histone variants, which contain specific changes in their amino acid sequence. Variant histones tend to be present at lower copy number than their canonical counterparts, and show differences in distribution or function. Their expression is not linked to the cell cycle, and the mRNAs are polyadenylated and do not always contain the RNA hairpin element.[10]

2.1.3 Post-Translational Modifications of Histones

The first histone post-translational modification (PTM) whose functional significance was investigated was acetylation in the mid-1960s.[11] Remarkably, only several years later were the internal acetylated residues identified as ε-acetyl lysine.[12] Over the 50 years since, the number of reported histone PTMs has increased to more than 100 (see HIstome: The Histone Infobase at http://www.iiserpune.ac.in/~coee/histome/ for a database containing just some of the known PTMs, along with some of the relevant modifying enzymes). PTM of histones can involve the covalent addition of chemical groups to selected amino acids, such as the acetylation of lysines, methylation of lysines and arginines, and phosphorylation of serines, threonines, tyrosines or histidines.[13] Some of the sites of histone acetylation, methylation and phosphorylation are shown in Figure 2.1. Histones can also be modified by attachment of the small peptides ubiquitin or small ubiquitin-like modifier (SUMO).[14–17] Alternatively, modifications may involve a chemical change to the amino acid, such as deimination of arginine to produce citrulline,[18,19] or a structural change, such as proline isomerisation, the conversion of proline between *cis* and *trans* forms.[20] Histones can also be modified by proteolytic cleavage, which has been reported to occur on the N-terminal side of alanine 21 in the H3 tail of both yeast and mammalian cells.[21,22]

Most histone modifications are dynamic, often turning over with half-lives of just a few minutes.[23,24] The level at which any particular histone amino

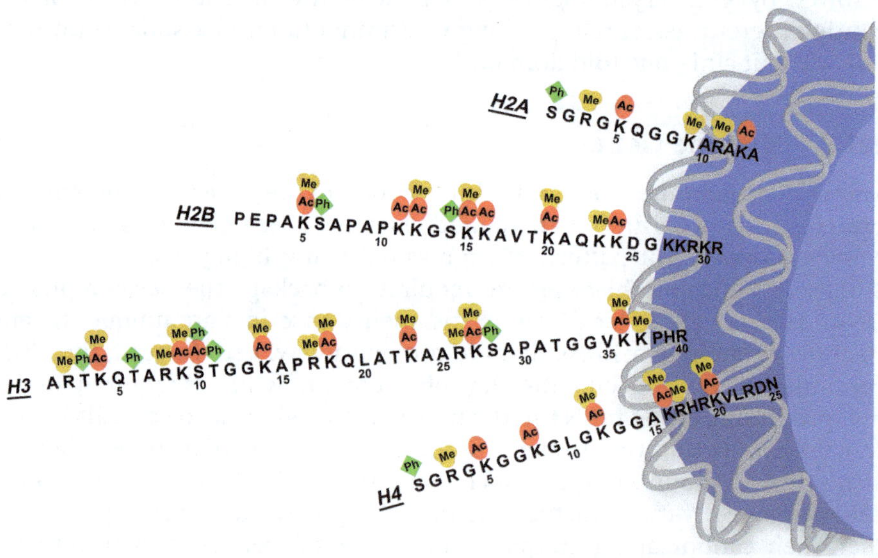

Figure 2.1 Acetylation, methylation and phosphorylation of the core histone tails. The amino acids on each histone tail that are thought to be subject to acetylation, methylation and phosphorylation are shown. Some residues, such as H3K9, can be either acetylated or methylated, so both modifications are shown, although they cannot be present at the same time. Modifications that occur in the internal globular region of each histone are not shown. The position along the protein chain at which each histone tail exits the nucleosome is consistent with experimental evidence (*e.g.* proteolytic cleavage), but their positioning relative to nucleosomal DNA is for illustration only. Every five amino acids in the histone tail sequences are numbered. Ac = acetylation; Me = methylation; Ph = phosphorylation.

acid is modified reflects a steady-state equilibrium between the activities of modifying and demodifying enzymes.

The distribution of histone marks in different cell types is one of the features of the human genome that is under investigation by the ENCODE (*Enc*yclopaedia *o*f *D*NA *E*lements) consortium, a multinational group consisting of hundreds of scientists established with the aim of understanding the organisation and control of the human genome. Much effort has been aimed at attempting to understand the ways in which histone PTMs regulate processes such as transcription and DNA repair, and it is hoped that the information on the distribution of histone modifications gained by ENCODE will aid such lines of enquiry. However, new PTM events are still being discovered, with 67 novel sites of modification revealed in a single screen using mass spectrometry.[25] These included two previously unknown types of modification; lysine crotonylation and tyrosine hydroxylation. The functional analysis of these new modifications promises new insights and research directions.

2.2 How Histone Modifications Influence Chromatin Function

2.2.1 Associations between Modifications and Chromatin Function

Different histone PTMs are known to be associated with specific chromatin-based processes. For example, acetylation events have long been associated with enhanced transcription,[11,30–32] as has the trimethylation of H3 lysine 4 (H3K4me3).[33,34] By contrast, H3K27me3, H3K9me3 and H4K20me3 are associated with repression of gene expression. H3K9me3 and H4K20me3 are found to be highly enriched in areas of constitutive heterochromatin, such as the peri-centromeric regions,[35,36] whereas H3K27me3 is located in regions of facultative heterochromatin, formed when genes are switched off during the course of development, as seen at *Hox* gene clusters.[37] Histone modification patterns can also be used to predict enhancers and enhancer activity,[38,39] and as a signature of transcriptional elongation occurring at a particular gene.[40,41] However, it is important to remember that these overall associations do not always apply in individual cases. Some silent genes are highly acetylated and not all regions enriched in H3K4me1 are enhancers.

2.2.2 Nucleosome Signalling and the Histone Code Hypothesis

The association of different histone modifications with distinct chromatin states is well documented, but how do histone PTMs contribute to the structure and function of chromatin? The effects can be direct, as may sometimes be the case for histone acetylation, which has been said to relax chromatin by reducing the net positive charge of histones,[42,43] or indirect, *via* the recruitment of specific non-histone proteins. The latter possibility, the nucleosome signalling hypothesis, was first suggested over 20 years ago, proposing that PTMs on the histone tails provide molecular receptors on the nucleosome surface which are recognised by dedicated binding proteins that then regulate chromatin function.[26,27] A protein with just such predicted binding properties was first identified in 1999, with the recognition that the bromodomain of an acetyltransferase enzyme bound selectively to acetylated lysine.[28] Since then, domains specific for a variety of other PTMs have been identified (Table 2.1). For example, proteins with an ankyrin domain recognise mono- or dimethylated H3K9, and 14-3-3 domain proteins recognise H3 phosphorylated at serine 10 or 28.[44–46] However, it is important to remember that the interaction between histone modifications and genome function is not one-directional. Although histone modifications and their binding proteins do affect chromatin-based processes the reverse is also true. For example, the trimethylation of H3K36 is catalysed during transcriptional elongation by the KMT3A enzyme, which associates with the RNA polymerase II complex. The transcription-dependent deposition of H3K36me3 is thought to prevent the initiation of aberrant transcripts in the gene body.[47]

A linguistic metaphor is often used to categorise components of the nucleosome signalling system, with the modifying enzymes referred to as 'writers' and the binding proteins as 'readers'.[29] The demodifying enzymes can be referred to as 'erasers'. Like most metaphors, this one is not perfect; for example it provides a static image and gives no indication of the often rapid turnover of modifications, but the terms provide a useful shorthand for categorising groups of proteins with shared functions. Protein families and domains which have been reported to function in the establishment, erasure and reading of histone acetylation, methylation and phosphorylation are listed in Table 2.1.

The idea that chromatin function is controlled by the activity of a variety of non-histone proteins, the binding of which is determined by combinations of histone modifications, led to the use of terms such as 'epigenetic code'[48] or 'histone code'.[29] This terminology, like the writer/reader metaphor, has descriptive value, but can also mislead. It is useful in that it emphasises the functional importance of combinations of PTMs, but misleading in that it indicates that histone modifications can function as a predictive code, analogous to and alongside the genetic code itself. This is an exciting possibility, but as yet there is little evidence to support its existence.[49] Even if we restrict the purpose of the code to defining the probability that a specific protein will bind to a specific region of a histone tail on the nucleosome surface, whether or not such binding actually happens can be influenced by modification of the binding protein itself, its association with protein partners, binding of other proteins to the nucleosome surface or reconfiguration of the histone tails, amongst other things. Such complexity will make the translation of any code based on histone modifications highly context dependent, thereby limiting its predictive value.

Table 2.1 Proteins that read, write and erase histone acetylation, methylation and phosphorylation.[a]

Activity	Enzyme families/ protein domains	Histone modifications
Acetylation	NAT	Histone lysines
Deacetylation	Rpd3/Hda1	Acetylated histone lysines
	Sir2	H3K9ac, H3K18ac, H3K56ac, H4K16ac
Acetyl binding	Bromodomain	Acetylated lysines of core histones
	DPH	H3K56ac
	DPF	Acetylated lysines of H3 and H4
Methylation	Dot1	H3K79
	PRMT	Histone arginines
	SET	Histone lysines
Demethylation	JmjC	H3K4me2,3, H3K9me1,2,3, H3K36me2,3, H3K27me2,3, H4K20me1
	Amine oxidase	H3K4me1,2, H3K9me1,2
	PAD (?)	Arginine methylation
Methyl binding	ADD	H3K9me3, H4R3me2s
	Ankyrin	H3K9me1,2
	BAH	H4K20me2
	Chromodomain and chromobarrel	H3K4me1, H3K9me2,3, H3K27me2,3, H3K36me2,3, H4K20me1
	DCD	H3K4me1,2,3
	MBT	H3Kme1,2, H4Kme1,2
	PHD	H3K4me0,2,3, H3K9me3
	PWWP	H3K36me3, H3K79me3, H4K20me1,3
	TTD	H3K4me0,3, H3K9me3, H4K20me2,3
	Tudor	H3K36me2,3, H3R2me3a/s, H3R17me2a, H4R3me2a
	WD40	H1K26me3, H3K9me3, H3K27me3, H3R2me2s, H4K20me3
Phosphorylation	H kinase	H4H18, H4H75
	STY kinase	Histone serines, threonines and tyrosines
Dephosphorylation	PPI and PPII	Tyrosine phosphorylation
	PP1-5 and PP2C-like	Serine and threonine phosphorylation
Phospho binding	14-3-3	H3S10ph, H3S28ph
	Tandem BRCT	H2AXS139ph
	BIR	H3T3ph

[a]Abbreviations: ADD, ATRX-DNMT3-DNMT3L; BAH, bromo adjacent homology; BIR, Baculovirus IAP repeat; BRCT, BRCA1 C-terminus; DCD, double chromodomain; DPF, double PHD finger; DPH, double pleckstrin homology; JmjC, Jumonji C; NAT, *N*-acetyltransferase; PAD, peptidyl arginine deiminase; PHD, plant homeodomain; PRMT, protein arginine methyltransferase; PWWP, proline-tryptophan-tryptophan-proline; STY, serine threonine tyrosine; PP, protein phosphatase; TTD, tandem Tudor domain. See text for further discussion and references.

2.3 Enzymes that Write Histone Modifications and Protein Domains that Read them

Enzymes that establish, erase and recognise histone acetylation, methylation and phosphorylation are listed in Table 2.1. These are the most highly characterised and frequently occurring PTMs, although they represent only a small proportion of the different types of histone modification that are known to exist. Some modifications (*e.g.* crotonylation) are rare and of uncertain function while others (*e.g.* attachment of the small peptides ubiquitin and SUMO) are dealt with elsewhere. Although different kinds of histone modification are discussed separately here, it is important to recognise that this is not the situation in the cell, where there will be multiple modifications on each of the histone tails in a nucleosome. These modifications are dynamic, and there is also likely to be substantial crosstalk between them, resulting in a highly complex, changing system. The crosstalk between histone modifications will be discussed later.

2.3.1 Histone Acetylation

Most histone acetylation occurs on lysine residues (Figure 2.2), but it has also very recently been reported to affect several serine, threonine and tyrosine residues in the H3 tail of mammalian cells.[50,51] Lysine acetylation is commonly associated with active transcription. It has also been found to be important in the activation of replication origins during S-phase[52] and in the repair of DNA double-strand breaks.[53] What these processes have in common is the requirement for an open chromatin structure to allow access of various protein complexes to DNA. The core histones become acetylated at multiple (but specific) lysine residues (Figure 2.1). The linker histones also undergo acetylation,[54,55] and acetylation at lysine 34 of the most highly conserved linker histone variant, H1.4, has been shown to stimulate transcription both by recruiting the general transcription factor TFIID and by increasing H1 mobility.[56]

Histone lysine acetylation is a highly dynamic modification, with an overall half-life of only a few minutes,[23,24] although this varies between cell types and is likely to differ from one part of the genome to another, depending on the enzymes present. Acetylation is put in place by histone lysine acetyltransferases (KATs), which catalyse the transfer of an acetyl group from the cofactor acetyl coenzyme A to the lysine ε-amino group. The activity of KATs is opposed by the action of histone deacetylases (HDACs), which remove the acetyl group (Figure 2.2). Both KATs and HDACs tend to localise to actively transcribed genes.[57]

There are three main families of KATs – the GNAT, MYST and CBP proteins,[58] all of which contain a catalytic *N*-acetyltransferase (NAT) domain. Members of the GNAT and MYST families are conserved from yeast to humans, but the CBP proteins (CBP and p300, also named KAT3A and KAT3B, respectively) are only found in multicellular organisms. Whilst a

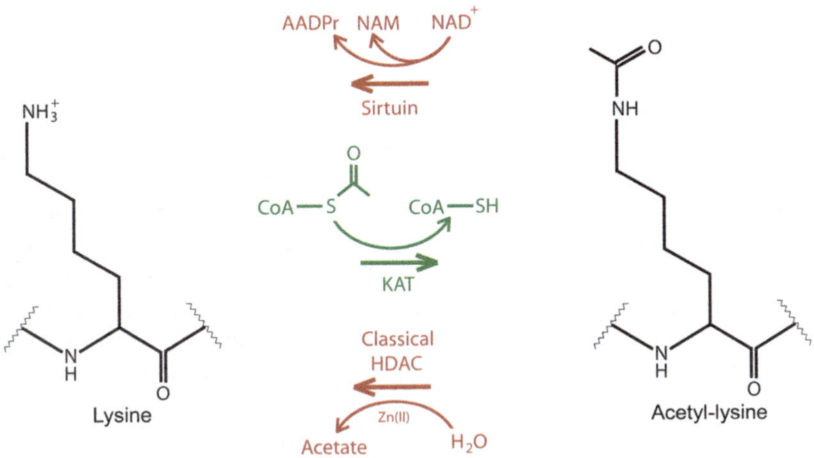

Figure 2.2 Histone lysine acetylation and deacetylation. Acetylation of histone lysine residues is catalysed by lysine acetyl transferases (KATs), which transfer the acetyl group from acetyl coenzyme A (reaction shown in green). Deacetylation of histone lysines can be carried out in two ways (red reactions). One reaction is catalysed by 'classical' histone deacetylases (HDACs), which require the presence of Zn(II) as a cofactor. This mechanism is water-promoted, and produces acetate. The second reaction is catalysed by sirtuins, which couple lysine deacetylation with NAD^+ hydrolysis. Deacetylation by sirtuins produces nicotinamide (NAM) and acetyl-ADP-ribose (AADPr).

subset of KATs show acetylation activity which is specific for one particular histone lysine residue, the majority of these enzymes act on multiple lysines in the core histones and even on non-histone proteins. For example, KAT6A and KAT6B are specific for H3K14, whereas KAT3A and KAT3B acetylate lysines on both H2A and H2B.[59] In humans there are 18 separate HDAC enzymes, grouped into four classes.[60] The 'classical' HDACs belonging to classes I, II and IV require Zn^{2+} as a cofactor, whereas the class III HDACs (also called the sirtuins) require NAD^+ (Figure 2.2). HDACs have largely overlapping substrate specificities and, despite their name, target both histone and non-histone proteins. Like the acetyltransferases, some HDACs occur as part of large multi-protein complexes that crucially affect both their catalytic activity and susceptibility to certain inhibitors (discussed further below).

Histone acetylation has the potential to disrupt the electrostatic interactions which would otherwise maintain the compact structure of chromatin, by neutralising the positive charge of lysine residues. Initially it was proposed that acetylation may decrease interactions which occur between histone proteins and the negatively-charged DNA,[61] however the modification could also conceivably disturb the interactions *between* nucleosomes, by reducing the interaction of the H4 tail with an acidic patch on an H2A/H2B dimer of the adjacent nucleosome.[2] *In vitro*, the level of acetylation of nucleosomal arrays affects transcription mediated by RNA polymerase III, with higher levels of acetylation causing a significant increase in transcription initiation.[42]

Evidence from mutational experiments in yeast suggests that there is a cumulative effect of lysine acetylation of the H4 tail.[62] Yeast strains were created carrying combinations of mutations of H4 lysines 5, 8, 12 and 16 to arginine, mimicking the deacetylated state. When looking at mutations of K5, K8 and K12 it was found that over 20% of genes examined showed cumulative effects, with expression changing incrementally with the number of lysines mutated. These results are consistent with the idea that the charge-neutralising effects of acetylation may have a role in the regulation of gene expression. Work on the H3 histone tail also suggests that the acetylatable lysines K4, K9, K14, K18, K23 and K27 have redundant roles in transcription.[63] However, the same genetic approach showed that H4K16ac had effects on transcription that were independent of those exerted by acetylation of lysines 5, 8 and 12. In this respect, it is noteworthy that H4K16 is selectively acetylated on the transcriptionally hyperactive male X chromosome in *D. melanogaster* larvae.[26] Work from a number of labs has since revealed that the tail of H4 (specifically the region containing amino acids 14–19) is necessary for the formation of higher-order chromatin structure, and that acetylation of H4K16 counteracts this function.[64–66]

Thus, it seems that both charge-mediated effects due to overall changes in histone tail acetylation, and acetylation of individual lysines, play roles in regulating chromatin function. For some genes the acetylation of histone tails *per se* does not alter DNA–histone interactions, but is a prerequisite for nucleosome remodelling.[67] The chromatin remodelling complexes SWI/SNF and RSC contain bromodomains, and both of these complexes can be recruited to acetylated histones; the presence of acetylated lysines actually enhances their ability to mobilise nucleosomes,[68] thus leading to a more open chromatin structure, facilitating access to the DNA.

2.3.2 Histone Methylation

Methylation of histones occurs at selected lysines or arginines and, in contrast to acetylation, does not affect the net charge of these residues. Methylation is also distinct from acetylation in that each modifiable amino acid in the histone tail can carry multiple methyl groups; lysine residues can be mono-, di- or trimethylated (Figure 2.3A), and arginines can be mono- or dimethylated, with dimethylated arginine existing in symmetric and asymmetric forms (Figure 2.3B). The effects of histone methylation are mediated through specific reader proteins, and methyl readers can distinguish between mono-, di- or trimethylation, with readers of arginine dimethylation additionally discriminating between symmetric and asymmetric forms. Methylation is linked to opposing transcriptional outcomes depending on the residue which is methylated. For example, trimethylation of H3K4 is linked to positive transcriptional outcomes, whereas trimethylation of H3K27 usually marks silent genes. The interpretation of lysine methylation is also affected by the degree to which the residue is

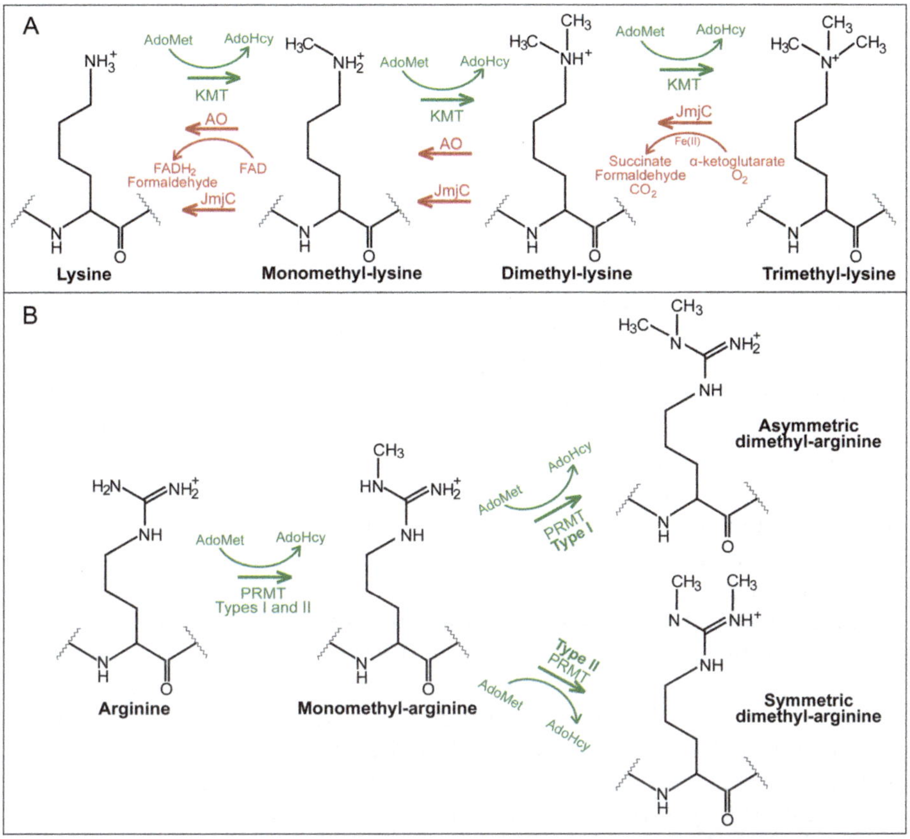

Figure 2.3 Chemistry of lysine and arginine methylation and demethylation reactions. (A) Lysine methylation and demethylation. The methylation and demethylation steps of H3K4 are illustrated. Lysine can be sequentially mono-, di- and trimethylated by lysine methyltransferases (KMTs). These enzymes transfer a methyl group from *S*-adenosyl-L-methionine (AdoMet), producing *S*-adenosyl-L-homocysteine (AdoHcy) (green reactions). Histone lysine demethylases contain either a JmjC domain or an amine oxidase (AO) domain (reactions shown in red). JmjC domain proteins can demethylate mono-, di- and trimethyl forms of H3K4, but AO proteins can only demethylate H3K4me1 and H3K4me2. JmjC enzymes require the presence of Fe(II) and α-ketoglutarate for their mechanism of action, which also uses O_2. Oxidative decarboxylation of α-ketoglutarate produces succinate, CO_2 and ferryl, and the latter oxidises a C–N bond in a methyl group, forming an unstable intermediate which breaks down to release formaldehyde and cause the loss of the methyl group. AO enzymes require FAD to demethylate lysines, *via* an amine oxidase reaction which produces an imine intermediate. This is hydrolysed to produce an unstable carbinolamine that degrades to produce formaldehyde and demethylated lysine. (B) Arginine methylation. The methylation of histone arginines is catalysed by protein arginine methyltransferases (PRMTs). These enzymes catalyse the transfer of a methyl group from *S*-adenosyl-L-methionine (AdoMet), producing *S*-adenosyl-L-homocysteine (AdoHcy) (green reactions). Both Type I and Type II PRMTs can mono- and dimethylate arginine, but they differ with respect to the kind of dimethylated product that is produced. Type I enzymes generate asymmetrically dimethylated arginine, whereas Type II enzymes generate symmetrically dimethylated arginine.

methylated. For example, relatively high levels of H3K4 monomethylation are found at gene enhancers whilst trimethylation is enriched at promoters.

As is the case for histone acetylation there are multiple enzymes which can methylate histones. There are 33 known human lysine methyltransferases (KMTs),[69] and the majority of these enzymes contain a SET catalytic domain. SET domains catalyse the transfer of a methyl group from the methyl donor *S*-adenosyl-L-methionine (SAM or AdoMet), resulting in the production of a methylated protein and *S*-adenosyl-L-homocysteine (SAH or AdoHcy).[70] In fact, of the 33 human KMTs only one, KMT4, contains an alternative catalytic domain, the Dot1 domain, which also uses AdoMet as the methyl donor.[59,71] Methylation of histone lysines is reversible, and the first histone lysine demethylase (KDM1A) was reported in 2004.[72] There are two groups of enzymes that are known to catalyse the demethylation of histone lysines: one group, including KDM1A, contains an amine oxidase catalytic domain which depends on the presence of FAD for its function. The second group of KDM enzymes contains a JmjC domain, and they require Fe(II) and α-ketoglutarate for their activity (Figure 2.3A).

Histone arginines are methylated by protein arginine methyltransferases (PRMTs) which, like the KMT enzymes, use AdoMet as the methyl donor. There are nine known PRMT enzymes (PRMT1–9), of which a subset has activity against histone substrates.[73] The PRMTs are divided into Type I and Type II enzymes. Both classes are competent in the monomethylation of arginine residues. Type I enzymes (*e.g.* PRMT1 and 3) also catalyse asymmetric arginine dimethylation, whereas Type II enzymes (*e.g.* PRMT 5 and 7) yield the symmetrically dimethylated isoform[74] (Figure 2.3B).

The demethylation of histone arginine is not as well understood as the demethylation of histone lysines. JMJD6 was reported to demethylate both H3R2 and H4R3,[75] with activity against both asymmetric and symmetric dimethylated forms. However, experimental work[76] and structural analysis[77] indicate that JMJD6 is not an arginine demethylase. The chemical conversion of arginine residues to citrulline, mediated by peptidylarginine deiminases (PADIs), has also been put forward as a potential method of removing arginine methylation.[78,79] However this mechanism of demethylation is still under debate, as PADI4 activity has since been reported to be inhibited even by monomethylation of the arginine substrate.[80,81]

2.3.3 The Multiple Readers of Histone Methylation

There are several protein domains that bind specifically to methylated histones, including the plant homeodomain (PHD), WD40, Tudor, Chromo and malignant brain tumour (MBT) domains (Table 2.1). The binding of these proteins is sensitive to the methylation state of specific residues in the histone tails, and they act as platforms to recruit effectors of processes such as transcription or gene silencing. Some illustrative examples are given below.

The human genome encodes around 150 PHD domain proteins.[82] These proteins can 'read' the H3 histone tail, and they are sensitive to the methylation state of H3K4 and H3R2. PHD-containing proteins often act to recruit multi-protein complexes which can regulate the processes of gene transcription. There is some amino acid sequence variation between PHD domains, and although all of these proteins can interact with the H3 tail there is variation in their specificity with respect to methylation status.[83] For example the TAF3 PHD binds to H3K4me3,[84] whereas the PHD of AIRE or the DNA methylation co-regulator DNMT3L will only bind to H3 tails which are unmethylated at H3K4.[85,86]

H3K4me3 is frequently associated with active genes,[33,34] and the interaction of TAF3 with H3K4me3 provides an example of how the histone modification might be functionally linked with gene expression. TAF3 is a subunit of the transcription factor TFIID; one of the general transcription factors (TFs) that make up the RNA polymerase II pre-initiation complex. In the processes that lead up to gene expression, TFIID is the first general TF to bind gene promoters. It was reported in 2007 that TAF3 could bind directly to nucleosomes which were trimethylated at H3K4 *via* its PHD.[84] Reduction in the levels of H3K4me3 led to reduced expression of some genes, which also showed reduced TAF3 association. The fact that not all genes were affected led the authors to suggest that H3K4me3 was particularly important in the binding of TFIID to promoters which lack canonical core promoter elements, such as a TATA box. Of course, there are many other PHD-containing proteins that can interact with H3K4me3 and which will have different roles in regulating chromatin function.

Chromodomain (CHD)-containing proteins are involved in the remodelling of chromatin and the repression of gene transcription.[87] The CHD is sensitive to the methylation state of histone lysines, and proteins belonging to this group include the HP1 family, and the chromatin recruitment modules of Polycomb repressive complex 1 (PRC1), which has a role in orchestrating the silencing of developmental genes. Both of these protein families will be discussed in more detail below. As with the PHD family, slight variations in amino acid sequence of the CHD result in a group of proteins with a range of different affinities for methylated histone lysines.[88]

The chromatin binding of the HP1α, β and γ proteins is sensitive to the methylation state of H3K9. These proteins will bind H3K9me2 and H3K9me3 with high affinity.[89,90] The canonical function of HP1 proteins is in the formation and maintenance of heterochromatin. In both mammalian and *Drosophila* cells, the artificial recruitment of these proteins to reporter transgenes leads to gene repression.[91,92] The general model of heterochromatin formation by HP1 involves N-terminal CHD-mediated binding of the protein to methylated H3K9, followed by dimerisation of HP1 proteins *via* their C-terminal chromoshadow domains.[93,94] Less specific binding to DNA and linker histones *via* the HP1 hinge region may also play a role.[95] This process is thought to draw adjacent nucleosomes together, compacting chromatin. In addition, HP1 recruits an H3K9 methyltransferase, which will catalyse further H3K9 methylation, stimulating the binding of further HP1 proteins.[96] HP1α and β have also been found to interact with the lamin B receptor, located in the nuclear lamina.[97] This interaction links H3K9 methylation with the localisation of silenced chromatin at the nuclear periphery.

Despite characterisation of HP1 proteins as markers of heterochromatin, additional work suggests a role for H3K9 methylation in active transcription as well. Although the presence of H3K9me2 and me3 at gene promoters is linked to gene repression,[98,99] H3K9me2/3 in the gene body is correlated with expression.[100] HP1γ associates with methylated H3K9 in transcribed chromatin, and the enrichment of HP1γ and H3K9 methylation is dependent on transcription,[100] illustrating how processes such as transcription can alter local histone modifications. The functional impact of H3K9 methylation on transcribed genes is not known.

H3K27me3 is crucial for gene silencing by the Polycomb complexes PRC1 and PRC2. H3K27me3, deposited by PRC2, is read by CHD proteins within the PRC1 complex. Multiple versions of the PRC1 and PRC2 complexes exist, with different combinations of subunits.[101] The chromodomain proteins CBX2, 4, 6, 7 and 8 act as chromatin recruitment modules for PRC1, by targeting it to H3K27me3. The binding of PRC1 complexes is associated with the compaction of the chromatin structure[102] and PRC1 also catalyses mono-ubiquitination of H2AK119 at PRC1 target sites.[103,104] Both PRC1-associated chromatin compaction and H2AK119ub have been reported to function independently in the silencing of target genes; Eskeland *et al.*[105] report that compaction of chromatin by PRC1 at murine *Hoxb* and *Hoxd* loci leads to gene repression even in the absence of histone ubiquitination, and Endoh *et al.*[106] have shown that for other PRC1 targets H2AK112ub is not required for chromatin compaction but *is* necessary for PRC1-mediated gene silencing. Thus, the PRC1 complex appears to contribute to transcriptional silencing in several different ways. Questions remain, as the links between H2AK119ub and transcriptional silencing are not yet fully understood. It should be noted that H3K27me3 binding is not the only means by which PRC1 recognises genomic targets. It has also been found be targeted by interaction with specific TFs[107–109] and by the interaction of KDM2B (a component of a subset of PRC1 complexes) with CpG islands in embryonic stem (ES) cells.[110]

2.3.4 Histone Phosphorylation

The phosphorylation of histone tails occurs at serines, threonines, tyrosines and histidines, and although it is found at fewer sites than acetylation or methylation it still has important consequences for genome function, particularly in DNA repair, transcription and the compaction of DNA as cells enter mitosis. Like acetylation, phosphorylation changes the charge of modified residues, but by introducing a negative charge rather than neutralising a positive one. This modification therefore has the potential to create a charge repulsion effect between histone tails and DNA.[111] However, many of its effects *in vivo* are known to be coordinated by the action of specific binding proteins. Much of the function of histone phosphorylation is mediated *via* crosstalk with other histone modifications, and this will be discussed later in the chapter.

The phosphorylation of histone residues is catalysed by a variety of cellular kinases in response to upstream signalling pathways. Such kinases belong to a large protein family containing a conserved catalytic region which catalyses the transfer of a phosphate group from ATP to target proteins. They are often important components of cell signalling pathways, and examples include JAK2, PKB and PKCα, β or δ.[112] These kinases have a wide range of target proteins, and are also known to phosphorylate chromatin modifiers and TFs, potentially allowing them to coordinate chromatin responses to external signals. Kinase activity is opposed by phosphatases, which remove the phosphate group. Phosphatases involved in the dephosphorylation of histones include PP1, PP1γ, PP2A and PP4.[113–116]

There are three protein domains which are known to interact with phosphorylated histones; the 14-3-3 domain, the BRCT domain and the BIR domain.[46] These readers are specific for different phosphorylated histone residues. For example, the BRCT domain recognises the histone variant H2AX phosphorylated at S139. This histone modification (often referred to as γH2AX) is a marker of DNA damage and is established rapidly by the ATM, ATR and DNA-PK enzymes after DNA damage has been induced.[117] γH2AX recruits components of the DNA repair pathway, many of which contain BRCT domains, to the site of DNA damage: these proteins, such as MRN, can then act as a molecular scaffold to assemble repair proteins.[118–121] There is also evidence to suggest that the outcome of the DNA damage response is further modulated by phosphorylation at tyrosine 142 of histone H2A. When this residue is unmodified serine 139 phosphorylation is less stable,[122] and it has been suggested that modification of the tyrosine may mediate a cellular decision between DNA repair and apoptosis.[123]

2.4 Interactions between Histone Modifications – Crosstalk

In reality, no histone modification functions in isolation. Figure 2.1 shows that histone tails can be extensively modified, even when only three possible types of modification are considered. Crosstalk between histone modifications results from interactions between the components of histone reading and modifying systems. Considered broadly, crosstalk between histone modifications can be thought to occur at three stages (Figure 2.4A): (i) the deposition of a histone PTM, (ii) the interaction of the PTM with a reader protein, and (iii) the removal of the PTM. Many examples of crosstalk have been documented in the literature. Given the huge number of modifications that histones undergo, and the number of amino acids in histone proteins which can be modified, cataloguing and understanding the possible interactions represents an immense challenge.

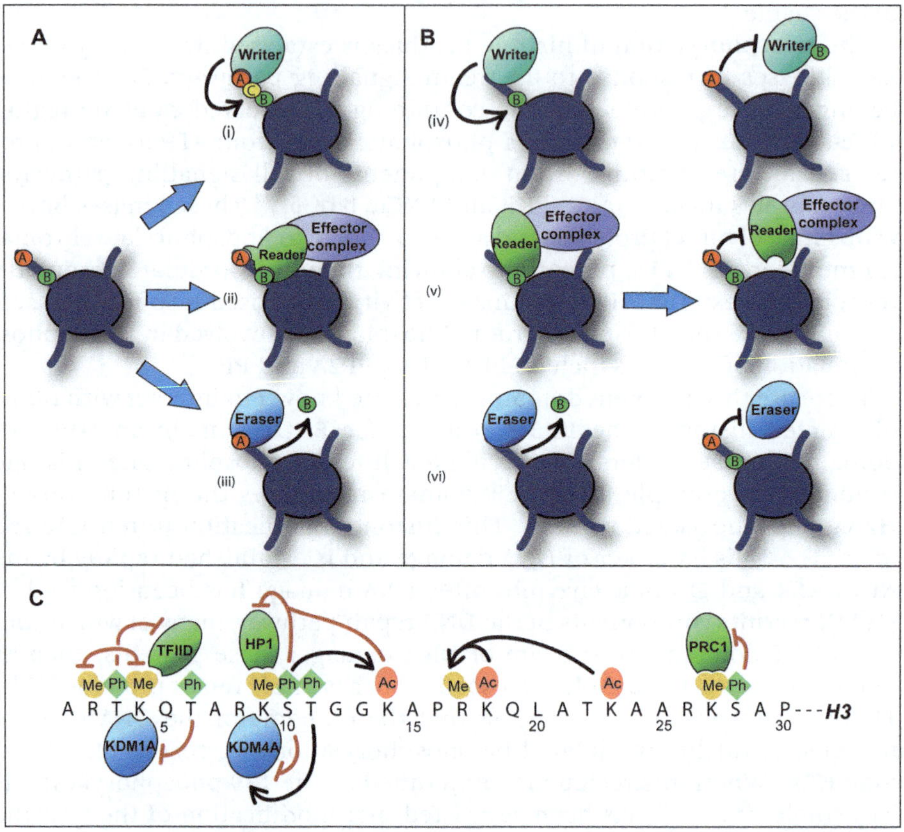

Figure 2.4 Mechanisms of histone crosstalk. Interactions between histones can have both positive and negative effects on the writing, reading and erasing processes. (A) Modification A (red mark) stimulates interactions with writers, readers and erasers of nearby modifications. (i) Modification A recruits the writer of mark C (yellow). (ii) Modification A stabilises the association of a reader protein with mark B (green), allowing recruitment of effector complexes. (iii) Modification A recruits an eraser of B. (B) Modification A inhibits the binding of writers, readers and erasers of nearby histone marks. (iv) The presence of modification A (red mark) prevents a writer protein from establishing mark B (green). (v) When modification A (red) is added the reader of B (green) can no longer recognise the mark. (vi) Modification A inhibits the binding of an eraser protein specific to B (green). (C) Schematic illustrating some of the crosstalk between histone modifications that occurs on the H3 tail, and which is referred to in the main text. The first 30 amino acids of the histone are shown. Readers of histone modification are shown in green, erasers shown in blue. Ac = acetylation; Me = methylation; Ph = phosphorylation.

2.4.1 Effects on Further Modification of the Same Histone Tail

Histone modifications interact, and the idea that histone modifications can alter how a histone tail is further modified is self-evident. For example, if a histone is acetylated at a specific lysine (*e.g.* H3K9 or K27), it cannot also be methylated at the same position. However, more indirect mechanisms of interference also exist. For example, the modification of one residue can negatively affect the ability of a 'writer' enzyme to modify an adjacent or nearby residue [Figure 2.4B, example (iv)]. This sort of interaction occurs between H3R2 and H3K4, where asymmetric dimethylation of the arginine prevents trimethylation of the lysine, by preventing binding of the methylase (Figure 2.4C). H3R2me2a was found to be required for silencing of heterochromatin in yeast; it inhibits the binding of the Set1 complex component Spp1, which is required for deposition of the active chromatin mark H3K4me3.[124] The same interaction occurs in mammalian cells, where it has additionally been found that H3K4me3 prevents asymmetric H3R2 dimethylation (Figure 2.4C).[125,126] The more recent finding that symmetrical dimethylation of H3R2 is tightly correlated with H3K4me3 at active promoters in mouse suggests that the switch between symmetric and asymmetric methylation of this arginine may have a role in determining gene activity.[127] Another example of one histone modification interfering with the deposition of another has been reported in yeast, where the isomerisation of H3 proline 38 inhibits methylation of H3K36 and *vice versa*.[128]

A number of histone modifications have also been found to facilitate the deposition of other histone marks. For example, phosphorylation of H3S10 enhances the acetylation of lysine 14 (Figure 2.4C), with the HAT GCN5 showing up to 10-fold preference for substrates which are phosphorylated at H3S10. Computer modelling of yeast Gcn5 bound to histone tails suggests that H3S10ph stabilises the interaction of Gcn5 with its substrate.[129,130] A similar facilitation event occurs further along the H3 histone tail, where methylation of H3R17 by CARM1 is augmented by the presence of H3K18ac, and to a more limited extent by H3K23 acetylation (Figure 2.4C).[131] Both of these interactions occur within one histone tail; however such facilitation can also take place between histone tails within a nucleosome (Figure 2.5), as demonstrated by the role of H2B ubiquitination in stimulating the methylation of H3K4 and H3K79.[71,132–135]

2.4.2 Effects on Recognition of Nearby Modifications

There are lots of examples of one histone modification affecting the interpretation of a nearby modification. This can occur due to the presence of multiple binding domains in proteins which interact with chromatin. For example, KMT2A (a lysine methyltransferase) and BPTF (a component of the NURF nucleosomal remodelling complex) both contain PHD fingers and a bromodomain. In fact, the separate chromatin-binding modules of BPTF are

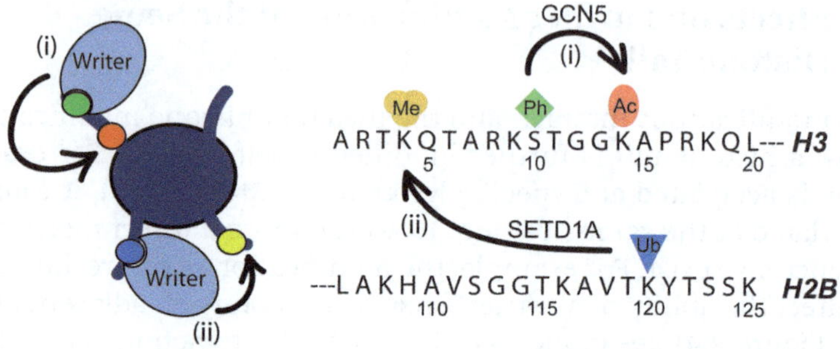

Figure 2.5 Interactions between histone modifications can occur within or between histone tails. (i) Interactions can occur between histone modifications located on the same histone tail. In the left figure one modification (green circle) recruits the writer of a second modification (red circle). An example of this is shown on the right, where phosphorylation of serine 10 of H3 facilitates acetylation of H3K14 by the Gcn5 enzyme. (ii) Similar interactions can occur between histone tails. In the left figure a writer that interacts with a modification on one tail (blue circle) can modify histones on another tail (yellow circle). An example is shown in the right figure, where ubiquitination of lysine 120 of H2A allows methylation of H3K4 by SETD1A.

known to interact with different histone tails in a nucleosome simultaneously.[136] The assembly of such proteins into complexes which interact with chromatin will bring together a range of proteins with specificity for different chromatin modifications, the binding of which will be affected by multiple histone modifications in the local chromatin environment.

There are also cases where a histone modification affects the binding of a reader protein to a nearby histone modification [Figure 2.4B, example (v)]. One of the most widely cited examples of this sort of interaction is the role of H3S10ph in the eviction of HP1 from sites of H3K9me2/3 as a cell condenses its DNA during entry into mitosis.[137,138] In contrast to interphase cells, where H3S10ph is located at distinct genomic locations, typically associated with active genes, mitotic cells show genome-wide enrichment for the mark. This begins at the peri-centromeric heterochromatin late in G2 phase, subsequently spreading to encompass the entire chromosome by metaphase,[139] and is thought to be necessary for the initiation of chromosome condensation.[140] A similar cell cycle-dependent change in distribution is observed for H3S28ph.[141] The mechanism by which H3S10ph causes loss of HP1 binding, even in the presence of a steady level of H3K9me3, is unknown. Over-expression of a catalytically inactive version of the relevant kinase, Aurora B, prevents phosphorylation of the two serines as cells enter mitosis, and causes incomplete chromosome condensation, as well as affecting the alignment of chromosomes on the metaphase plate.[142] However, data derived from both experimental work and computer modelling suggests that H3S10ph in itself is insufficient to cause the loss of HP1 binding, and it has been suggested that the acetylation of H3K14ac (a process which is itself facilitated by H3S10ph), in conjunction with H3S10ph, is necessary for dissociation of HP1 (Figure 2.4C).[143,144]

The initial association of H3S10ph with the loss of HP1 binding led to the suggestion of a 'phospho-methyl switch' method of histone modification crosstalk.[145] There are multiple positions in the histone tails where a serine or threonine residue which has the potential to be phosphorylated is located adjacent to a lysine which is known to be methylated. The 'phospho-methyl switch' proposes that phosphorylation and dephosphorylation act to modulate access to the adjacent modified lysine by its binding proteins. Such a switch could thereby regulate gene activity without affecting the longer-term epigenetic memory of gene expression mediated by lysine methylation. In addition to the reported H3S10ph/H3K9me3 interaction there is also evidence that the phosphorylation of H3T3 has a role in chromatin function during mitosis, inhibiting gene expression by causing TFIID to dissociate from H3K4me3-marked gene promoters (Figure 2.4C).[146] During interphase, interactions between phosphorylated H3S10 and S28 and the adjacent H3K9me2 and H3K27me3 are associated with transcriptional activation. Phosphorylation of H3S10 at the *HDAC1* promoter is reported to lead to loss of the HP1γ–H3K9me2 interaction and the recruitment of 14-3-3ζ, followed by transcriptional activation.[147] A similar interaction between H3S28ph and Polycomb group proteins has been reported at a subset of Polycomb target loci (Figure 2.4C).[148]

2.4.3 Effects on Erasure of Nearby Modifications

A further method of crosstalk is derived from the potential for one type of histone modification to alter the propensity of a nearby modification to be removed. For a number of histone modification pairings, one member of the pair has been found to protect the other from erasure [Figure 2.4B, example (vi)]. Such an interaction has been reported between H3T6 and H3K4; during the activation of androgen receptor (AR)-responsive genes H3T6 becomes phosphorylated by PKCβ. This prevents KDM1A from demethylating H3K4me3 (Figure 2.4C). Reduction in PKCβ levels causes a concomitant decrease in H3T6ph, reduced H3K4me3 levels and is related to reduced transcriptional activation.[149] This sort of 'protective' crosstalk is also evident at the H3S10ph/H3K9me pairing (discussed in the previous section), where the phosphorylation event has been found to prevent demethylation of the lysine residue by KDM4A (Figure 2.4C),[150] creating a potential mechanism for the maintenance of the repressive mark in the absence of HP1 binding during late G2 and mitosis. Interestingly the demethylation of H3K9me3 is also affected by phosphorylation of H3T11 (Figure 2.4C), although in this case the phosphorylation event *enhances* demethylation of H3K9 by KDM4C, allowing transcriptional activation.[151]

These are just a few examples of the varied ways in which histone modifications have the potential to undergo complex crosstalk. It is likely that further research will reveal many novel and interesting interactions, especially with regard to modifications such as crotonylation, propionylation or butyrylation, on which research is still in its infancy.

2.5 The Importance of Protein Complexes

Many, probably most, of the enzymes discussed in this chapter operate as components of multi-protein complexes. In attempting to unravel the links between histone-modifying enzymes and chromatin function, we must deal with the fact that the same enzyme can exert very different functional effects depending on the proteins with which it associates. This is well illustrated by recent studies on histone deacetylases (HDACs).

2.5.1 HDACs and their Inhibitors

Recent work has shown that class I/IIa HDACs are catalytically active only when complexed with specific partner proteins. Class IIb HDACs show little activity, even in complexes, and their role may be primarily structural. The gene-regulatory complexes, of which catalytically active HDACs are key components, have been purified and subjected to detailed structural analysis.[152] There are four such complexes, namely CoRest, NuRD, Sin3 and NCoR/SMRT. The first three contain HDAC1 and HDAC2, while NCoR contains just HDAC3. The complexes are responsible for the up- and/or down-regulation of specific genes and are necessary for the correct differentiation of various cell types.

A second important and unexpected insight from structural studies is the demonstration that for both the NCoR and NuRD complexes, HDAC activity requires the incorporation of a molecule of D-myo-inositol 1,4,5,6 tetrakisphosphate [ins(1,4,5,6)P_4].[153] Inositol phosphates are crucial mediators of intracellular signalling pathways. It is significant that the K_d for $InsP_4$ incorporation into the NuRD complex is around 5 μM, within the intracellular concentration range, suggesting that varying the concentrations of this important signalling molecule might be another means by which the epigenetic effects of HDACs are regulated.

Remarkably, the sensitivity of a particular HDAC to a specific inhibitor seems to be dependent on the complex of which it is part. Using an affinity purification/inhibition approach coupled to state-of-the-art mass spectrometry, it has been shown that HDAC1 and HDAC2, when part of the CoRest complex, are inhibited by the short-chain fatty acid valproic acid (VPA, as the sodium salt) with a K_d of around 0.65 mM.[154] By contrast, the same two enzymes are inhibited with a K_d of about 1.4 mM in the NuRD complex, and 13 mM in the Sin3 complex. Thus, there is a 20-fold difference in the sensitivity of these HDACs to the HDAC inhibitor (HDACi) VPA, depending on the complex of which the enzymes are part. HDAC3 in the NCoR complex is inhibited with a K_d of around 4 mM. All this matters because VPA is widely used clinically, both as an antiepileptic and in clinical trials against various cancers, and 0.65 mM is just below the maximum plasma concentration achievable by oral administration. These findings also raise the possibility of differentially affecting the function of specific HDAC complexes, and the genes they regulate, by adjusting the dosage of the HDACi. Intriguingly, a second HDACi, suberoyl anilide hydroxamic acid (SAHA, Vorinostat), which was recently FDA approved for treatment of cutaneous T-cell lymphoma, inhibits HDAC activity similarly irrespective of the complex, with K_d values ranging from about 0.2 to 0.4 μM. Thus, SAHA acts as a pan-specific HDACi and, as things stand, does not offer the potential of dose-specific effects.[154]

2.6 Conclusions

Nucleosomes are not simply a convenient DNA packaging system for eukaryotic cells, but rather act as signalling modules which can be read by effector protein complexes. These complexes mediate genomic functions such as transcription, DNA synthesis and DNA repair, with further modifications occurring as these processes are carried out. The result is a highly dynamic but controlled system, allowing both plasticity of responses (to various cell signalling pathways, for example), as well as the maintenance of the *status quo* where this is necessary, such as in the maintenance of cellular identity or the silencing of repetitive genomic regions.

The reading, writing and erasing of histone PTMs is affected by the local histone modification environment. The number of different histone modifications that can occur and the dynamic nature of these modifications combine to create a complex environment which is likely to be context dependent, differing between cell types and in response to cell signalling and metabolic processes. This intricacy certainly presents both intellectual and technical challenges, but provides a vast number of potential drug targets, offering unparalleled opportunities for the design and delivery of effective epigenetic therapies for a variety of human diseases.

Acknowledgements

Research in the authors' laboratory is supported by Cancer Research UK.

References

1. A. Klug, D. Rhodes, J. Smith, J. T. Finch and J. O. Thomas, *Nature*, 1980, **287**, 509.
2. K. Luger, A. W. Mader, R. K. Richmond, D. F. Sargent and T. J. Richmond, *Nature*, 1997, **389**, 251.
3. G. Arents, R. W. Burlingame, B. C. Wang, W. E. Love and E. N. Moudrianakis, *Proc. Natl. Acad. Sci. U. S. A.*, 1991, **88**, 10148.
4. G. Arents and E. N. Moudrianakis, *Proc. Natl. Acad. Sci. U. S. A.*, 1995, **92**, 11170.

5. H. E. Kasinsky, J. D. Lewis, J. B. Dacks and J. Ausio, *FASEB J.*, 2001, **15**, 34.
6. N. Heintz, H. L. Sive and R. G. Roeder, *Mol. Cell. Biol.*, 1983, **3**, 539.
7. W. F. Marzluff, *Gene Expression*, 1992, 2, 93.
8. N. B. Pandey, A. S. Williams, J. H. Sun, V. D. Brown, U. Bond and W. F. Marzluff, *Mol. Cell. Biol.*, 1994, **14**, 1709.
9. D. R. Gallie, N. J. Lewis and W. F. Marzluff, *Nucleic Acids Res.*, 1996, **24**, 1954.
10. P. B. Talbert and S. Henikoff, *Nat. Rev. Mol. Cell Biol.*, 2010, **11**, 264.
11. V. G. Allfrey, R. Faulkner and A. E. Mirsky, *Proc. Natl. Acad. Sci. U. S. A.*, 1964, **51**, 786.
12. G. Vidali, E. L. Gershey and V. G. Allfrey, *J. Biol. Chem.*, 1968, **243**, 6361.
13. S. R. Bhaumik, E. Smith and A. Shilatifard, *Nat. Struct. Mol. Biol.*, 2007, **14**, 1008.
14. I. L. Goldknopf, C. W. Taylor, R. M. Baum, L. C. Yeoman, M. O. Olson, A. W. Prestayko and H. Busch, *J. Biol. Chem.*, 1975, **250**, 7182.
15. S. I. Matsui, B. K. Seon and A. A. Sandberg, *Proc. Natl. Acad. Sci. U. S. A.*, 1979, **76**, 6386.
16. M. H. West and W. M. Bonner, *Nucleic Acids Res.*, 1980, **8**, 4671.
17. Y. Shiio and R. N. Eisenman, *Proc. Natl. Acad. Sci. U. S. A.*, 2003, **100**, 13225.
18. T. Hagiwara, K. Nakashima, H. Hirano, T. Senshu and M. Yamada, *Biochem. Biophys. Res. Commun.*, 2002, **290**, 979.
19. K. Nakashima, T. Hagiwara and M. Yamada, *J. Biol. Chem.*, 2002, **277**, 49562.
20. C. J. Nelson, H. Santos-Rosa and T. Kouzarides, *Cell*, 2006, **126**, 905.
21. E. M. Duncan, T. L. Muratore-Schroeder, R. G. Cook, B. A. Garcia, J. Shabanowitz, D. F. Hunt and C. D. Allis, *Cell*, 2008, **135**, 284.
22. H. Santos-Rosa, A. Kirmizis, C. Nelson, T. Bartke, N. Saksouk, J. Cote and T. Kouzarides, *Nat. Struct. Mol. Biol.*, 2009, **16**, 17.
23. V. Jackson, A. Shires, R. Chalkley and D. K. Granner, *J. Biol. Chem.*, 1975, **250**, 4856.
24. A. Chestier and M. Yaniv, *Proc. Natl. Acad. Sci. U. S. A.*, 1979, **76**, 46.
25. M. Tan, H. Luo, S. Lee, F. Jin, J. S. Yang, E. Montellier, T. Buchou, Z. Cheng, S. Rousseaux, N. Rajagopal, Z. Lu, Z. Ye, Q. Zhu, J. Wysocka, Y. Ye, S. Khochbin, B. Ren and Y. Zhao, *Cell*, 2011, **146**, 1016.
26. B. M. Turner, A. J. Birley and J. Lavender, *Cell*, 1992, **69**, 375.
27. B. M. Turner, *Cell*, 1993, **75**, 5.
28. C. Dhalluin, J. E. Carlson, L. Zeng, C. He, A. K. Aggarwal and M. M. Zhou, *Nature*, 1999, **399**, 491.
29. B. D. Strahl and C. D. Allis, *Nature*, 2000, **403**, 41.
30. T. R. Hebbes, A. W. Thorne and C. Crane-Robinson, *EMBO J.*, 1988, 7, 1395.
31. D. Y. Lee, J. J. Hayes, D. Pruss and A. P. Wolffe, *Cell*, 1993, **72**, 73.
32. A. Eberharter and P. B. Becker, *EMBO Rep.*, 2002, **3**, 224.
33. D. Schübeler, D. M. MacAlpine, D. Scalzo, C. Wirbelauer, C. Kooperberg, F. van Leeuwen, D. E. Gottschling, L. P. O'Neill, B. M. Turner, J. Delrow, S. P. Bell and M. Groudine, *Genes Dev.*, 2004, **18**, 1263.

34. B. E. Bernstein, M. Kamal, K. Lindblad-Toh, S. Bekiranov, D. K. Bailey, D. J. Huebert, S. McMahon, E. K. Karlsson, E. J. Kulbokas III, T. R. Gingeras, S. L. Schreiber and E. S. Lander, *Cell*, 2005, **120**, 169.
35. A. H. Peters, D. O'Carroll, H. Scherthan, K. Mechtler, S. Sauer, C. Schofer, K. Weipoltshammer, M. Pagani, M. Lachner, A. Kohlmaier, S. Opravil, M. Doyle, M. Sibilia and T. Jenuwein, *Cell*, 2001, **107**, 323.
36. G. Schotta, M. Lachner, K. Sarma, A. Ebert, R. Sengupta, G. Reuter, D. Reinberg and T. Jenuwein, *Genes Dev.*, 2004, **18**, 1251.
37. R. Cao, L. Wang, H. Wang, L. Xia, H. Erdjument-Bromage, P. Tempst, R. S. Jones and Y. Zhang, *Science*, 2002, **298**, 1039.
38. N. D. Heintzman, R. K. Stuart, G. Hon, Y. Fu, C. W. Ching, R. D. Hawkins, L. O. Barrera, S. Van Calcar, C. Qu, K. A. Ching, W. Wang, Z. Weng, R. D. Green, G. E. Crawford and B. Ren, *Nat. Genet.*, 2007, **39**, 311.
39. G. E. Zentner, P. J. Tesar and P. C. Scacheri, *Genome Res.*, 2011, **21**, 1273.
40. A. J. Bannister, R. Schneider, F. A. Myers, A. W. Thorne, C. Crane-Robinson and T. Kouzarides, *J. Biol. Chem.*, 2005, **280**, 17732.
41. C. R. Vakoc, M. M. Sachdeva, H. Wang and G. A. Blobel, *Mol. Cell. Biol.*, 2006, **26**, 9185.
42. C. Tse, T. Sera, A. P. Wolffe and J. C. Hansen, *Mol. Cell. Biol.*, 1998, **18**, 4629.
43. S. M. Gorisch, M. Wachsmuth, K. F. Toth, P. Lichter and K. Rippe, *J. Cell Sci.*, 2005, **118**, 5825.
44. S. D. Taverna, H. Li, A. J. Ruthenburg, C. D. Allis and D. J. Patel, *Nat. Struct. Mol. Biol.*, 2007, **14**, 1025.
45. T. G. Kutateladze, *Cell*, 2011, **146**, 842.
46. C. A. Musselman, M. E. Lalonde, J. Cote and T. G. Kutateladze, *Nat. Struct. Mol. Biol.*, 2012, **19**, 1218.
47. E. J. Wagner and P. B. Carpenter, *Nat. Rev. Mol. Cell Biol.*, 2012, **13**, 115.
48. B. M. Turner, *BioEssays*, 2000, **22**, 836.
49. B. M. Turner, *Nat. Cell Biol.*, 2007, **9**, 2.
50. R. Y. Tweedie-Cullen, A. M. Brunner, J. Grossmann, S. Mohanna, D. Sichau, P. Nanni, C. Panse and I. M. Mansuy, *PLoS One*, 2012, **7**, e36980.
51. L. M. Britton, A. Newhart, N. V. Bhanu, R. Sridharan, M. Gonzales-Cope, K. Plath, S. M. Janicki and B. A. Garcia, *Epigenetics*, 2013, **8**, 1101.
52. A. Unnikrishnan, P. R. Gafken and T. Tsukiyama, *Nat. Struct. Mol. Biol.*, 2010, **17**, 430.
53. Y. Xu and B. D. Price, *Cell Cycle*, 2011, **10**, 261.
54. B. A. Garcia, S. A. Busby, C. M. Barber, J. Shabanowitz, C. D. Allis and D. F. Hunt, *J. Proteome Res.*, 2004, **3**, 1219.
55. A. Vaquero, M. Scher, D. Lee, H. Erdjument-Bromage, P. Tempst and D. Reinberg, *Mol. Cell*, 2004, **16**, 93.
56. K. Kamieniarz, A. Izzo, M. Dundr, P. Tropberger, L. Ozretić, J. Kirfel, E. Scheer, P. W. Tropel J. R., L. Tora, S. Viville, R. Buettner and R. Schneider, *Genes Dev.*, 2012, **26**, 797.
57. Z. Wang, C. Zang, K. Cui, D. E. Schones, A. Barski, W. Peng and K. Zhao, *Cell*, 2009, **138**, 1019.

58. K. K. Lee and J. L. Workman, *Nat. Rev. Mol. Cell Biol.*, 2007, **8**, 284.
59. C. D. Allis, S. L. Berger, J. Cote, S. Dent, T. Jenuwien, T. Kouzarides, L. Pillus, D. Reinberg, Y. Shi, R. Shiekhattar, A. Shilatifard, J. Workman and Y. Zhang, *Cell*, 2007, **131**, 633.
60. B. Barneda-Zahonero and M. Parra, *Mol. Oncol.*, 2012, **6**, 579.
61. L. Hong, G. P. Schroth, H. R. Matthews, P. Yau and E. M. Bradbury, *J. Biol. Chem.*, 1993, **268**, 305.
62. M. F. Dion, S. J. Altschuler, L. F. Wu and O. J. Rando, *Proc. Natl. Acad. Sci. U. S. A.*, 2005, **102**, 5501.
63. A. M. Martin, D. J. Pouchnik, J. L. Walker and J. J. Wyrick, *Genetics*, 2004, **167**, 1123.
64. B. Dorigo, T. Schalch, K. Bystricky and T. J. Richmond, *J. Mol. Biol.*, 2003, **327**, 85.
65. M. Shogren-Knaak, H. Ishii, J. M. Sun, M. J. Pazin, J. R. Davie and C. L. Peterson, *Science*, 2006, **311**, 844.
66. P. J. Robinson, W. An, A. Routh, F. Martino, L. Chapman, R. G. Roeder and D. Rhodes, *J. Mol. Biol.*, 2008, **381**, 816.
67. H. Reinke, P. D. Gregory and W. Horz, *Mol. Cell*, 2001, **7**, 529.
68. N. Chatterjee, D. Sinha, M. Lemma-Dechassa, S. Tan, M. A. Shogren-Knaak and B. Bartholomew, *Nucleic Acids Res.*, 2011, **39**, 8378.
69. J. C. Black, C. Van Rechem and J. R. Whetstine, *Mol. Cell*, 2012, **48**, 491.
70. S. C. Dillon, X. Zhang, R. C. Trievel and X. Cheng, *Genome Biol.*, 2005, **6**, 227.
71. H. H. Ng, Q. Feng, H. Wang, H. Erdjument-Bromage, P. Tempst, Y. Zhang and K. Struhl, *Genes Dev.*, 2002, **16**, 1518.
72. Y. Shi, F. Lan, C. Matson, P. Mulligan, J. R. Whetstine, P. A. Cole, R. A. Casero and Y. Shi, *Cell*, 2004, **119**, 941.
73. A. Di Lorenzo and M. T. Bedford, *FEBS Lett.*, 2011, **585**, 2024.
74. B. C. Smith and J. M. Denu, *Biochim. Biophys. Acta*, 2009, **1789**, 45.
75. B. Chang, Y. Chen, Y. Zhao and R. K. Bruick, *Science*, 2007, **318**, 444.
76. C. J. Webby, A. Wolf, N. Gromak, M. Dreger, H. Kramer, B. Kessler, M. L. Nielsen, C. Schmitz, D. S. Butler, J. R. Yates 3rd, C. M. Delahunty, P. Hahn, A. Lengeling, M. Mann, N. J. Proudfoot, C. J. Schofield and A. Bottger, *Science*, 2009, **325**, 90.
77. X. Hong, J. Zang, J. White, C. Wang, C. H. Pan, R. Zhao, R. C. Murphy, S. Dai, P. Henson, J. W. Kappler, J. Hagman and G. Zhang, *Proc. Natl. Acad. Sci. U. S. A.*, 2010, **107**, 14568.
78. Y. Wang, J. Wysocka, J. Sayegh, Y. H. Lee, J. R. Perlin, L. Leonelli, L. S. Sonbuchner, C. H. McDonald, R. G. Cook, Y. Dou, R. G. Roeder, S. Clarke, M. R. Stallcup, C. D. Allis and S. A. Coonrod, *Science*, 2004, **306**, 279.
79. G. L. Cuthbert, S. Daujat, A. W. Snowden, H. Erdjument-Bromage, T. Hagiwara, M. Yamada, R. Schneider, P. D. Gregory, P. Tempst, A. J. Bannister and T. Kouzarides, *Cell*, 2004, **118**, 545.
80. Y. Hidaka, T. Hagiwara and M. Yamada, *FEBS Lett.*, 2005, **579**, 4088.
81. R. Raijmakers, A. J. Zendman, W. V. Egberts, E. R. Vossenaar, J. Raats, C. Soede-Huijbregts, F. P. Rutjes, P. A. van Veelen, J. W. Drijfhout and G. J. Pruijn, *J. Mol. Biol.*, 2007, **367**, 1118.

82. M. Bienz, *Trends Biochem. Sci.*, 2006, **31**, 35.
83. R. Sanchez and M. M. Zhou, *Trends Biochem. Sci.*, 2011, **36**, 364.
84. M. Vermeulen, K. W. Mulder, S. Denissov, W. W. Pijnappel, F. M. van Schaik, R. A. Varier, M. P. Baltissen, H. G. Stunnenberg, M. Mann and H. T. Timmers, *Cell*, 2007, **131**, 58.
85. T. Org, F. Chignola, C. Hetenyi, M. Gaetani, A. Rebane, I. Liiv, U. Maran, L. Mollica, M. J. Bottomley, G. Musco and P. Peterson, *EMBO Rep.*, 2008, **9**, 370.
86. S. K. Ooi, C. Qiu, E. Bernstein, K. Li, D. Jia, Z. Yang, H. Erdjument-Bromage, P. Tempst, S. P. Lin, C. D. Allis, X. Cheng and T. H. Bestor, *Nature*, 2007, **448**, 714.
87. K. L. Yap and M. M. Zhou, *Biochemistry*, 2011, **50**, 1966.
88. B. J. Blus, K. Wiggins and S. Khorasanizadeh, *Crit. Rev. Biochem. Mol. Biol.*, 2011, **46**, 507.
89. A. J. Bannister, P. Zegerman, J. F. Partridge, E. A. Miska, J. O. Thomas, R. C. Allshire and T. Kouzarides, *Nature*, 2001, **410**, 120.
90. M. Lachner, D. O'Carroll, S. Rea, K. Mechtler and T. Jenuwein, *Nature*, 2001, **410**, 116.
91. K. Ayyanathan, M. S. Lechner, P. Bell, G. G. Maul, D. C. Schultz, Y. Yamada, K. Tanaka, K. Torigoe and F. J. Rauscher III, *Genes Dev.*, 2003, **17**, 1855.
92. Y. Li, J. R. Danzer, P. Alvarez, A. S. Belmont and L. L. Wallrath, *Development*, 2003, **130**, 1817.
93. R. Aasland and A. F. Stewart, *Nucleic Acids Res.*, 1995, **23**, 3168.
94. N. P. Cowieson, J. F. Partridge, R. C. Allshire and P. J. McLaughlin, *Curr. Biol.*, 2000, **10**, 517.
95. R. R. Meehan, C. F. Kao and S. Pennings, *EMBO J.*, 2003, **22**, 3164.
96. G. Schotta, A. Ebert, V. Krauss, A. Fischer, J. Hoffmann, S. Rea, T. Jenuwein, R. Dorn and G. Reuter, *EMBO J.*, 2002, **21**, 1121.
97. Q. Ye and H. J. Worman, *J. Biol. Chem.*, 1996, **271**, 14653.
98. A. Barski, S. Cuddapah, K. Cui, T. Y. Roh, D. E. Schones, Z. Wang, G. Wei, I. Chepelev and K. Zhao, *Cell*, 2007, **129**, 823.
99. T. S. Mikkelsen, M. Ku, D. B. Jaffe, B. Issac, E. Lieberman, G. Giannoukos, P. Alvarez, W. Brockman, T. K. Kim, R. P. Koche, W. Lee, E. Mendenhall, A. O'Donovan, A. Presser, C. Russ, X. Xie, A. Meissner, M. Wernig, R. Jaenisch, C. Nusbaum, E. S. Lander and B. E. Bernstein, *Nature*, 2007, **448**, 553.
100. C. R. Vakoc, S. A. Mandat, B. A. Olenchock and G. A. Blobel, *Mol. Cell*, 2005, **19**, 381.
101. J. A. Simon and R. E. Kingston, *Nat. Rev. Mol. Cell Biol.*, 2009, **10**, 697.
102. N. J. Francis, R. E. Kingston and C. L. Woodcock, *Science*, 2004, **306**, 1574.
103. M. de Napoles, J. E. Mermoud, R. Wakao, Y. A. Tang, M. Endoh, R. Appanah, T. B. Nesterova, J. Silva, A. P. Otte, M. Vidal, H. Koseki and N. Brockdorff, *Dev. Cell*, 2004, **7**, 663.
104. H. Wang, L. Wang, H. Erdjument-Bromage, M. Vidal, P. Tempst, R. S. Jones and Y. Zhang, *Nature*, 2004, **431**, 873.

105. R. Eskeland, M. Leeb, G. R. Grimes, C. Kress, S. Boyle, D. Sproul, N. Gilbert, Y. Fan, A. I. Skoultchi, A. Wutz and W. A. Bickmore, *Mol. Cell*, 2010, **38**, 452.
106. M. Endoh, T. A. Endo, T. Endoh, K. Isono, J. Sharif, O. Ohara, T. Toyoda, T. Ito, R. Eskeland, W. A. Bickmore, M. Vidal, B. E. Bernstein and H. Koseki, *PLoS Genet.*, 2012, **8**, e1002774.
107. X. Ren and T. K. Kerppola, *Mol. Cell. Biol.*, 2011, **31**, 2100.
108. N. Dietrich, M. Lerdrup, E. Landt, S. Agrawal-Singh, M. Bak, N. Tommerup, J. Rappsilber, E. Sodersten and K. Hansen, *PLoS Genet.*, 2012, **8**, e1002494.
109. M. Yu, T. Mazor, H. Huang, H. T. Huang, K. L. Kathrein, A. J. Woo, C. R. Chouinard, A. Labadorf, T. E. Akie, T. B. Moran, H. Xie, S. Zacharek, I. Taniuchi, R. G. Roeder, C. F. Kim, L. I. Zon, E. Fraenkel and A. B. Cantor, *Mol. Cell*, 2012, **45**, 330.
110. A. M. Farcas, N. P. Blackledge, I. Sudbery, H. K. Long, J. F. McGouran, N. R. Rose, S. Lee, D. Sims, A. Cerase, T. W. Sheahan, H. Koseki, N. Brockdorff, C. P. Ponting, B. M. Kessler and R. J. Klose, *Elife*, 2012, **1**, e00205.
111. J. A. North, S. Javaid, M. B. Ferdinand, N. Chatterjee, J. W. Picking, M. Shoffner, R. J. Nakkula, B. Bartholomew, J. J. Ottesen, R. Fishel and M. G. Poirier, *Nucleic Acids Res.*, 2011, **39**, 6465.
112. S. H. Baek, *Mol. Cell*, 2011, **42**, 274.
113. J. R. Paulson, J. S. Patzlaff and A. J. Vallis, *J. Cell Sci.*, 1996, **109**(6), 1437.
114. M. Shimada, M. Haruta, H. Niida, K. Sawamoto and M. Nakanishi, *EMBO Rep.*, 2010, **11**, 883.
115. D. Chowdhury, M. C. Keogh, H. Ishii, C. L. Peterson, S. Buratowski and J. Lieberman, *Mol. Cell*, 2005, **20**, 801.
116. S. Nakada, G. I. Chen, A. C. Gingras and D. Durocher, *EMBO Rep.*, 2008, **9**, 1019.
117. E. P. Rogakou, D. R. Pilch, A. H. Orr, V. S. Ivanova and W. M. Bonner, *J. Biol. Chem.*, 1998, **273**, 5858.
118. G. S. Stewart, B. Wang, C. R. Bignell, A. M. Taylor and S. J. Elledge, *Nature*, 2003, **421**, 961.
119. I. M. Ward, K. Minn, K. G. Jorda and J. Chen, *J. Biol. Chem.*, 2003, **278**, 19579.
120. Z. Shao, F. Li, S. M. Sy, W. Yan, Z. Zhang, D. Gong, B. Wen, M. S. Huen, Q. Gong, J. Wu and Y. Shi, *J. Struct. Biol.*, 2012, **177**, 459.
121. T. T. Paull, E. P. Rogakou, V. Yamazaki, C. U. Kirchgessner, M. Gellert and W. M. Bonner, *Curr. Biol.*, 2000, **10**, 886.
122. A. Xiao, H. Li, D. Shechter, S. H. Ahn, L. A. Fabrizio, H. Erdjument-Bromage, S. Ishibe-Murakami, B. Wang, P. Tempst, K. Hofmann, D. J. Patel, S. J. Elledge and C. D. Allis, *Nature*, 2009, **457**, 57.
123. T. Banerjee and D. Chakravarti, *Mol. Cell. Biol.*, 2011, **31**, 4858.
124. A. Kirmizis, H. Santos-Rosa, C. J. Penkett, M. A. Singer, M. Vermeulen, M. Mann, J. Bahler, R. D. Green and T. Kouzarides, *Nature*, 2007, **449**, 928.
125. E. Guccione, C. Bassi, F. Casadio, F. Martinato, M. Cesaroni, H. Schuchlautz, B. Luscher and B. Amati, *Nature*, 2007, **449**, 933.

126. D. Hyllus, C. Stein, K. Schnabel, E. Schiltz, A. Imhof, Y. Dou, J. Hsieh and U. M. Bauer, *Genes Dev.*, 2007, **21**, 3369.
127. C. C. Yuan, A. G. Matthews, Y. Jin, C. F. Chen, B. A. Chapman, T. K. Ohsumi, K. C. Glass, T. G. Kutateladze, M. L. Borowsky, K. Struhl and M. A. Oettinger, *Cell. Rep.*, 2012, **1**, 83.
128. C. J. Nelson, H. Santos-Rosa and T. Kouzarides, *Cell*, 2006, **126**, 905.
129. P. Cheung, K. G. Tanner, W. L. Cheung, P. Sassone-Corsi, J. M. Denu and C. D. Allis, *Mol. Cell*, 2000, **5**, 905.
130. W. S. Lo, R. C. Trievel, J. R. Rojas, L. Duggan, J. Y. Hsu, C. D. Allis, R. Marmorstein and S. L. Berger, *Mol. Cell*, 2000, **5**, 917.
131. S. Daujat, U. M. Bauer, V. Shah, B. M. Turner, S. Berger and T. Kouzarides, *Curr. Biol.*, 2002, **12**, 2090.
132. J. Dover, J. Schneider, M. A. Tawiah-Boateng, A. Wood, K. Dean, M. Johnston and A. Shilatifard, *J. Biol. Chem.*, 2002, **277**, 28368.
133. S. D. Briggs, T. Xiao, Z. W. Sun, J. A. Caldwell, J. Shabanowitz, D. F. Hunt, C. D. Allis and B. D. Strahl, *Nature*, 2002, **418**, 498.
134. Z. W. Sun and C. D. Allis, *Nature*, 2002, **418**, 104.
135. J. Kim, M. Guermah, R. K. McGinty, J. S. Lee, Z. Tang, T. A. Milne, A. Shilatifard, T. W. Muir and R. G. Roeder, *Cell*, 2009, **137**, 459.
136. A. J. Ruthenburg, H. Li, D. J. Patel and C. D. Allis, *Nat. Rev. Mol. Cell Biol.*, 2007, **8**, 983.
137. W. Fischle, B. S. Tseng, H. L. Dormann, B. M. Ueberheide, B. A. Garcia, J. Shabanowitz, D. F. Hunt, H. Funabiki and C. D. Allis, *Nature*, 2005, **438**, 1116.
138. T. Hirota, J. J. Lipp, B. H. Toh and J. M. Peters, *Nature*, 2005, **438**, 1176.
139. M. J. Hendzel, Y. Wei, M. A. Mancini, A. Van Hooser, T. Ranalli, B. R. Brinkley, D. P. Bazett-Jones and C. D. Allis, *Chromosoma*, 1997, **106**, 348.
140. A. Van Hooser, D. W. Goodrich, C. D. Allis, B. R. Brinkley and M. A. Mancini, *J. Cell Sci.*, 1998, **111**(23), 3497.
141. H. Goto, Y. Tomono, K. Ajiro, H. Kosako, M. Fujita, M. Sakurai, K. Okawa, A. Iwamatsu, T. Okigaki, T. Takahashi and M. Inagaki, *J. Biol. Chem.*, 1999, **274**, 25543.
142. H. Goto, Y. Yasui, E. A. Nigg and M. Inagaki, *Genes Cells*, 2002, **7**, 11.
143. B. Mateescu, P. England, F. Halgand, M. Yaniv and C. Muchardt, *EMBO Rep.*, 2004, **5**, 490.
144. G. V. Papamokos, G. Tziatzos, D. G. Papageorgiou, S. D. Georgatos, A. S. Politou and E. Kaxiras, *Biophys. J.*, 2012, **102**, 1926.
145. W. Fischle, Y. Wang and C. D. Allis, *Nature*, 2003, **425**, 475.
146. R. A. Varier, N. S. Outchkourov, P. de Graaf, F. M. van Schaik, H. J. Ensing, F. Wang, J. M. Higgins, G. J. Kops and H. T. Timmers, *EMBO J.*, 2010, **29**, 3967.
147. S. Winter, E. Simboeck, W. Fischle, G. Zupkovitz, I. Dohnal, K. Mechtler, G. Ammerer and C. Seiser, *EMBO J.*, 2008, **27**, 88.
148. S. S. Gehani, S. Agrawal-Singh, N. Dietrich, N. S. Christophersen, K. Helin and K. Hansen, *Mol. Cell*, 2010, **39**, 886.

149. E. Metzger, A. Imhof, D. Patel, P. Kahl, K. Hoffmeyer, N. Friedrichs, J. M. Muller, H. Greschik, J. Kirfel, S. Ji, N. Kunowska, C. Beisenherz-Huss, T. Gunther, R. Buettner and R. Schule, *Nature*, 2010, **464**, 792.
150. S. S. Ng, K. L. Kavanagh, M. A. McDonough, D. Butler, E. S. Pilka, B. M. Lienard, J. E. Bray, P. Savitsky, O. Gileadi, F. von Delft, N. R. Rose, J. Offer, J. C. Scheinost, T. Borowski, M. Sundstrom, C. J. Schofield and U. Oppermann, *Nature*, 2007, **448**, 87.
151. E. Metzger, N. Yin, M. Wissmann, N. Kunowska, K. Fischer, N. Friedrichs, D. Patnaik, J. M. Higgins, N. Potier, K. H. Scheidtmann, R. Buettner and R. Schule, *Nat. Cell Biol.*, 2008, **10**, 53.
152. R. D. Kelly and S. M. Cowley, *Biochem. Soc. Trans.*, 2013, **41**, 741.
153. C. J. Millard, P. J. Watson, I. Celardo, Y. Gordiyenko, S. M. Cowley, C. V. Robinson, L. Fairall and J. W. Schwabe, *Mol. Cell*, 2013, **51**, 57.
154. M. Bantscheff, C. Hopf, M. M. Savitski, A. Dittmann, P. Grandi, A. M. Michon, J. Schlegl, Y. Abraham, I. Becher, G. Bergamini, M. Boesche, M. Delling, B. Dumpelfeld, D. Eberhard, C. Huthmacher, T. Mathieson, D. Poeckel, V. Reader, K. Strunk, G. Sweetman, U. Kruse, G. Neubauer, N. G. Ramsden and G. Drewes, *Nat. Biotechnol.*, 2011, **29**, 255.

CHAPTER 3

Epigenetics and Disease

ADELE MURRELL*[a]

[a]Centre for Regenerative Medicine, Department of Biology and Biochemistry, University of Bath, Claverton Down, Bath BA2 7AY, UK
*E-mail: amm95@bath.ac.uk

3.1 Introduction

Only a small proportion of the genome consists of DNA coding for genes. The remainder of the genome consists of non-coding DNA, some of which contains regulatory sequences that direct functional properties of gene transcription (promoters, enhancers), DNA replication (origins of replication) and the organisational structure of chromosomes (satellite repeats, centromeres, telomeres). The genome is packed into chromatin that is organised and structured in three-dimensional space. Non-coding DNA and chromatin structure make up the major components of the epigenome. While mutations within single genes leading to Mendelian inheritance of disease are well understood, complex diseases such as cancer, cardiovascular and neurological diseases could be due to mutations in pleiotropic transcription modifiers and the consequential deregulation of several genes. In these diseases, even normal polymorphic variations within genes encoding for transcription factors, components or regulators of chromatin structure can influence the severity of the phenotype. In addition to genetic mutations, epigenetic mechanisms that can further drive and/or propagate aberrant gene expression patterns underpin the aetiology of complex diseases. Because the epigenome is sensitive to environmental factors, it is proposed that lifestyle influences the

RSC Drug Discovery Series No. 48
Epigenetics for Drug Discovery
Edited by Nessa Carey

Published by the Royal Society of Chemistry, www.rsc.org

penetrance and severity of adult-onset and acquired complex diseases, with genetics being a predisposing factor.

The individual components that embody the epigenome act in concert to drive cellular gene expression programmes in response to developmental and environmental cues. The epigenome is also influenced by cellular metabolism, nutrition and oxidative stress. This responsive nature of the epigenome makes it difficult to distinguish cause and effect. Thus an epigenetic mark such as a gain of DNA methylation may reflect a historical event of gene silencing and not necessarily reflect a gene expression change driving pathological disease. Mutations that lead to Mendelian inherited syndromes have been reported in several of the genes that encode chromatin-modifying proteins, indicating that epigenetic factors can indeed drive disease. The study of such diseases (Mendelian epigenetic diseases (MEDs)) will lead to a better understanding of how individual components within the epigenome influence specific development programmes. It will also lead to the identification of direct and indirect target genes of epigenetic modifiers, which may have roles in more complex late-onset and acquired diseases.

Congenital non-Mendelian epigenetic diseases (NMEDs) include genomic imprinting disorders, where the aetiology involves aberrant DNA methylation (or other epigenetic lesion) rather than a genetic mutation. DNA methylation is reprogrammed (erased and reset) during germ cell and early embryonic development and therefore diseases due to aberrant DNA methylation are likely to be sporadic rather than familial. Indeed, apart from genomic imprinting, only isolated examples of transgenerational epigenetic inheritance have been reported. Even in imprinting disorders, DNA sequence mutations may affect the acquisition or maintenance of an epigenetic mark and thus the underlying cause could still be predominantly genetic rather than epigenetic. In contrast to congenital syndromes, acquired genetic mutations as well as epimutations (*e.g.* aberrant DNA methylation profiles) can drive late-onset diseases such as cancer. This review will examine the balance between genetic and epigenetic interactions, highlighting examples of congenital disease such as genomic imprinting disorders, mutations within genes encoding chromatin-modifying proteins, and acquired epimutations in disease. Where therapeutic intervention strategies have been identified, these will be highlighted and further considered.

3.2 Congenital Mendelian Inherited Epigenetic Diseases

Mendelian inheritance of mutations within genes that encode chromatin or DNA methylation modifiers comprises a broad (and growing) category of congenital epigenetic diseases. An excellent and extensive review of these genetic disorders has recently been published by Berdasco and Esteller.[1] These authors have comprehensively catalogued and tabulated Mendelian disorders associated with all known mutations in genes encoding DNA methyltransferases and chromatin modifiers and therefore only a few diseases will be discussed here to exemplify selected epigenetic aetiologies such as DNA methylation, histone modifications, nucleosomal remodelling and chromatin topography (Figure 3.1).

Epigenetic modifiers often have tissue restricted expression and regularly function as part of a transcription factor and/or DNA replication complex. Additionally in many cases a particular chromatin modifier may be ubiquitous but its interacting factors may be tissue-specific. While most MEDs show a degree of phenotypic variation between individuals bearing similar mutations within a gene, each of the MEDs has distinctive characteristic phenotypes. Phenotypic variation is expected when multiple genes are regulated by one mutated gene, but the distinctive phenotypic outcome is indicative of the importance of epigenetic regulation of embryonic and foetal developmental gene expression programmes. The field is now amassing data from genome-wide studies that will identify the full complement of aberrantly expressed genes associated with altered epigenetic profiles in these congenital MEDs. Several MEDs are neurodegenerative disorders, which could suggest that an accumulation of epigenetic lesions underlie the pathophysiology. An alternative explanation for the late onset of these diseases is that different isoforms or paralogous family members of these proteins may initially functionally compensate during early development.

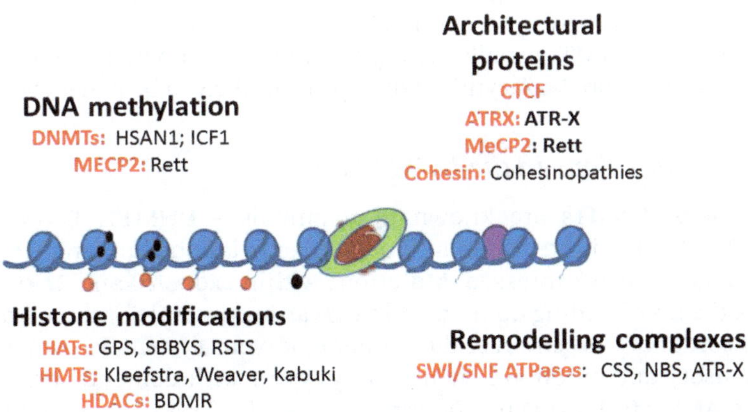

Figure 3.1 Mendelian routes to epigenetic disease. Mutations in genes that encode proteins with DNA methylation functions (DNA methyltransferases, DNMTs or DNA methylation readers such as MECP2), or chromatin modifying function such as histone modifications, including histone acetylases (HATs), histone deacetylases (HDACs) and histone methyltransferases (HTMs), or proteins that either shape chromatin conformation (architectural proteins) or remodel nucleosomes (remodelling complexes). Note overlap between epigenetic functions, for instance MECP2, which leads to Rett syndrome, is a DNA methylation reader and also contributes to chromatin architecture.

3.2.1 Mendelian Inherited Congenital Syndromes Involving Mutations in Genes that Encode DNA Methylation Readers and Modifiers

DNA methylation readers include proteins that bind to methylated DNA sequences, such as transcription factors with methyl-binding domains (including *MBD1*, *MBD2*, *MBD3* and *MeCP2*), while the modifiers include DNA methyltransferases (DNMTs) that have the ability to place DNA methylation marks. Although these proteins all have DNA-binding domains, consensus DNA-binding motifs have not been identified. Indeed, this seems to be the case with most chromatin modifiers, suggesting that their recruitment to DNA sequences is reliant on additional factors. Removal of DNA methylation is believed to be either through passive lack of maintenance of the methylation mark during cell division, or *via* an active demethylation process. Current knowledge points to demethylation involving DNA repair pathways and/or oxidisation by the ten-eleven translocation (TET) family of enzymes with 5-hydroxymethylcytosine being an intermediate step (reviewed in Williams *et al.*[2]).

DNMTs are targets for DNA methylation intervention therapies. 5-azacytidine-based therapies target the catalytic domain of DNMTs and prevent maintenance of DNA methylation in cell lines (see Chapter 4). Interestingly, the CpGs across the whole genome are not homogeneously susceptible to demethylation by 5-azacytidine. Future therapies directed at specific genomic regions will be possible once we understand more about the mechanisms that direct the *de novo* DNMTs to their target sites and the additional roles that DNMTs play during cell cycle progression and DNA replication and repair. Congenital diseases involving DNMTs usually display loss of DNMT function and hypomethylation and therefore 5-azacytidine therapy is unlikely to be beneficial.

3.2.1.1 Mutations in the DNMT Genes

Three major DNMTs are known in mammals – DNMT1, DNMT3A and DNMT3B. DNMT1 is predominantly a DNA methylation maintenance enzyme that is ubiquitously expressed. Mutations within exons 20 and 21 of *DNMT1* disrupt the DNA-binding domain and cause an autosomal dominant, degenerative disorder of the central and peripheral nervous system, known as hereditary sensory and autonomic neuropathy type 1 with dementia and hearing loss (HSAN1; MIM #614116). *In vitro* studies with DNMT1 exons 20 and 21 mutations show reduction of DNA methylation at heterochromatin and satellite repeats. However, there is also unexpected hypermethylation at specific loci. This disease is a typical example of how compromised function of a ubiquitous epigenetic maintenance enzyme results in a specific phenotype with tissue-specific effects. An unanswered question is whether the maintenance of DNA methylation in these patients is *via* the normal DNMT1 homologue or whether the *de novo* DNMTs can compensate for reduced DNMT1 activity.

Mutations in *DNMT3B*, a *de novo* DNA methyltransferase, account for 60% of the rare autosomal recessive, immunodeficiency centromeric instability facial syndrome 1 (ICF1, MIM #242860). This syndrome is characterised by chromatin decondensation and chromosome instability. Several chromosomal abnormalities with a particularly high incidence of somatic recombination between chromosomes 1, 9 and 16 are present. The peri-centromeric inactive heterochromatin shows polymorphic variation at these chromosomes. ICF1 patients have hypomethylation at these centromeres as well as at other non-coding repetitive sequences such as satellite repeats, Alu repeats and subtelomeric sequences. In addition to hypomethylation, ICF1 patients have altered expression of more than 700 genes, which may explain the immunodeficiency and characteristic developmental facial phenotype. Curiously, although *DNMT3B* is often aberrantly expressed in cancer, ICF1 patients do not have an increased susceptibility to cancer. This is even more surprising given their immunodeficiency, chromosomal instability and cell cycle defects.

3.2.1.2 Mutations within DNA Methylation Readers

The complexity of the interactions between different epigenetic components is highlighted by Rett syndrome (MIM #312750), a progressive X-linked disorder caused by mutations in methyl-CpG binding protein 2 gene (*MeCP2*). *MeCP2* is widely expressed in several tissues, with the highest expression levels in brain. It was initially identified as a DNA methylation binding protein that recruits histone deacetylases (HDACs) and Sin3 to repress gene expression.[3] However it also binds at unmethylated gene promoters associated with active genes.[4] Several MeCP2 target genes have been identified including *UBE23A*, *DLX5*, *BDNF* and *PRODH*. There is a significant overlap between symptoms of Rett syndrome and Angelman syndrome, suggesting that some of the MeCP2 target genes may lie within the imprinted Angelman/Prader–Willi domain.[5] Imprinting defects have not been found in Rett syndrome patients, which is probably not surprising if MeCP2 has a role as an epigenetic reader rather than as a modifier. These methylation-dependent reader functions seem to include contributions to chromatin conformation,[6] RNA splicing,[7] chromosome condensation[8] and silencing of repetitive elements.[9]

Mouse models for Rett syndrome have shown that the Rett phenotype can be partially rescued by restoration of MeCP2 expression.[10–12] These are very promising results for future therapeutic intervention, especially considering that, even when mice are treated at late time-points after developing Rett syndrome symptoms, their neurodegenerative phenotypes improve when MeCP2 expression is restored.

3.3 Mendelian Inherited Congenital Syndromes Involving Mutations in Genes that Encode Histone Modifiers

Post-translational histone modifications have combinatorial effects on gene expression. Thus active or repressive 'signatures' have been identified which predict whether the associated genes are expressed or silenced. Acetyl groups at lysine residues of histones are added by a large family of histone acetyl transferases (HATs) and removed by histone deacetylases (HDACs). Histone acetylation is considered an activating modification and can be targeted by HDAC inhibitors such as acetyl butyrate, trichostatin A and valproic acid, among others (see Chapter 5). Diseases due to mutations in HATs or HDACs show phenotypic features that overlap with several other congenital diseases involving post-translational modifiers of histones, which could be due to their involvement in similar transcription pathways. 'Crosstalk' has been reported between histone deacetylation and DNMTs, presumably because they are both epigenetic silencing mechanisms. However wider networks of epigenetic interactions can be affected when one component of the epigenome is disrupted, possibly because HDAC inhibitors can also target acetylation on non-histone proteins, including some transcription factors and chromatin modifiers.

HATs have been implicated in genitopatellar syndrome (GPS, MEM #606170) and Say–Barber–Biesecker–Young–Simpson syndrome (SBBYS, MEM #603736) caused by mutations in the *KAT6B* gene. *KAT6B* is a member of the MYST family of HAT proteins (reviewed in Berdasco and Esteller[1]). The genetic disruption of the CREB binding protein (CBP) and its homologue p300 has been implicated in 45–55% of Rubinstein–Taybi syndrome (RSTS, MIM #180849) patients (reviewed in Berdasco and Esteller[1]). RSTS is a candidate for therapeutic use of HDAC inhibitors. The rationale is that mutations in HAT enzymes slow down/reduce the amount of acetylation and therefore the prevention of deacetylation should maintain the steady states of acetylated histones required to keep genes active. In mouse models of RSTS, treatment with SAHA or TSA has been shown to ameliorate cognitive impairment.[13] However there is evidence that HDAC inhibitors effectively enhance synaptic plasticity and memory only if some CREB or CBP function remains. More work is still needed to identify the critical genes whose expression is regulated by histone acetylation during memory storage before RSTS can be treated with HDAC inhibitors.

Mutations in a histone deacetylase HDAC4 have been described in brachydactyly-mental retardation syndrome (BDMR, also known as Albright hereditary osteodystrophy, MIM #600430) patients.[14,15] Phenotypic symptoms of BDMR, such as skeletal and neuromuscular defects, are due to the role of HDAC4 in regulating genes important in bone, muscle, neurological and cardiac development.[14]

Histone methyltransferases such as EHMT1 specific for H3K9 methylation, EZH2 specific for H3K27 trimethylation, and MLL2 specific for H3K4 methylation are associated with Kleefstra (MIM #610253),[16] Weaver2 (WVS2, MIM #614421) and Kabuki-1 syndromes (MIM #147920), respectively (reviewed in Berdasco and Esteller[1]). Kleefstra syndrome is characterised by microcephaly with distinctive facial appearance, childhood hypotonia, urogenital defects, developmental delay and features of autism (MIM #610253).[16] The large number of target genes and the context-dependent interactions of histone methyltransferases with other epigenetic modifiers account for some of the phenotypic variation within Kleefstra syndrome, as well as for the overlap of clinical features with other MEC syndromes. For instance, features of Kleefstra syndrome are found with mutations in another histone methyltransferase MLL3 and in a nucleosomal remodelling protein SMARCB1.[17] Thus in modelling terms there is a 'many to many' relationship between the phenotypes and mutations in histone modifiers.

3.4 Mutations in Genes Encoding Nucleosomal Remodelling and Architectural Proteins

Compelling evidence indicates that alterations in chromatin remodelling proteins play an important role in the pathogenesis of human disease. Members of the SWI/SNF chromatin remodelling complex, such as the ATPases SMARCB1, SMARCA2 and SMARCA4 and their interacting subunits ARID1A and ARID1B, which control nucleosomal positioning and the accessibility of DNA to transcription factors, have been shown to be mutated in Coffin–Siris syndrome (CSS, MIM #135900).[18] Mutations in *SMARCA2* have been reported in patients with Nicolaides–Baraitser syndrome (NBS, MIM #601358).[19,20] Another example of the SWI/SNF remodelling complex involved in congenital disease is alpha-thalassaemia X-linked mental retardation, a neurodevelopmental syndrome with varying degrees of gonadal dysgenesis (ATR-X (MIM #301040)). The *ATRX* gene encodes a SWI2/SNF2-like ATPase protein at its C-terminal and a cysteine-rich N-terminal domain named ATRX-DMNT3-DNMT3L (ADD domain) that can interact with H3.[21,22] Changes in DNA methylation profiles have been reported in ATR-X syndrome.[23] ATRX also interacts with MECP2.[24] In addition to ATR-X syndrome, mutations in *ATRX* also contribute to mental retardation-hypotonic facies (MIM #309580) and alpha-thalassaemia myelodysplasia (MIM #300448) syndromes. Although ATRX was initially described as a transcriptional regulator of β-globin gene expression, further epigenetic regulatory functions, including control of centromere and telomere stability, and chromosome cohesion, have been reported (reviewed in de la Fuente *et al.*[25]).

The genome is structurally organised into chromatin looping domains (also termed topologically associated domains). These domains are characterised by boundaries that comprise active transcription start sites as well as chromatin insulator binding sites for CTCF.[26] CTCF sites are present in clusters at approximately every 40–50 kb along the genome and at most CTCF sites, cohesin has been shown to co-localise.[27–29] Cohesin also accumulates at large transcription factor complexes such as mediator at promoters and enhancers.[30,31] *In vitro* experiments using RNAi-mediated depletion of cohesin have indicated that cohesin has a function in stabilising chromatin loops.[32–34] Cohesin also has an indirect role in regulating transcription by determining the amount of elongating RNA polymerase on genes.[35]

In addition to CTCF and cohesin, chromatin loops have been shown to be orchestrated by MeCP2, ATRX, Polycomb complexes, long non-coding RNAs and several transcription factors. It is feasible that there is a hierarchical order of stable and dynamic looping structures mediated by several interacting factors which have multiple context-dependent roles in gene activation and silencing. The architectural proteins that shape the topography of the loops are sensitive to DNA methylation and we have recently shown that overall chromatin conformation structure can be restored at the imprinted *IGF2* locus by epigenetic intervention of DNA methylation using 5-azacytidine.[36]

Recently a small number of patients with mutations in *CTCF* have been identified.[37] These patients have congenital growth defects including microcephaly. Interestingly the *CTCF* mutations result in functional haplo-insufficiency, and transcriptome data from affected patients identified several gene expression changes. Preliminary examination of ChIA–PET interaction databases indicate that gene expression changes in these patients could feasibly be due to changes in chromatin topology.[37]

An overlapping spectrum of human syndromes can be attributed to mutations in cohesin subunits, or to regulators of cohesin loading and unloading from chromosomes. These are known as cohesinopathies and are characterised by a variety of developmental defects, including growth and mental retardation, limb deformities and craniofacial anomalies. The cohesinopathies are reviewed extensively elsewhere.[38] Cohesinopathies have not yet been shown to be caused by defects in gene regulation as a result of destabilised chromatin conformation.

The best known cohesinopathy is Cornelia de Lange syndrome (CdLS (OMIM 122470)), also termed Brachmann de Lange syndrome (BdLS), a broad spectrum disorder with multiple developmental and cognitive abnormalities (reviewed in Liu and Baynam[39]). All patients within the CdLS spectrum have neurodevelopmental delay and highly variable mental retardation. More than half of CdLS cases (~65%) are dominantly inherited, and caused by mutations in the NIPBL gene[40,41] (OMIM 608667). However, heterozygous missense mutations in genes for cohesin subunits *SMC1A* (OMIM 300040) and *SMC3* (OMIM 606062) also give rise to syndromes that fall within the CdLS spectrum, and account for about 5% of CdLS cases.[42–44] Although there is no direct evidence for aberrantly structured chromatin looping domains in CdLS patients, Nolen *et al.* have recently demonstrated, using fluorescence *in situ* hybridisation, that cells from CdLS patients with NIPBL mutations have noticeably compacted chromatin in gene-rich regions of the genome.[45]

The recessive cohesinopathy, Roberts/SC phocomelia syndrome RBS (OMIM 268300), caused by mutations in the *ESCO2* gene, is of interest from an epigenetic point of view. Most *ESCO2* mutations are truncating, but at least two mutations disrupt its acetyltransferase activity.[46,47] *ESCO2* is expressed in human embryonic tissues in a pattern that is consistent with the systems and organs affected in individuals with Roberts syndrome.[46] A unique and invariant feature of the chromosomes in cells from RBS patients is a characteristic puffing of heterochromatic chromatin regions.[48] ESCO2 may function in replication-coupled cohesion, and it has been shown that the processivity of DNA replication forks is reduced in cells from RBS patients.[49] Future studies still need to address how replication defects contribute to the disease aetiology in RBS patients, as well as whether chromatin structure and gene regulation downstream of *ESCO2* is responsible for RBS pathology.

3.5 Congenital NMEDs Exemplified by Syndromes Due to Loss of Genomic Imprinting

NMEDs can be thought of as diseases that have aberrant DNA methylation and which strictly speaking do not have an associated genetic mutation. Congenital NMEDs could arise if there were transgenerational epigenetic inheritance, and genomic imprinting disorders may be the best example of such diseases. Genomic imprinting is a phenomenon whereby a gene is gender-specifically marked in the embryonic germline and subsequently silenced in a parent-of-origin manner in the next generation. Several caveats apply to the inclusion of imprinting as transgenerational epigenetic inheritance. These include the observation that most imprinted diseases are *de novo* rather than familial, and usually when the disease is familial this is associated with a chromosomal abnormality. Even when it appears that the underlying aetiology of the disease is epigenetic it is not certain whether the localised aberrant DNA methylation is due to an undetected underlying genetic mutation within or near the locus. Despite intensive research efforts, a DNA sequence element that characterises and identifies an imprinted locus has not been identified to date. Comparative studies of allele-specific gene expression between imprinted and non-imprinted genes have revealed that the sequence context is important for allelic expression and acquisition of DNA methylation. Small changes in sequence can influence whether a gene is expressed or silenced and SNPs associated with random monoallelic expression have been identified at non-imprinted genes.[50] Interestingly the absence of SNP-associated changes in allelic expression at imprinted genes suggests that parent-of-origin epigenetic marks and the resultant inherent haploinsufficiency exert a strong selection against SNPs that may potentially silence an allele that has been marked for activation in the germline.

Congenital imprinted diseases have been identified at several imprinted gene clusters. Haploinsufficiency and gene dosage generally accounts for the phenotype. The earliest observations of imprinted disease were indications of structural chromosomal deletions that had varying phenotypes depending upon whether the mutation was of maternal or paternal origin such as Prader–Willi or Angelman syndrome (reviewed in Murrell[51]).

While mutations (deletions) within genes or regulatory regions in a cluster can be the underlying cause of the disease, epigenetic lesions such as aberrant DNA methylation at imprint control regions seem to account for Beckwith–Wiedemann syndrome (BWS, MIM #130650). The spectrum of clinical symptoms varies within BWS patients. Genotype–phenotype correlations have been made, which can inform clinical decisions and genetic counselling as well as predicting susceptibility to childhood cancer (reviewed in Demars and Gicquel,[52] Weksberg *et al.*[53] and Greer *et al.*[54]).

Most of the imprinted genes identified thus far function in signalling pathways that are involved in foetal growth and development. This is perhaps not surprising when the earliest screens for imprinted genes involved generating androgenic and gynogenic embryos which resulted in growth defects during early embryogenesis.[55] Nevertheless, loss of imprinting at several loci associated with increased cellular growth has been suggested as an early epigenetic event in cancer. Additionally several embryonal cancers have multiple genes with aberrant imprinting. An interesting correlation of germline tumour origin and imprinted status was described in a review by Oosterhuis and Looijenga in 2005.[56] These authors highlighted the similar imprinted states of several germline tumours and gonadal stem cells where imprints are erased and re-established.[56] Correlations such as these suggest that epigenetic profiles reflect the maturation state of their cells of origin, rather than loss of imprinting.

Aside from the function in cell growth and hence the obvious targets for therapeutic intervention, genomic imprinting research has increased our understanding of gene regulation. This has led to several paradigms of epigenetic mechanisms that are widely applicable to all loci. Of particular relevance is the role of DNA methylation at imprinting control regions (ICRs), which is usually acquired in the germline (reviewed in Demars and Gicquel[52]). Following fertilisation, imprinted regions resist genome-wide demethylation and *de novo* methylation. Thus DNA methylation at ICRs seems to be remarkably resistant to post-fertilisation epigenetic demethylation events, and such allele-specific methylation remains stable throughout development in all tissues, even where the imprinted genes are not expressed. *De novo* methylation at ICRs is mediated by DNMT3A and its cofactor DNMT3L. These methyltransferases are expressed in oocytes (reviewed in Abramowitz and Bartolomei[57]). In addition, the establishment of the imprint at the ICRs requires the H3K4me2 demethylase KDM1B (LSD2) (expressed in the growing oocyte), the KRAB zinc finger protein ZFP57 (expressed in oocytes, sperm and zygote), the C6ORF221 and caterpillar proteins NLP2 and NLP7 (expressed in oocytes and the zygote) (reviewed in Abramowitz and Bartolomei[57]). These proteins have been discovered in congenital imprinted diseases involving several imprinted loci (such as multi-locus hypomethylation syndrome[58] and familial biparental hydatidiform mole[59]). Proteins involved in the maintenance of imprinting include DNMT1, CTCF, NuRD complex proteins, MBD3, MTA2 and PGC7/STELLA (reviewed in Abramowitz and Bartolomei[57]).

The allele-specific methylation at ICRs pivotally regulates allele-specific expression of several genes within a cluster and also affects epigenetic modifications at a distance. The ICRs seem to regulate genomic imprinting by two mechanisms (which are not necessarily mutually exclusive). The first mechanism is that it may contain binding sites for CTCF and be involved in the topological organisation of the locus, such that each of the parental chromosomes has a differential looping domain.[34,36,60–62] However genetic defects at the ICR are uncommon and have only been identified in BWS patients with gain of methylation at the ICR. Mutations/small deletions of OCT4 and SOX2 binding sites within the ICR have been identified in BWS patients with gain of methylation at the ICR.[63,64] In these case reports, the BWS phenotype segregated with transmission of the mutation through the female germline with no phenotype under paternal transmission.[63,64] It is therefore likely that binding of pluripotency factors at the ICR prevents methylation of the maternal ICR. OCT4 and SOX2 binding sites have also been identified at other imprinted ICRs.[65] Loss of CTCF binding at the *IGF2-H19* ICR in mouse models led to increased methylation on the maternal allele (reviewed in Lewis and Murrell[66]). Further work is required to establish how pluripotency factors affect the establishment and maintenance of imprinting and how these influence CTCF function.

The second mechanism involves the ICRs being promoter regions for lncRNAs that can regulate several genes *in cis*, either by transcriptional interference or by acting to recruit chromatin modifiers (reviewed in Pauler *et al.*[67]). The most well-characterised ICR regulated by lncRNAs is the *Airn* transcript at the mouse *Igf2r* locus. *Airn* silences *Igf2r* by transcriptional overlap of the promoter.[68] It is also likely that similar transcriptional interference mechanisms, without the involvement of a mature RNA, may regulate imprinted gene expression at the *GNAS*, and *COPG2* loci (reviewed in Pauler *et al.*[67]). To date the evidence suggesting that imprinted genes are regulated by lncRNA recruiting chromatin factors includes reports of *Airn* recruiting histone methyltransferase G9a to the *Igf2r* locus;[69] *KCNQ1OT1* recruiting G9a and PRC2 to the *KCNQ1* locus.[70] In the last 5 years it has been shown that lncRNA is abundant within the genome and plays a prominent part in regulating gene expression and modifying chromatin (reviewed in Hu *et al.*[71]).

It is likely that imprinting is regulated by a combination of the above mechanisms, not just at the establishment stage but also at the maintenance stages. One useful feature of imprinted genes is that the DNA methylation patterns at the ICRs are very stable and therefore a useful marker for aberrant epigenetic response in acquired disease, as detailed below.

3.6 Complex Disease and Epigenetics

Multifactorial diseases, from cancer to rheumatoid arthritis, potentially start with a genetic component for susceptibility, followed by acquired genetic mutations and epigenetic changes during the lifetime of an individual. The contribution of epigenetic changes to multifactorial disease is not understood. In particular it is not understood whether epigenetic changes are the result of disease or essential drivers of disease. Driver events are defined as mutations/epimutations in critical pathways that lead to the development of any of the hallmarks of cancer that appear during early tumorigenesis.[72,73] It is likely that epigenetic changes may be a response in some instances and drivers in others. The challenge is to understand which of the observed epigenetic changes are indeed pathological and what the consequences are for gene expression.

Identifying epigenetic mutations that are present early on during tumorigenesis is not always practicable, and 'driver epigenetic mutations' have been inferred when similar mutations are present in congenital conditions that predispose to cancer. For example, somatic mutations in the histone methyltransferase *NSD1* have been identified in haematological malignancies, whereas germline mutations cause Sotos syndrome.[74] It is fairly well established that in sporadic colorectal cancer patients, driver epigenetic mutations such as DNA methylation of the *hMLH1* gene causes microsatellite instability. Genetic mutations of mismatch DNA repair genes are known to be an underlying cause of hereditary non-polyposis colorectal cancer (HNPCC). Unusually, in some familial cases of HNPCC, germline acquisition and even transgenerational acquisition of DNA methylation that silences *hMLH1* and *hMSH2* genes have been reported. Epigenetic mutations can also be considered drivers if they globally affect genome integrity by inhibiting DNA repair, or inhibit cellular events such as apoptosis and senescence. Thus epigenetic silencing of *BRCA1*, *WRN*, *FANCF* and *CHFR* are drivers in the sense that they contribute to genome instability. It is increasingly being found that epigenetic and genetic mutations cluster together, for example DNA methylation at the *MGMT* gene is often found together with *K-RAS* and *p53* mutations.

Genome-wide profiling of DNA methylation and histone modifications of cancer tissues compared to normal tissues are beginning to yield data that inform on the interrelatedness of epigenetic marks and have highlighted epigenetic features of cancer. Thus it has been known for a long time that DNA methylation is globally reduced in tumours, leading to overall reduction in heterochromatin and genome instability. DNA methylation reprogramming is accompanied by localised hypermethylation at gene-specific regions. This is presumed to cause silencing of tumour suppressor genes and to promote prolonged survival and replication of the cancer cell. More recent studies have refined these analyses and analysed DNA methylation relative to CpG density, and also within the CpG islands and regions flanking the CpG islands (shores). Large genomic regions susceptible to DNA methylation change have been identified. These have been labelled as hypermethylation and hypomethylation blocks. As yet it has not been shown whether these 'blocks' are subsequently more susceptible to DNA damage or to chromosome aberrations such as duplications or deletions.

For many years, attempts have been made to correlate global DNA methylation states in cancer with expression of the DNMTs. Varying reports have implicated expression changes of the *de novo* DNMT3a and DNMT3b with methylation changes in cancer. The rediscovery of 5-hydroxymethylcytosine (5-hmC), and the fact that it is produced by TET enzymes, have led to the formulation of new theories regarding the mechanisms whereby DNA can be actively demethylated. Unfortunately it is not as straightforward as these theories propose, mainly because the TET enzymes seem to be selective for the regions that they demethylate. In all tissues studied so far, 5-hmC has been shown to accumulate within gene bodies and gene promoters of active genes.[75–78] Indeed the regions that are usually hypomethylated in cancer, such as heterochromatin and satellite repeats, have no 5-hmC.[75–78] In several haematological malignancies, the TET genes are inactivated due to chromosomal translocations, and it has been proposed that hypermethylation in these cancers is due to the lack of TET enzyme activity.[79,80]

Further exciting discoveries in the field revealed that TET enzymes are inhibited by the oncometabolite D-2-hydroxyglutarate, which accumulates when the isocitrate dehydrogenase genes *IDH1/2* are mutated. Several cancers have mutations in *IDH1/2*, with a particularly high incidence in acute myeloid leukaemia (35%) and glioblastomas (80%) (reviewed in Xu *et al.*[81]). As expected these cancers have very low levels of 5-hmC. An increasing number of reports indicate that reduced levels of 5-hmC are a hallmark of cancer in general. In melanomas it has been observed that benign naevi have normal levels of 5-hmC while malignant tumours have low 5-hmC levels.[82] Most melanomas have mutations in *TET2* or *IDH1/2*. Cells isolated from these melanomas produce subcutaneous tumours in mice. However if the cells are treated with ectopic TET2 or IDH1/2, they are less efficient at producing tumours in the *in vivo* mouse models.[82] These results indicate that increasing TET enzyme activity is a likely future therapeutic avenue for cancer. However, it seems that some cancers, such as colorectal cancer, have reduced 5-hmC levels that cannot be explained by mutations in TETs or IDH1/2. Indeed when primary cells are placed into cell culture and start growing, 5-hmC production is stalled. It is likely that 5-hmC is reduced in rapidly dividing cells such as tumours both due to failure to maintain the mark and due to further oxidative demethylation steps. The TETs have roles as transcription factors and may also regulate gene expression independently of their demethylation function.[2] The TETs have been shown to influence differentiation of embryonic stem cells.[78,83–85] However during epithelial differentiation of the colon, we have found that there is no temporal change in TET expression, despite very low levels of 5-hmC in the stem cell compartment of the crypts and the transit amplifying cells, but high levels of 5-hmC in terminally differentiated epithelium.

DNA methylation and demethylation are still considered strong epigenetic intervention targets. DNA methyltransferase inhibitors are the longest-standing epigenetic drugs, identified as anti-proliferative agents before their epigenetic mechanism of action had been determined (see Chapter 4). The global changes of DNA methylation observed in cancer reflect the large scale of epigenetic reprogramming that occurs during the neoplastic process. Strong evidence that *de novo* methylation in cancer may be an instructive rather than a random process is the finding that gene promoters that are hypermethylated in a number of cancers are marked by bivalent chromatin signatures in embryonic stem cells.[86–88] These bivalent marks manifest as histone methylation at lysine 4 (H3K4me3), together with a silent mark of histone methylation at lysine 27 (H3K27me3).[89,90] Genes with bivalent chromatin at their promoters are PcG target genes that are expressed at low levels and are poised for active transcription upon lineage commitment.[89,90] More recently it has been shown that the subset of bivalent genes that become hypermethylated in cancer are developmental regulators and their silencing could be the reason that cancer cells have a stem-like state.[91]

3.7 Conclusions

Epigenetic factors probably contribute to all diseases to varying extents. It is becoming easier to identify DNA methylation and histone modifications in multiple disease states, but these are mostly the epigenetic marks that are left behind by previous pathological events. The occurrence of congenital disorders caused by Mendelian transmittance of mutations in genes that encode epigenetic modifiers provides strong evidence for epigenetic mechanisms being drivers and initiators of disease. A feature of many congenital epigenetic diseases is a susceptibility to cancer, which further suggests that epigenetic mechanisms also drive multifactorial diseases. All complex and multifactorial disease processes are potentially due to aberrant gene expression programmes caused by deregulation of transcription factor binding to DNA regulatory elements such as enhancers or promoters. Genetic and epigenetic barriers need to be overcome for malignant gene expression (Figure 3.2). Targeting epigenetic mechanisms has strong potential for therapeutic avenues. Particularly promising is the prospect that even developmentally-induced neurodegeneration might be reversed in the future by epigenetic drugs.

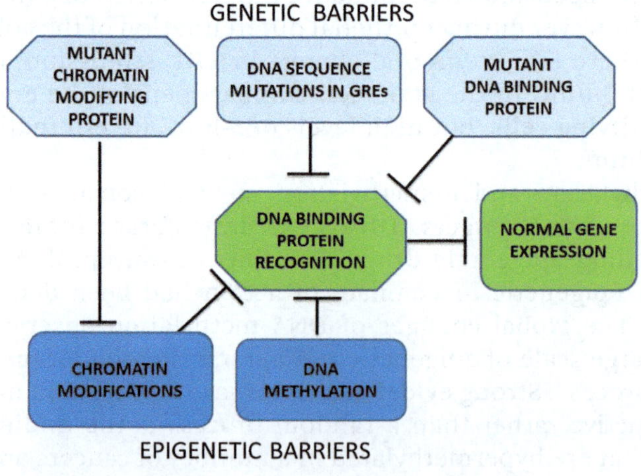

Figure 3.2 Genetic and epigenetic barriers prevent aberrant interactions between DNA and DNA binding proteins at sequence elements such as promoters and enhancers to enable normal gene expression. Complex diseases such as cancer may have an accumulation of both genetic and epigenetic mutations.

References

1. M. Berdasco and M. Esteller, Genetic syndromes caused by mutations in epigenetic genes, *Hum. Genet.*, 2013, **132**, 359–383.
2. K. Williams, J. Christensen and K. Helin, DNA methylation: TET proteins-guardians of CpG islands? *EMBO Rep.*, 2012, **13**, 28–35.
3. P. Jones, G. Veenstra, P. Wade, D. Vermaak, S. Kass, N. Landsberger, J. Strouboulis and A. Wolffe, Methylated DNA and MeCP2 recruit histone deacetylase to repress transcription, *Nat. Genet.*, 1998, **19**, 187–191.
4. D. Yasui, S. Peddada, M. Bieda, R. Vallero, A. Hogart, R. Nagarajan, K. Thatcher, P. Farnham and J. Lasalle, Integrated epigenomic analyses of neuronal MeCP2 reveal a role for long-range interaction with active genes, *Proc. Natl. Acad. Sci. U. S. A.*, 2007, **104**, 19416–19421.
5. R. Samaco, R. Nagarajan, D. Braunschweig and J. LaSalle, Multiple pathways regulate MeCP2 expression in normal brain development and exhibit defects in autism-spectrum disorders, *Hum. Mol. Genet.*, 2004, **13**, 629–639.
6. S.-i. Horike, S. Cai, M. Miyano, J.-F. Cheng and T. Kohwi-Shigematsu, Loss of silent-chromatin looping and impaired imprinting of DLX5 in Rett syndrome, *Nat. Genet.*, 2005, **37**, 31–40.
7. J. Young, E. Hong, J. Castle, J. Crespo-Barreto, A. Bowman, M. Rose, D. Kang, R. Richman, J. Johnson, S. Berget, *et al.*, Regulation of RNA splicing by the methylation-dependent transcriptional repressor methyl-CpG binding protein 2, *Proc. Natl. Acad. Sci. U. S. A.*, 2005, **102**, 17551–17558.
8. T. Ishibashi, A. Thambirajah and J. Ausió, MeCP2 preferentially binds to methylated linker DNA in the absence of the terminal tail of histone H3 and independently of histone acetylation, *FEBS Lett.*, 2008, **582**, 1157–1162.
9. A. Muotri, M. Marchetto, N. Coufal, R. Oefner, G. Yeo, K. Nakashima and F. Gage, L1 retrotransposition in neurons is modulated by MeCP2, *Nature*, 2010, **468**, 443–446.
10. E. Giacometti, S. Luikenhuis, C. Beard and R. Jaenisch, Partial rescue of MeCP2 deficiency by postnatal activation of MeCP2, *Proc. Natl. Acad. Sci. U. S. A.*, 2007, **104**, 1931–1936.
11. D. Tropea, E. Giacometti, N. Wilson, C. Beard, C. McCurry, D. Fu, R. Flannery, R. Jaenisch and M. Sur, Partial reversal of Rett Syndrome-like symptoms in MeCP2 mutant mice, *Proc. Natl. Acad. Sci. U. S. A.*, 2009, **106**, 2029–2034.
12. J. Guy, J. Gan, J. Selfridge, S. Cobb and A. Bird, Reversal of neurological defects in a mouse model of Rett syndrome, *Science*, 2007, **315**, 1143–1147.
13. T. Hallam and R. Bourtchouladze, Rubinstein-Taybi syndrome: molecular findings and therapeutic approaches to improve cognitive dysfunction. Cellular and molecular life sciences, *CMLS*, 2006, **63**, 1725–1735.
14. B. Morris, C. Etoubleau, S. Bourthoumieu, S. Reynaud-Perrine, C. Laroche, A. Lebbar, C. Yardin and S. Elsea, Dose dependent expression of

HDAC4 causes variable expressivity in a novel inherited case of brachydactyly mental retardation syndrome, *Am. J. Med. Genet., Part A*, 2012, **158A**, 2015–2020.

15. S. Williams, M. Aldred, V. Der Kaloustian, F. Halal, G. Gowans, D. McLeod, S. Zondag, H. Toriello, R. Magenis and S. Elsea, Haploinsufficiency of HDAC4 causes brachydactyly mental retardation syndrome, with brachydactyly type E, developmental delays, and behavioral problems, *Am. J. Hum. Genet.*, 2010, **87**, 219–228.
16. T. Kleefstra, W. van Zelst-Stams, W. Nillesen, V. Cormier-Daire, G. Houge, N. Foulds, M. van Dooren, M. Willemsen, R. Pfundt, A. Turner, *et al.*, Further clinical and molecular delineation of the 9q subtelomeric deletion syndrome supports a major contribution of EHMT1 haploinsufficiency to the core phenotype, *J. Med. Genet.*, 2009, **46**, 598–606.
17. T. Kleefstra, J. Kramer, K. Neveling, M. Willemsen, T. Koemans, L. Vissers, W. Wissink-Lindhout, M. Fenckova, W. van den Akker, N. Kasri, *et al.*, Disruption of an EHMT1-associated chromatin-modification module causes intellectual disability, *Am. J. Hum. Genet.*, 2012, **91**, 73–82.
18. Y. Tsurusaki, N. Okamoto, H. Ohashi, T. Kosho, Y. Imai, Y. Hibi-Ko, T. Kaname, K. Naritomi, H. Kawame, K. Wakui, *et al.*, Mutations affecting components of the SWI/SNF complex cause Coffin-Siris syndrome, *Nat. Genet.*, 2012, **44**, 376–378.
19. D. Wolff, S. Endele, S. Azzarello-Burri, J. Hoyer, M. Zweier, I. Schanze, B. Schmitt, A. Rauch, A. Reis and C. Zweier, In-Frame Deletion and Missense Mutations of the C-Terminal Helicase Domain of SMARCA2 in Three Patients with Nicolaides-Baraitser Syndrome, *Mol. Syndromol.*, 2012, **2**, 237–244.
20. J. K. Van Houdt, B. A. Nowakowska, S. B. Sousa, B. D. van Schaik, E. Seuntjens, N. Avonce, A. Sifrim, O. A. Abdul-Rahman, M. J. van den Boogaard, A. Bottani, *et al.*, Heterozygous missense mutations in SMARCA2 cause Nicolaides-Baraitser syndrome, *Nat. Genet.*, 2012, **44**, 445–449.
21. Y. Xue, R. Gibbons, Z. Yan, D. Yang, T. L. McDowell, S. Sechi, J. Qin, S. Zhou, D. Higgs and W. Wang, The ATRX syndrome protein forms a chromatin-remodeling complex with Daxx and localizes in promyelocytic leukemia nuclear bodies, *Proc. Natl. Acad. Sci. U. S. A.*, 2003, **100**, 10635–10640.
22. A. Argentaro, J. C. Yang, L. Chapman, M. S. Kowalczyk, R. J. Gibbons, D. R. Higgs, D. Neuhaus and D. Rhodes, Structural consequences of disease-causing mutations in the ATRX-DNMT3-DNMT3L (ADD) domain of the chromatin-associated protein ATRX, *Proc. Natl. Acad. Sci. U. S. A.*, 2007, **104**, 11939–11944.
23. R. J. Gibbons, T. L. McDowell, S. Raman, D. M. O'Rourke, D. Garrick, H. Ayyub and D. R. Higgs, Mutations in ATRX, encoding a SWI/SNF-like protein, cause diverse changes in the pattern of DNA methylation, *Nat. Genet.*, 2000, **24**, 368–371.
24. X. Nan, J. Hou, A. Maclean, J. Nasir, M. J. Lafuente, X. Shu, S. Kriaucionis and A. Bird, Interaction between chromatin proteins MECP2 and ATRX

is disrupted by mutations that cause inherited mental retardation, *Proc. Natl. Acad. Sci. U. S. A.*, 2007, **104**, 2709–2714.
25. R. De La Fuente, C. Baumann and M. M. Viveiros, Role of ATRX in chromatin structure and function: implications for chromosome instability and human disease, *Reproduction*, 2011, **142**, 221–234.
26. J. R. Dixon, S. Selvaraj, F. Yue, A. Kim, Y. Li, Y. Shen, M. Hu, J. S. Liu and B. Ren, Topological domains in mammalian genomes identified by analysis of chromatin interactions, *Nature*, 2012, **485**, 376–380.
27. V. Parelho, S. Hadjur, M. Spivakov, M. Leleu, S. Sauer, H. C. Gregson, A. Jarmuz, C. Canzonetta, Z. Webster, T. Nesterova, *et al.*, Cohesins functionally associate with CTCF on mammalian chromosome arms, *Cell*, 2008, **132**, 422–433.
28. E. D. Rubio, D. J. Reiss, P. L. Welcsh, C. M. Disteche, G. N. Filippova, N. S. Baliga, R. Aebersold, J. A. Ranish and A. Krumm, CTCF physically links cohesin to chromatin, *Proc. Natl. Acad. Sci. U. S. A.*, 2008, **105**, 8309–8314.
29. K. S. Wendt and J. M. Peters, How cohesin and CTCF cooperate in regulating gene expression, *Chromosome Res.*, 2009, **17**, 201–214.
30. M. H. Kagey, J. J. Newman, S. Bilodeau, Y. Zhan, D. A. Orlando, N. L. van Berkum, C. C. Ebmeier, J. Goossens, P. B. Rahl, S. S. Levine, *et al.*, Mediator and cohesin connect gene expression and chromatin architecture, *Nature*, 2010, **467**, 430–435.
31. D. Schmidt, P. C. Schwalie, C. S. Ross-Innes, A. Hurtado, G. D. Brown, J. S. Carroll, P. Flicek and D. T. Odom, A CTCF-independent role for cohesin in tissue-specific transcription, *Genome Res.*, 2010, **20**, 578–588.
32. R. Chien, W. Zeng, S. Kawauchi, M. A. Bender, R. Santos, H. C. Gregson, J. A. Schmiesing, D. A. Newkirk, X. Kong, A. R. Ball Jr, *et al.*, Cohesin mediates chromatin interactions that regulate mammalian beta-globin expression, *J. Biol. Chem.*, 2011, **286**, 17870–17878.
33. S. Hadjur, L. M. Williams, N. K. Ryan, B. S. Cobb, T. Sexton, P. Fraser, A. G. Fisher and M. Merkenschlager, Cohesins form chromosomal cis-interactions at the developmentally regulated IFNG locus, *Nature*, 2009, **460**, 410–413.
34. R. Nativio, K. S. Wendt, Y. Ito, J. E. Huddleston, S. Uribe-Lewis, K. Woodfine, C. Krueger, W. Reik, J. M. Peters and A. Murrell, Cohesin is required for higher-order chromatin conformation at the imprinted IGF2-H19 locus, *PLoS Genet.*, 2009, **5**, e1000739.
35. A. Fay, Z. Misulovin, J. Li, C. A. Schaaf, M. Gause, D. S. Gilmour and D. Dorsett, Cohesin selectively binds and regulates genes with paused RNA polymerase, *Curr. Biol.*, 2011, **21**, 1624–1634.
36. Y. Ito, R. Nativio and A. Murrell, Induced DNA demethylation can reshape chromatin topology at the IGF2-H19 locus, *Nucleic Acids Res.*, 2013, **41**, 5290–5302.
37. A. Gregor, M. Oti, E. N. Kouwenhoven, J. Hoyer, H. Sticht, A. B. Ekici, S. Kjaergaard, A. Rauch, H. G. Stunnenberg, S. Uebe, *et al.*, De novo mutations in the genome organizer CTCF cause intellectual disability, *Am. J. Hum. Genet.*, 2013, **93**, 124–131.

38. S. Remeseiro, A. Cuadrado and A. Losada, Cohesin in development and disease, *Development*, 2013, **140**, 3715–3718.
39. J. Liu and G. Baynam, Cornelia de Lange syndrome, *Adv. Exp. Med. Biol.*, 2010, **685**, 111–123.
40. I. D. Krantz, J. McCallum, C. DeScipio, M. Kaur, L. A. Gillis, D. Yaeger, L. Jukofsky, N. Wasserman, A. Bottani, C. A. Morris, *et al.*, Cornelia de Lange syndrome is caused by mutations in NIPBL, the human homolog of Drosophila melanogaster Nipped-B, *Nat. Genet.*, 2004, **36**, 631–635.
41. E. T. Tonkin, T. J. Wang, S. Lisgo, M. J. Bamshad and T. Strachan, NIPBL, encoding a homolog of fungal Scc2-type sister chromatid cohesion proteins and fly Nipped-B, is mutated in Cornelia de Lange syndrome, *Nat. Genet.*, 2004, **36**, 636–641.
42. M. A. Deardorff, M. Kaur, D. Yaeger, A. Rampuria, S. Korolev, J. Pie, C. Gil-Rodriguez, M. Arnedo, B. Loeys, A. D. Kline, *et al.*, Mutations in cohesin complex members SMC3 and SMC1A cause a mild variant of cornelia de Lange syndrome with predominant mental retardation, *Am. J. Hum. Genet.*, 2007, **80**, 485–494.
43. L. Mannini, J. Liu, I. D. Krantz and A. Musio, Spectrum and consequences of SMC1A mutations: the unexpected involvement of a core component of cohesin in human disease, *Hum. Mutat.*, 2010, **31**, 5–10.
44. A. Musio and I. D. Krantz, Cohesin biology and the cohesinopathies: Abstracts from the Second Biennial Conference, Pontignano, Italy, 2009, *Am. J. Med. Genet., Part A*, 2010, **152A**, 1630–1640.
45. L. D. Nolen, S. Boyle, M. Ansari, E. Pritchard and W. A. Bickmore, Regional chromatin decompaction in Cornelia de Lange syndrome associated with NIPBL disruption can be uncoupled from cohesin and CTCF, *Hum. Mol. Genet.*, 2013, **22**, 4180–4193.
46. H. Vega, A. H. Trainer, M. Gordillo, M. Crosier, H. Kayserili, F. Skovby, M. L. Uzielli, R. E. Schnur, S. Manouvrier, E. Blair, *et al.*, Phenotypic variability in 49 cases of ESCO2 mutations, including novel missense and codon deletion in the acetyltransferase domain, correlates with ESCO2 expression and establishes the clinical criteria for Roberts syndrome, *J. Med. Genet.*, 2010, **47**, 30–37.
47. M. Gordillo, H. Vega, A. H. Trainer, F. Hou, N. Sakai, R. Luque, H. Kayserili, S. Basaran, F. Skovby, R. C. Hennekam, *et al.*, The molecular mechanism underlying Roberts syndrome involves loss of ESCO2 acetyltransferase activity, *Hum. Mol. Genet.*, 2008, **17**, 2172–2180.
48. D. Tomkins, A. Hunter and M. Roberts, Cytogenetic findings in Roberts-SC phocomelia syndrome(s), *Am. J. Med. Genet.*, 1979, **4**, 17–26.
49. P. van der Lelij, B. C. Godthelp, W. van Zon, D. van Gosliga, A. B. Oostra, J. Steltenpool, J. de Groot, R. J. Scheper, R. M. Wolthuis, Q. Waisfisz, *et al.*, The cellular phenotype of Roberts syndrome fibroblasts as revealed by ectopic expression of ESCO2, *PLoS One*, 2009, **4**, e6936.
50. A. Chess, Mechanisms and consequences of widespread random monoallelic expression, *Nat. Rev. Genet.*, 2012, **13**, 421–428.

51. A. Murrell, Genomic imprinting and cancer: from primordial germ cells to somatic cells, *Sci. World J.*, 2006, **6**, 1888–1910.
52. J. Demars and C. Gicquel, Epigenetic and genetic disturbance of the imprinted 11p15 region in Beckwith-Wiedemann and Silver-Russell syndromes, *Clin. Genet.*, 2012, **81**, 350–361.
53. R. Weksberg, C. Shuman and J. B. Beckwith, Beckwith-Wiedemann syndrome. European journal of human genetics, *EJHG*, 2010, **18**, 8–14.
54. K. J. Greer, S. J. Kirkpatrick, R. Weksberg and R. M. Pauli, Beckwith-Wiedemann syndrome in adults: observations from one family and recommendations for care, *Am. J. Med. Genet. A*, 2008, **146A**, 1707–1712.
55. M. A. Surani, S. C. Barton and M. L. Norris, Development of reconstituted mouse eggs suggests imprinting of the genome during gametogenesis, *Nature*, 1984, **308**, 548–550.
56. J. W. Oosterhuis and L. H. Looijenga, Testicular germ-cell tumours in a broader perspective, *Nat. Rev. Cancer*, 2005, **5**, 210–222.
57. L. K. Abramowitz and M. S. Bartolomei, Genomic imprinting: recognition and marking of imprinted loci, *Curr. Opin. Genet. Dev.*, 2012, **22**, 72–78.
58. D. J. Mackay, J. L. Callaway, S. M. Marks, H. E. White, C. L. Acerini, S. E. Boonen, P. Dayanikli, H. V. Firth, J. A. Goodship, A. P. Haemers, *et al.*, Hypomethylation of multiple imprinted loci in individuals with transient neonatal diabetes is associated with mutations in ZFP57, *Nat. Genet.*, 2008, **40**, 949–951.
59. C. M. Wang, P. H. Dixon, S. Decordova, M. D. Hodges, N. J. Sebire, S. Ozalp, M. Fallahian, A. Sensi, F. Ashrafi, V. Repiska, *et al.*, Identification of 13 novel NLRP7 mutations in 20 families with recurrent hydatidiform mole; missense mutations cluster in the leucine-rich region, *J. Med. Genet.*, 2009, **46**, 569–575.
60. S. Kurukuti, V. K. Tiwari, G. Tavoosidana, E. Pugacheva, A. Murrell, Z. Zhao, V. Lobanenkov, W. Reik and R. Ohlsson, CTCF binding at the H19 imprinting control region mediates maternally inherited higher-order chromatin conformation to restrict enhancer access to Igf2, *Proc. Natl. Acad. Sci. U. S. A.*, 2006, **103**, 10684–10689.
61. A. Murrell, S. Heeson and W. Reik, Interaction between differentially methylated regions partitions the imprinted genes Igf2 and H19 into parent-specific chromatin loops, *Nat. Genet.*, 2004, **36**, 889–893.
62. R. Nativio, A. Sparago, Y. Ito, R. Weksberg, A. Riccio and A. Murrell, Disruption of genomic neighbourhood at the imprinted IGF2-H19 locus in Beckwith-Wiedemann syndrome and Silver-Russell syndrome, *Hum. Mol. Genet.*, 2011, **20**, 1363–1374.
63. S. Berland, M. Appelback, O. Bruland, J. Beygo, K. Buiting, D. J. Mackay, I. Karen Temple and G. Houge, Evidence for anticipation in Beckwith-Wiedemann syndrome. European journal of human genetics, *EJHG*, 2013, **21**, 1344–1348.
64. N. Hori, M. Yamane, K. Kouno and K. Sato, Induction of DNA demethylation depending on two sets of Sox2 and adjacent Oct3/4 binding sites

(Sox-Oct motifs) within the mouse H19/insulin-like growth factor 2 (Igf2) imprinted control region, *J. Biol. Chem.*, 2012, **287**, 44006–44016.
65. Y. Kaufman, M. Heled, J. Perk, A. Razin and R. Shemer, Protein-binding elements establish in the oocyte the primary imprint of the Prader-Willi/Angelman syndromes domain, *Proc. Natl. Acad. Sci. U. S. A.*, 2009, **106**, 10242–10247.
66. A. Lewis and A. Murrell, Genomic imprinting: CTCF protects the boundaries, *Curr. Biol.*, 2004, **14**, R284–R286.
67. F. M. Pauler, D. P. Barlow and Q. J. Hudson, Mechanisms of long range silencing by imprinted macro non-coding RNAs, *Curr. Opin. Genet. Dev.*, 2012, **22**, 283–289.
68. P. A. Latos, F. M. Pauler, M. V. Koerner, H. B. Senergin, Q. J. Hudson, R. R. Stocsits, W. Allhoff, S. H. Stricker, R. M. Klement, K. E. Warczok, *et al.*, Airn transcriptional overlap, but not its lncRNA products, induces imprinted Igf2r silencing, *Science*, 2012, **338**, 1469–1472.
69. T. Nagano, J. A. Mitchell, L. A. Sanz, F. M. Pauler, A. C. Ferguson-Smith, R. Feil and P. Fraser, The Air noncoding RNA epigenetically silences transcription by targeting G9a to chromatin, *Science*, 2008, **322**, 1717–1720.
70. R. R. Pandey, T. Mondal, F. Mohammad, S. Enroth, L. Redrup, J. Komorowski, T. Nagano, D. Mancini-Dinardo and C. Kanduri, Kcnq1ot1 antisense noncoding RNA mediates lineage-specific transcriptional silencing through chromatin-level regulation, *Mol. Cell*, 2008, **32**, 232–246.
71. W. Hu, J. R. Alvarez-Dominguez and H. F. Lodish, Regulation of mammalian cell differentiation by long non-coding RNAs, *EMBO Rep.*, 2012, **13**, 971–983.
72. S. Kalari and G. P. Pfeifer, Identification of driver and passenger DNA methylation in cancer by epigenomic analysis, *Adv. Genet.*, 2010, **70**, 277–308.
73. D. Hanahan and R. A. Weinberg, Hallmarks of cancer: the next generation, *Cell*, 2011, **144**, 646–674.
74. N. Kurotaki, K. Imaizumi, N. Harada, M. Masuno, T. Kondoh, T. Nagai, H. Ohashi, K. Naritomi, M. Tsukahara, Y. Makita, *et al.*, Haploinsufficiency of NSD1 causes Sotos syndrome, *Nat. Genet.*, 2002, **30**, 365–366.
75. W. A. Pastor, U. J. Pape, Y. Huang, H. R. Henderson, R. Lister, M. Ko, E. M. McLoughlin, Y. Brudno, S. Mahapatra, P. Kapranov, *et al.*, Genome-wide mapping of 5-hydroxymethylcytosine in embryonic stem cells, *Nature*, 2011, **473**, 394–397.
76. M. J. Booth, M. R. Branco, G. Ficz, D. Oxley, F. Krueger, W. Reik and S. Balasubramanian, Quantitative sequencing of 5-methylcytosine and 5-hydroxymethylcytosine at single-base resolution, *Science*, 2012, **336**, 934–937.
77. M. R. Branco, G. Ficz and W. Reik, Uncovering the role of 5-hydroxymethylcytosine in the epigenome, *Nat. Rev. Genet.*, 2012, **13**, 7–13.
78. G. Ficz, M. R. Branco, S. Seisenberger, F. Santos, F. Krueger, T. A. Hore, C. J. Marques, S. Andrews and W. Reik, Dynamic regulation of 5-hydroxymethylcytosine in mouse ES cells and during differentiation, *Nature*, 2011, **473**, 398–402.

79. Z. Li, X. Cai, C. L. Cai, J. Wang, W. Zhang, B. E. Petersen, F. C. Yang and M. Xu, Deletion of Tet2 in mice leads to dysregulated hematopoietic stem cells and subsequent development of myeloid malignancies, *Blood*, 2011, **118**, 4509–4518.
80. M. Ko, Y. Huang, A. M. Jankowska, U. J. Pape, M. Tahiliani, H. S. Bandukwala, J. An, E. D. Lamperti, K. P. Koh, R. Ganetzky, *et al.*, Impaired hydroxylation of 5-methylcytosine in myeloid cancers with mutant TET2, *Nature*, 2010, **468**, 839–843.
81. W. Xu, H. Yang, Y. Liu, Y. Yang, P. Wang, S. H. Kim, S. Ito, C. Yang, P. Wang, M. T. Xiao, *et al.*, Oncometabolite 2-hydroxyglutarate is a competitive inhibitor of alpha-ketoglutarate-dependent dioxygenases, *Cancer Cell*, 2011, **19**, 17–30.
82. C. G. Lian, Y. Xu, C. Ceol, F. Wu, A. Larson, K. Dresser, W. Xu, L. Tan, Y. Hu, Q. Zhan, *et al.*, Loss of 5-hydroxymethylcytosine is an epigenetic hallmark of melanoma, *Cell*, 2012, **150**, 1135–1146.
83. H. Wu, A. C. D'Alessio, S. Ito, K. Xia, Z. Wang, K. Cui, K. Zhao, Y. E. Sun and Y. Zhang, Dual functions of Tet1 in transcriptional regulation in mouse embryonic stem cells, *Nature*, 2011, **473**, 389–393.
84. K. Iqbal, S. G. Jin, G. P. Pfeifer and P. E. Szabo, Reprogramming of the paternal genome upon fertilization involves genome-wide oxidation of 5-methylcytosine, *Proc. Natl. Acad. Sci. U. S. A.*, 2011, **108**, 3642–3647.
85. S. Ito, A. C. D'Alessio, O. V. Taranova, K. Hong, L. C. Sowers and Y. Zhang, Role of Tet proteins in 5mC to 5hmC conversion, ES-cell self-renewal and inner cell mass specification, *Nature*, 2010, **466**, 1129–1133.
86. J. E. Ohm, K. M. McGarvey, X. Yu, L. Cheng, K. E. Schuebel, L. Cope, H. P. Mohammad, W. Chen, V. C. Daniel, W. Yu, *et al.*, A stem cell-like chromatin pattern may predispose tumor suppressor genes to DNA hypermethylation and heritable silencing, *Nat. Genet.*, 2007, **39**, 237–242.
87. Y. Schlesinger, R. Straussman, I. Keshet, S. Farkash, M. Hecht, J. Zimmerman, E. Eden, Z. Yakhini, E. Ben-Shushan, B. E. Reubinoff, *et al.*, Polycomb-mediated methylation on Lys27 of histone H3 pre-marks genes for de novo methylation in cancer, *Nat. Genet.*, 2007, **39**, 232–236.
88. M. Widschwendter, H. Fiegl, D. Egle, E. Mueller-Holzner, G. Spizzo, C. Marth, D. J. Weisenberger, M. Campan, J. Young, I. Jacobs, *et al.*, Epigenetic stem cell signature in cancer, *Nat. Genet.*, 2007, **39**, 157–158.
89. M. Ku, R. P. Koche, E. Rheinbay, E. M. Mendenhall, M. Endoh, T. S. Mikkelsen, A. Presser, C. Nusbaum, X. Xie, A. S. Chi, *et al.*, Genomewide analysis of PRC1 and PRC2 occupancy identifies two classes of bivalent domains, *PLoS Genet.*, 2008, **4**, e1000242.
90. A. Meissner, T. S. Mikkelsen, H. Gu, M. Wernig, J. Hanna, A. Sivachenko, X. Zhang, B. E. Bernstein, C. Nusbaum, D. B. Jaffe, *et al.*, Genome-scale DNA methylation maps of pluripotent and differentiated cells, *Nature*, 2008, **454**, 766–770.
91. H. Easwaran, S. E. Johnstone, L. Van Neste, J. Ohm, T. Mosbruger, Q. Wang, M. J. Aryee, P. Joyce, N. Ahuja, D. Weisenberger, *et al.*, A DNA hypermethylation module for the stem/progenitor cell signature of cancer, *Genome Res.*, 2012, **22**, 837–849.

CHAPTER 4

Targeting DNA Methylation

TOM D. HEIGHTMAN*[a] AND MICHAEL McCULLAR[b]

[a]Astex Pharmaceuticals, 436 Cambridge Science Park, Cambridge CB4 0QA, UK; [b]Astex Pharmaceuticals Inc., 4140 Dublin Boulevard, Suite 200, Dublin, CA 94568, USA
*E-mail: tom.heightman@astx.com

4.1 Introduction

Over the last three decades DNA methylation has become established as a dominant mechanism in the regulation of gene expression.[1] Among the many compelling findings supporting this statement, perhaps the most elegant evidence has been gathered from studies comparing pairs of monozygotic twins.[2,3] Despite having identical genomes, twin pairs frequently experience discordant susceptibility to a range of diseases including cancers,[4] autoimmune diseases[5] and psychological disorders.[6] Such discordance has been ascribed to differences in environment during the lifetime of each twin, and is typically more pronounced according to age and differences in environment. At the molecular level, these differences in susceptibility correlate with differences in gene silencing between twins and, importantly, silencing co-occurs with DNA methylation at specific gene promoter regions.[7] As a result of these and many other observations, our understanding of epigenetics has emerged as a complex mechanism for dynamic, semi-reversible biochemical responses to environmental conditions ranging from diet to psychological stress, or in other words, a developmental plasticity in which life experiences as early as *in utero*[8] influence disease susceptibility in later

RSC Drug Discovery Series No. 48
Epigenetics for Drug Discovery
Edited by Nessa Carey

Published by the Royal Society of Chemistry, www.rsc.org

life and even in subsequent generations. The fact that some epigenetic modifications including DNA methylation appear to be reversible makes them an attractive area in which to seek new medicines with the potential to halt or reverse disease progression. This chapter will focus on the biochemical mechanisms controlling DNA methylation, consequences of aberrant DNA methylation in complex chronic diseases, existing modulators of DNA methylation used in the clinic, and opportunities for new drugs targeting this central epigenetic mechanism.

4.2 DNA Methylation and Its Role in Gene Expression

DNA methylation occurs principally at the C-5 position of cytosine within DNA sequences containing the CpG motif (Figure 4.1), and is mediated by three catalytically active DNA methyltransferases DNMT1, DNMT3a and DNMT3b, which use *S*-adenosylmethionine (SAM) as cofactor.† The C-5 methyl group resides in the major groove of duplex DNA.

During gametogenesis, epigenetic marks are established by differential methylation of paternal and maternal alleles of genes essential for embryo growth, a process known as imprinting.[9] Shortly after fertilisation, much of the DNA from both paternal and maternal origin becomes demethylated. This creates a totipotent epigenetic 'clean slate', such that cells within the developing zygote can pursue programmes that allow them to differentiate into cells of distinct identity.[10] Cellular progression from totipotency through multipotency and into full differentiation correlates with both irreversible gene silencing and DNA methylation, introduced by the '*de novo*' DNA methyltransferases DNMT3a and DNMT3b (Figure 4.2a). These isoforms play similar but distinct roles in development; DNMT3a is associated with imprinting,[11] while DNMT3b plays a key role in early somatic development.[12] Both DNMT3a and DNMT3b require the presence of catalytically inactive DNMT3L as a co-activator.[13,14] During the subsequent replication and division of somatic cells, DNA methylation is replicated by DNMT1 copying DNA methylation marks from the parental strand onto the newly synthesised daughter strand during S-phase. DNMT1 is hence referred to as the 'maintenance' DNA methyltransferase, and provides a way of passing epigenetic information between cell generations. Interestingly, this occurs with imperfect fidelity: failure of DNA methylation maintenance has been estimated to occur with a frequency of ~5% per CpG site per cell division, such that after many cell generations, methylation of specific loci may be at a much lower level than in the originating cells.[15] Since DNMT1 only methylates hemi-methylated duplex DNA, selective recognition of MeC is required: this is effected by the MeC-binding

---

†The enzyme originally denoted as DNMT2 does not methylate DNA but instead methylates a specific cytosine in $tRNA^{Asp}$ and has been subsequently named tRNA aspartic methyltransferase 1 (TRDMT1).

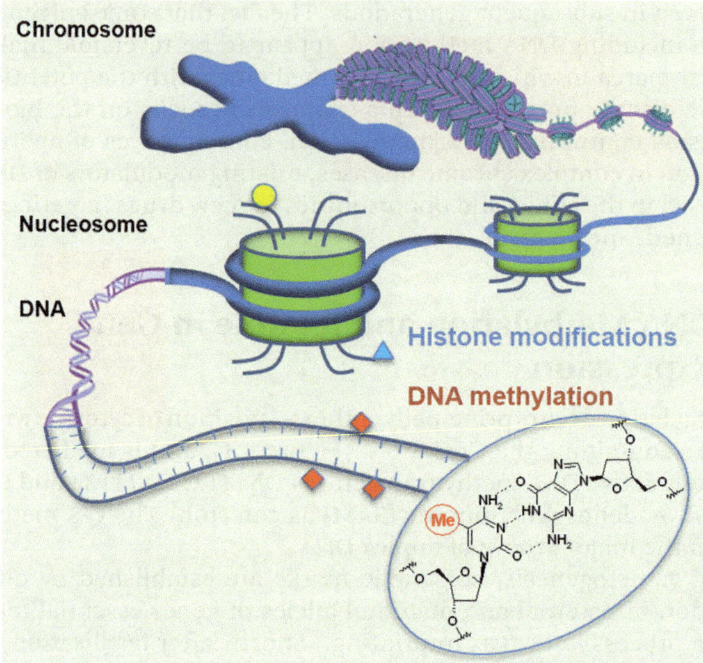

Figure 4.1 DNA methylation occurs principally at the C5 position of cytosine within DNA sequences containing the CpG motif.

protein UHRF1, a constituent of the multi-protein complex in which DNMT1 functions (Figure 4.2b).[16,17]

DNA methylation is typically associated with gene silencing, although the precise function appears to depend on context and many aspects remain poorly understood.[18] Genome-scale mapping has revealed DNA methylation at repetitive intergenic sequences such as long interspersed nuclear elements (LINEs), in gene bodies, and at transcriptional start sites containing CpG islands, loosely defined as tracts of at least 200 base pairs containing greater than 50% GC pairs and an observed : expected CpG ratio of greater than 60%.[19] In general, ~70% of gene promoters contain CpG islands, and these are usually unmethylated.[20] In contrast, methylation is more frequent in CpG islands within gene bodies.[21]

Unmethylated CpG islands are able to recruit a variety of histone-modifying enzyme complexes, including the SET1 H3K4 methyltransferase complex,[22] the PRC1 E3 ubiquitin ligase complex[23] and the histone demethylase KDM2A[24] to gene promoters, *via* the presence of Zn-finger CXXC domains in these complexes, which bind to the unmethylated CpG motif. Methylated CpG islands can mediate transcriptional repression both directly, by preventing transcription factor binding to specific CpG sequences,[25] and indirectly, by recruiting specific chromatin-modifying protein complexes containing methyl-DNA-binding proteins (MBDs).[26,27]

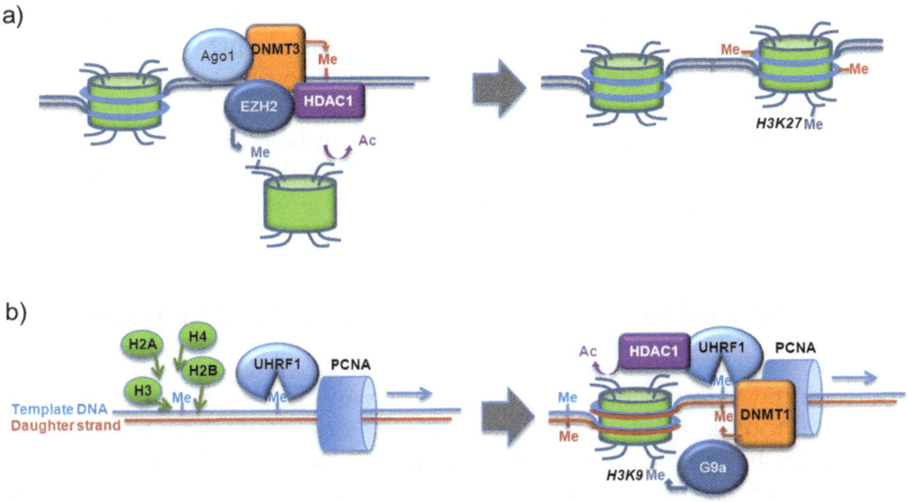

Figure 4.2 (a) Model for *de novo* methylation of DNA by DNMT3a/b: DNMT3a or b plus DNMT3L are recruited by complexes containing sequence-specific non-coding RNA associated with Argonaute1, repressive histone methyltransferases, *e.g.* EZH2 or G9a, and the histone deacetylase HDAC1. The removal of activating acetyl marks and introduction of repressive methylation marks on histones (at H3K9 and K36) is reinforced by DNA methylation to provide multiple layers of silencing. (b) Model for maintenance methylation of DNA by DNMT1: hemi-methylated CpG on newly synthesised DNA is recognised by the SRA domain of UHRF1, which binds to the processive PCNA. Histones are reassembled, and UHRF1 recruits DNMT1 to methylate the daughter DNA strand and thereby transfer methylation status. UHRF1 also recruits HDAC1 and G9a to deacetylate and methylate histones, respectively. UHRF1 also binds to methylated histone H3K9 *via* its PHD domain, implying that histone methylation may also precede DNA methylation.

Although the above observations strongly associate DNA methylation with gene silencing, it is important to note that DNA methylation is just one component of a complex mechanism and as such, there are contexts in which DNA methylation alone is either not sufficient or not required for gene silencing. The importance of DNA methylation in gene silencing was emphasised by early studies showing that for some tumour suppressor genes in which promoters are hypermethylated, silencing is reversed by drugs that reverse the hypermethylation (see below), but not by drugs that reverse histone marks associated with repression, such as HDAC inhibitors.[28] However, studies by other groups examining alternative tumour suppressor genes have observed that re-expression can be induced by HDAC inhibitor treatment *without* affecting DNA methylation levels.[29] These apparently conflicting findings have led to a model of gene silencing in which both DNA methylation and histone modifications are considered to play important roles, can act independently[30] or in concert,[31] and together define the degree of reversibility of silencing. Put simply, the histone state determines the immediate gene expression potential, while DNA methylation provides a long-term memory for gene silencing that can be temporarily overridden.[32] Support for this model is provided by the observation that the histone methyltransferase EZH2, which mediates transcriptional repression by methylation at H3K27, recruits DNA methyltransferases in the context of the Polycomb complexes 2 and 3, introducing DNA methylation at EZH2 target gene promoters and causing long-term silencing.[33]

4.3 Reversal and Modification of DNA Methylation

While there are complementary enzymes to add and remove many post-translational protein modifications, no enzyme has been identified that directly removes 5-methylcytosine marks. As described above, DNA methylation can be removed passively by successive rounds of DNA replication, due to the incomplete fidelity of DNA methylation inheritance (this mechanism may have evolved to introduce a degree of plasticity into epigenetic memory during the lifetime of an organism). Recently, a group of enzymes has been identified that convert 5-methylcytosine to 5-hydroxymethylcytosine. The TET oxygenases hydroxylate the 5-methyl group using iron and the cofactor 2-oxoglutarate.[34,35] These enzymes can also catalyse further oxidation to 5-formyl- and 5-carboxycytosine, which can then be excised by the base-excision repair enzyme TDG (thymine deglycosylase).[36,37] Post-fertilisation global DNA demethylation of the paternal genome in the zygote appears to be dependent on a combination of dilution of hydroxymethylated DNA during successive cell divisions and maternally-derived TET3 hydroxylation of the paternal genome[38] (while the maternal genome is protected from TET3-mediated hydroxymethylation/demethylation by the protein PGC7/STELLA).[39] Global demethylation during the development of primordial germ cells appears to be dependent on TET1/2.[40] Roles for TET proteins and hydroxymethylation in repressing or activating transcription of individual genes are still unclear. TET1 binds throughout the genome of embryonic stem cells, with the majority of binding sites located at transcription start sites (TSSs) of CpG-rich promoters and within genes.[41] The 5hmC modification is found in gene bodies and in contrast to 5mC is also enriched at CpG-rich TSSs. TET1 binds a significant proportion of Polycomb group target genes, and associates and co-localises with the SIN3A co-repressor complex.[42]

The recent identification of 5-hydroxymethyl-, 5-formyl- and 5-carboxycytosine binding proteins[43] suggests that these modifications may function in their own right as transcriptional regulators, rather than just acting as intermediates during demethylation. It is also worth noting here that traditional bisulfite-based methods for genomic mapping of 5-mC do not distinguish this modification from 5-hmC, so that many observations ascribed to 5-mC marks based on bisulfite-based sequencing may need to be revisited using methods capable of distinguishing the two marks.[44]

4.4 Aberrant DNA Methylation in Cancer

Aberrant methylation of DNA is a hallmark of many cancers,[45–47] although it may not always be causative in establishing gene silencing.[48] Widespread hypomethylation is commonly observed, consistent with activation of genes required for cellular proliferation, migration, adhesion, *etc.*; while hypermethylation occurs in CpG islands at the promoter regions of well-characterised tumour suppressors. Although a comprehensive review is beyond the scope of this chapter, a summary is provided in Table 4.1; the pattern is well illustrated using colon cancer as an example, where global hypomethylation reaches 10–15%, while localised promoter hypermethylation has been shown to contribute to loss of expression of the tumour suppressor APC.[49,50] More broadly, a 'CpG island methylator phenotype' (CIMP) has been defined whereby a subset of colon cancer patients show hypermethylation of a panel of genes including RUNX3, CACNA1G, IGF2 and hMLH1.[51,52] The CIMP has subsequently been used to describe CpG island promoter methylation in other tumour types, including bladder, breast, endometrial, gastric, glioblastoma (gliomas), hepatocellular, lung, ovarian, pancreatic, renal cell and prostate cancers, as well as for leukaemia, melanoma, duodenal adenocarcinomas, adrenocortical carcinomas and neuroblastomas.[53] In sporadic breast and ovarian cancers, methylation of the BRCA1 gene leads to its inactivation, mimicking the loss of function that occurs with BRCA1 germline mutations.[54,55] Other well-characterised examples of methylation-associated silencing include the oestrogen receptor (ER) in breast[56] and other cancers;[57] the CDK inhibitor CDKN2/p16/MTS1 across multiple cancers;[58] and Rb in retinoblastoma and nervous system tumours[59] (see Table 4.1). Expression of non-coding RNAs can also be regulated by DNA methylation, as exemplified by miR-124a. The miR-124a gene promoter is hypermethylated in a subset of acute lymphoblastic leukaemia patients, but treatment with a hypomethylating agent restored expression of miR-124a, leading to silencing of its target oncogene CDK6.[60,61] Hence, somewhat counter-intuitively, DNA hypermethylation can lead to an *increase* in gene expression, when it is silencing an inhibitory cascade.

Perhaps unsurprisingly, epigenetic changes including DNA methylation are also associated with acquired drug resistance in cancer. This has been exemplified by studies on ovarian cancer cells initially responsive to cisplatin or carboplatin, which become resistant after successive rounds of treatment, correlating with increases in CGI methylation; this chemoresistance can be restored by treatment with hypomethylating agents.[62,63]

Together, these properties show how aberrant methylation profiles fulfil many of the 'hallmarks of cancer',[64] and in this respect, bear functional resemblance to somatic mutations or translocations.[65,66] In fact, there is strong evidence for interplay between the two: new somatic mutations are being discovered leading to change of function of epigenetic regulators,[67] and conversely, differentially methylated regions of DNA often contain SNPs associated with disease through genome-wide association studies.[68] To date the

Table 4.1 Examples of tumour suppressor genes aberrantly methylated in cancer.

Gene	Tumour tissue (hypermethylation incidence)	Role in transformation	References
APC	Colon (17–28%), liver metastasis from colon (42%), breast, lung	Cell proliferation, migration, adhesion, chromosomal stability	49 and 50
BRCA1	Breast (13–15%), metaplastic breast (63%), ovarian (21%)	DNA repair, cell cycle checkpoint, genomic stability	54 and 55
CDKN2A (p16)	Colon (30–40%), head and neck (27–67%), breast (15–31%), lung (25–31%)	Loss of cell cycle checkpoint	58
CDKN2B (p15)	Acute myeloid leukaemia (71–94%), chronic myeloid leukaemia (24%)	Loss of cell cycle checkpoint	45, 141 and 142
E-cadherin	Thyroid (83%), breast, prostate	Cell proliferation, invasion, metastasis	143 and 144
ER	Breast (25% in oestrogen receptor-negative), prostate (80–95%), lung, leukaemia	Hormone resistance	56 and 57
miR-124a	ALL (%)	Loss of cell cycle checkpoint	60 and 61
MLH1	Endometrial (91%), gastric (95–100%), colon (9–15%), ovarian secondary tumours (56%)	DNA mismatch repair, gene mutations, microsatellite instability	145–148
RASSF1A	Lung (34–70%), breast (9%), ovarian (10–50%), kidney (23%)	Cell proliferation	149 and 150
Rb	Retinoblastoma (10–13%), nervous system tumours (19%)	Cell proliferation	59 and 151
VHL	Renal (10–19%)	Stability of oncogenic proteins	152 and 153

best characterised somatic mutations within the DNA methylation machinery are heterozygous DNMT3a mutations observed in a subset of myelodysplastic syndrome (MDS) and acute myeloid leukaemia (AML) patients, which correlate with poor prognosis.[69–71] Within AML patients, DNMT3a mutations have been associated with methylation and gene expression changes in the HOXB locus, which has been implicated in oncogenesis.[72,73] Interestingly, mutations to the 5-mC hydroxylating enzyme TET2 are also frequently observed in a subset of MDS patients, leading to hypermethylation.[74]

A further level of interplay is apparent when one considers that many epigenetic modulatory enzymes not only introduce post-translational modifications to histones and DNA, but also into other transcriptionally relevant proteins: for example, methylation of DNMT1 by SET7/9 leads to proteasomal degradation,[75] so that demethylation by LSD1 is required to maintain stable DNMT1,[76] providing a mechanism for crosstalk between the modifications and the modifiers themselves.

Overall, it is clear that genetic and epigenetic processes are inextricably intertwined, which will necessitate large-scale sequencing of both genome and epigenome in order to have a more complete understanding of cancer pathology. A number of pre-competitive consortia have been initiated to this end.[77]

4.5 DNA Methylation in Other Diseases

Aberrant DNA methylation profiles have been observed in a wide variety of disease areas beyond cancer, in particular in immune/inflammatory, psychiatric, neurological and metabolic diseases.[78] For many of these diseases it is not yet clear whether DNA methylation changes are causative, or simply an adaptive response to the disease state. However, in either case, DNA methylation profiling has the potential to be a powerful diagnostic and prognostic tool, and its importance is demonstrated by the initiation of a number of large-scale international consortia to collect epigenomic data from patient samples (*e.g.* Human Epigenome Project, Blueprint Epigenome). The collective findings are too numerous to cover in detail here; a small selection are described below, and a more detailed account of aberrant epigenetic profiles is given in Chapters 1–3.

The role of DNA methylation is particularly evident in autoimmune and inflammatory diseases such as rheumatoid arthritis (RA), asthma and lupus. Fibroblast-like synoviocytes (FLSs) obtained from RA and osteoarthritis patients have a significant decrease in DNMT1, 3a and 3b mRNA levels in response to low levels of IL-1 down to 1 pg mL^{-1}. The decreased levels of DNMTs result in a demethylation of the CHI3L1 locus that is hypomethylated in RA FLS.[79] DNA methylation has been implicated as an intermediary of genetic risk in RA. Two clusters have been identified within the MHC region, which have differential methylation profiles in RA cases relative to controls.[80] DNA methylation also plays a role in T-helper cell maturation. Differentially methylated loci linked to genes encoding proteins involved in T-helper cell polarisation and macrophage activation have been identified in peripheral blood mononuclear cells from obese asthmatic children.[81] Exposure to allergens can result in the reversible hypermethylation of the *IFNG*-promoter in $CD4^+$ cells which is associated with T_H1.[82] The hypermethylation of this locus correlates with a decrease in IFN-γ levels. Airway epithelial cells also show differential methylation patterns in asthmatic children. These include hypermethylation at CpG sites in the promoter region of the transcription factor STAT5 and demethylation of the promoter region of *CRIP1*.[83] T-cells in patients with systemic lupus erythematosus demonstrate global hypomethylation (15–20% reduction) and decreased DNMT1 expression, which is believed to drive production of autoreactive antibodies.[84]

4.6 Structure and Mechanism of DNMTs

Structures of the catalytic domains of mouse DNMT1 in complex with DNA and SAH (PDB code 3pt6)[85] and of human DNMT3a in complex with DNMT3L and SAH (PDB code 2qrv)[86] have been solved to ~3 Å resolution. The DNMT1 structure provides a double rationale for the prevention of methylation of unmethylated CpG sites (Figure 4.3). Firstly, the CXXC domain undergoes specific H-bonding interactions with the major groove at CpG sites that would be blocked in the presence of methylation: in this way unmethylated DNA is sequestered away from the active site of mDNMT1, while single-strand methylated DNA is available for maintenance methylation. Secondly, the autoinhibitory CXXC-BAH1 linker is positioned directly between the DNA and the active site and thereby prevents entrance of DNA into the catalytic pocket. Additional specificity for maintenance methylation is effected by the SRA domain of UHRF1 in the DNMT1 complex: the SRA domain flips bases out of the duplex DNA helix allowing the cytosine C-5 methyl group to be sensed by a specific hydrophobic pocket.[87]

The DNMT3a/DNMT3L structure confirms the requirement of DNMT3L to engender catalytic activity to DNMT3a. Two molecules of each protein form

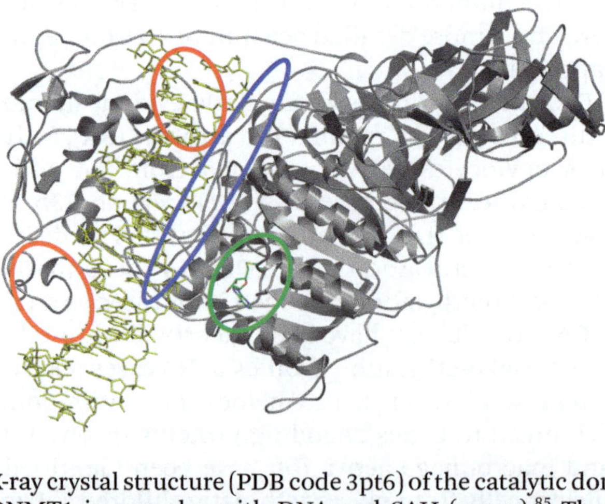

Figure 4.3 X-ray crystal structure (PDB code 3pt6) of the catalytic domains of mouse DNMT1 in complex with DNA and SAH (green).[85] The CXXC domain (red) undergoes specific H-bonding interactions with the major groove at CpG sites that would be blocked in the presence of methylation: in this way unmethylated DNA is sequestered away from the active site of mDNMT1, while single-strand methylated DNA is available for maintenance methylation. The autoinhibitory CXXC-BAH1 linker (blue) is positioned directly between the DNA and the active site and thereby prevents entrance of DNA into the catalytic pocket.

an active tetramer with two DNMT3a–DNMT3L interfaces and one DNMT3a–DNMT3a interface. Mutational studies showed that disruption of these interfaces led to loss of methyltransferase activity. The active tetramer contains two active sites separated by a distance equivalent to 8–10 base pairs on duplex DNA, similar to the periodicity observed for CpG methylation sites on imprinted genes. *In vitro*, on long DNA strands, the tetramer oligomerises and methylates CpG islands with a similar periodicity, suggesting a model of processive periodic methylation.[88] These conclusions are consistent with DNMT3a mutations in more aggressive forms of AML and MDS, which include three mutations at the tetramer interface and were associated with altered methylation patterns.[69,86]

Both structures confirm the presence of the catalytic residues required for methylation at C-5 of cytosine. Addition of the catalytic cysteine to the 6-position of the deoxycytidine generates an enamine with nucleophilic character at C-5 which, assisted by base catalysis from a neighbouring glutamate residue, attacks the methyl group from the SAM cofactor (Figure 4.4b). Elimination of the cysteine sulfhydryl group liberates the methylated product and, together with replacement of the reacted cofactor SAH with a new molecule of SAM, regenerates unoccupied enzyme for the next catalytic cycle.

4.7 DNMT Inhibitors

4.7.1 Nucleoside DNMT Inhibitors

The cytosine analogues decitabine and azacytidine are approved for the treatment of MDS and AML. Decitabine and azacytidine are both converted to deoxyazacytidine triphosphate *via* distinct metabolic pathways[89] (Figure 4.4a). Decitabine is phosphorylated successively by deoxycytidine kinase (DCK) to generate the monophosphate, and then nucleotide monophosphate and diphosphate kinases to generate the di- and triphosphates, respectively. Azacytidine is monophosphorylated by uridine-cytidine kinase and then converted to the diphosphate by nucleotide monophosphate kinase. At this point the majority of the Aza-DP is converted to the triphosphate where it becomes incorporated into RNA; a fraction (10–20%) is converted to decitabine diphosphate by ribonucleotide reductase, and is then converted to the triphosphate by nucleotide diphosphate kinase. Formation of deoxyazacytidine triphosphate from either drug leads to incorporation of deoxyazacytidine into DNA, where it becomes a suicide substrate for DNA methyltransferases. Addition of the catalytic cysteine to the 6-position of the deoxyazacytidine proceeds as for the natural deoxycytidine, followed by transfer of the methyl group from the SAM cofactor to the nitrogen at the 5-position of the azacytidine nucleobase (Figure 4.4b). The resulting intermediate is stable, lacking the beta-hydrogen present in cytidine that would allow elimination of the catalytic cysteine, thereby forming an irreversible covalent complex with DNMT1 which remains trapped on the DNA strand.[90] This triggers proteasome-mediated DNMT1 degradation and leads to cellular depletion.[91]

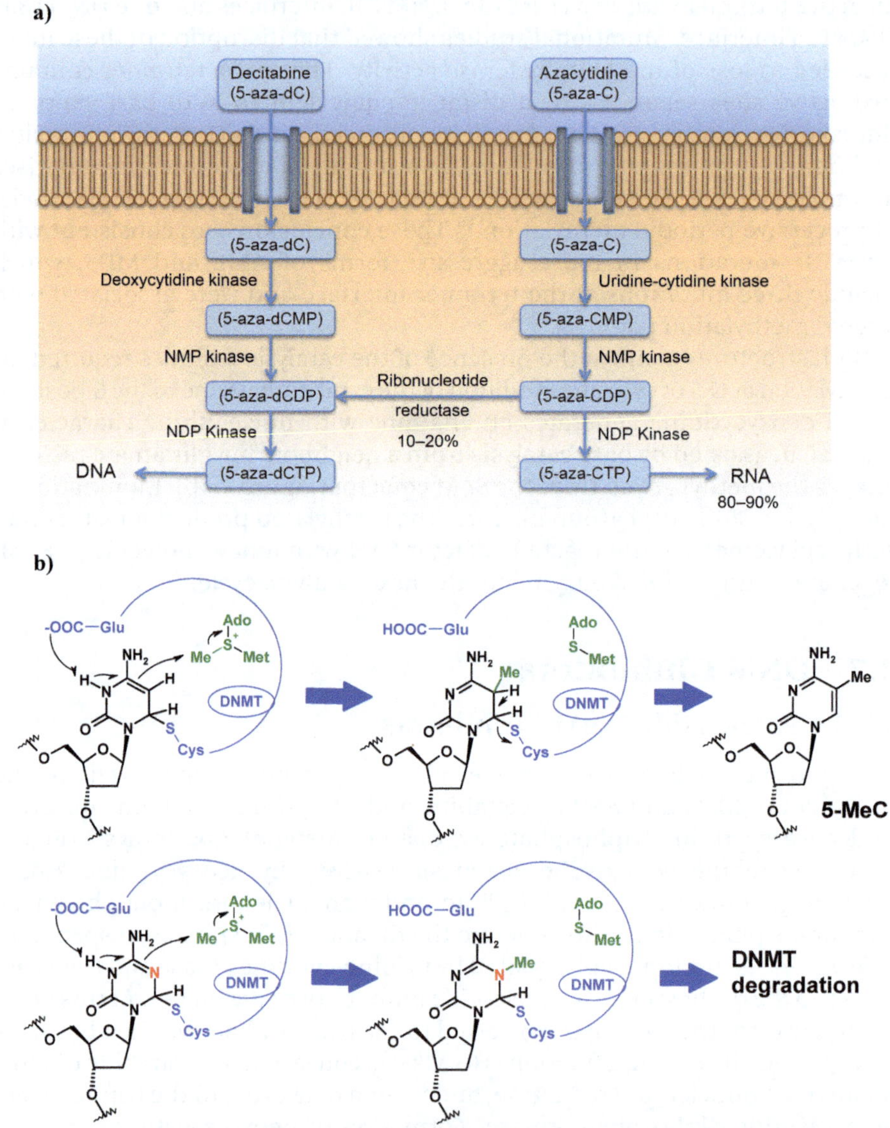

Figure 4.4 (a) Decitabine and azacytidine are both converted to deoxyazacytidine triphosphate *via* distinct metabolic pathways. (b) Mechanism for inhibition of DNMTs by deoxyazacytidine following its incorporation into DNA. Addition of the catalytic cysteine to the 6-position of the deoxyazacytidine proceeds as for the natural deoxycytidine, followed by transfer of the methyl group from the SAM cofactor to the nitrogen at the 5-position of the azacytidine nucleobase. The resulting intermediate is stable, lacking the beta-hydrogen present in cytidine that would allow elimination of the catalytic cysteine, thereby forming an irreversible covalent complex with DNMT1 which remains trapped on the DNA strand.

The precise mechanism of action of hypomethylating agents in MDS and AML remains the subject of some debate.[92] In AML cell lines, the effects of decitabine (DAC) and azacytidine (AZA) are highly concentration dependent. At low concentrations (30 nM DAC or 300 nM AZA) global DNA hypomethylation can be measured; at 100 nM DAC or 1 μM AZA DNMT1 levels are dramatically reduced, consistent with the proposed mechanism of action; while at higher concentrations (3–10 μM) DNA hypomethylation no longer occurs, presumably due to non-DNA methylation-dependent cytotoxicity which prevents incorporation of the nucleotide into DNA.[93,94] Clinical trials using low doses of AZA and DAC established suitable treatment schedules that allow efficient incorporation into DNA while minimising toxicity.

AZA was approved by the US Food and Drug Administration (FDA) in 2004 for the treatment of all MDS FAB subtypes, based on a primary randomised phase III trial where patients treated with AZA subcutaneously showed an improvement in objective response rates and conversion to transfusion independence relative to best supportive care. In the primary AZA registration study, the response rate in the AZA group was 16% (6% complete response and 10% partial response), and there were no responses in the best supportive care arm.[95–97] In addition, about 19% of the patients randomised to the AZA arm attained haematological improvement (not sufficient to attain a PR [partial response]) and approximately two-thirds of these patients attained transfusion independence. A subsequent European phase III trial showed significant improvement in median overall survival: 24.5 months with AZA *versus* 15.0 months with conventional care regimens.[98] AZA has been granted marketing authorisation by the European Commission for patients with high-risk MDS and AML. DAC was also approved for MDS treatment (IPSS High and Intermediate Risk) in 2006, based on objective response rate and transfusion independence in a randomised phase III study where DAC plus best supportive care was compared to best supportive care alone. In the DAC registration study, patients in the DAC treatment arm showed an objective response rate of 16% (15% complete response rate and 1% partial response).[99] Using preclinical findings that showed optimal DNA hypomethylation with low-dose prolonged exposure to DAC,[100] subsequent clinical trials demonstrated 39% complete responses,[101] and an overall improvement rate (CR + PR + HI) of 51%.[102] In a recent phase II study DAC demonstrated a complete response rate of 27% in older AML patients.[103] The clinical benefits of DAC appear to be mechanism specific, as clinical response in MDS has been shown to correlate strongly with DNA hypomethylation.[104]

DAC and AZA are the subject of a range of clinical trials both alone and in combination (Table 4.2). The majority of ongoing trials of DAC and AZA are in haematological malignancies as part of combination regimens.

Currently, AZA is approved for either subcutaneous (s.q.) or intravenous (i.v.) injection, while DAC is approved for i.v. administration. Both drugs have very short *in vivo* half-lives; patients treated with i.v. DAC demonstrate >90% plasma clearance of drug within an hour following its infusion[105] while those treated with s.q. AZA show a similarly rapid clearance.[106,107] Elimination

Table 4.2 Selected clinical trials involving AZA or DAC as single agent or in combination.

	DAC	
	Single agent	Combination
Haematological malignancies	• DAC in older or unfit AML patients • DAC in refractory immune thrombocytopenia • DAC *versus* AZA in low or int-1 risk MDS patients • DAC in patients with high-risk sickle cell disease	• DAC + dasatinib in CML patients • DAC + clofarabine in higher-risk MDS patients • DAC + ATRA + valproic acid in AML patients • DAC followed by clofarabine, idarubicin and cytarabine in AML • DAC + cladribine + cytarabine in patients with AML or high-risk MDS • DAC + vosaroxin in older patients with high-risk MDS and AML • DAC + mitoxantrone + etoposide + cytarabine in patients with relapsed/refractory AML or high-risk MDS • DAC + sapecitabine in newly diagnosed elderly AML patients
Solid malignancies		• DAC + temozolomide + panobinostat in patients with resistant metastatic melanoma • DAC + LHB589 + tamoxifen: re-expression of ER in triple negative breast cancers • DAC + genistein in advanced solid tumours and NSCLC
	AZA	
	Single agent	Combination
Haematological malignancies	• AZA in CML patients with minimal residual disease • AZA in CMML patients	• AZA + lenalidomide in elderly AML patients ≥65 years of age • AZA + lenalidomide in high-risk MDS and AML with Del (5q) • AZA + eltrombopag in MDS/AML patients
Solid malignancies		• AZA + entinostat in treating patients with recurrent advanced NSCLC

of these drugs occurs both by hydrolysis of the triazine ring and through N-4 deamination by the action of cytidine deaminase (CDA). These features have prompted a number of modifications to improve hypomethylating agent exposure (Table 4.3). SGI-110 is a dinucleotide pro-drug of DAC in which the 5-aza-deoxycytosine is linked through a 5′-phosphodiester to

Table 4.3 Nucleoside-based DNMT inhibitors.

Inhibitor	Structure	Phase	References
Azacytidine	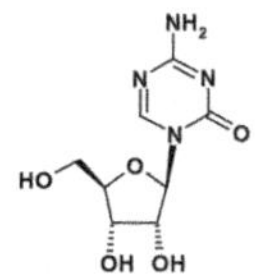	Approved (MDS, AML)	95–97
Decitabine	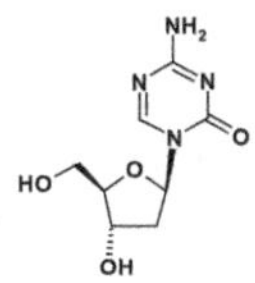	Approved (MDS, AML)	99–103
SGI-110	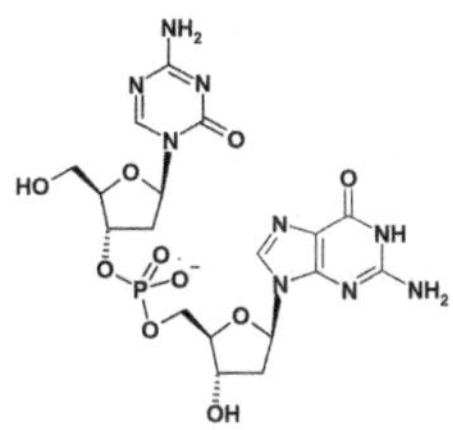	Phase II	154 and 155
NPEOC-DAC	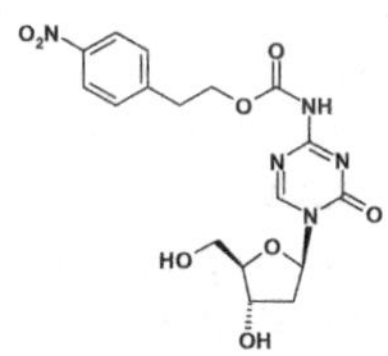	Preclinical	109
CP-4200	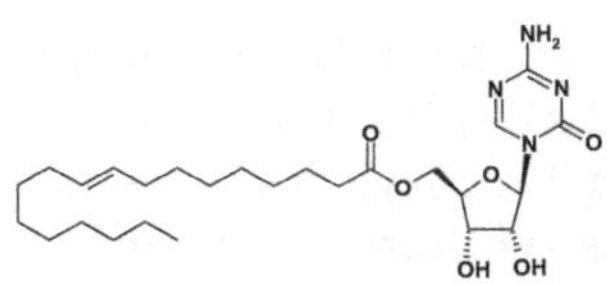	Preclinical	110
5,6-Dihydroazacytidine	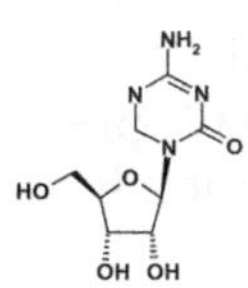	Preclinical	112
5-Fluoro-2′-deoxycytidine	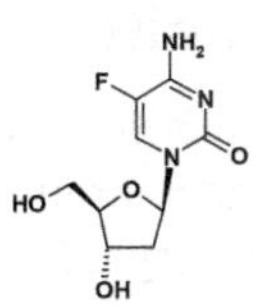	Preclinical	114 and 115
Zebularine	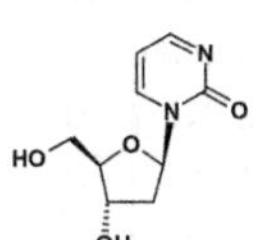	Preclinical	116

deoxyguanosine. SGI-110 is relatively stable to hydrolysis by CDA, and s.q. administration gives an altered exposure compared with i.v. administered DAC, with a lower C_{max} but significantly longer half-life, giving an overall similar AUC. SGI-110 is being tested against AML and MDS in both phase I and II clinical trials. Results from a portion of the phase I trial in MDS/AML suggest evidence of efficacy, notably in some patients who have failed prior DAC or AZA treatment.[108] The ability of SGI-110 to epigenetically restore sensitivity to carboplatin is also being investigated as part of a phase I/II trial in relapsed or refractory ovarian cancer in combination with carboplatin. SGI-110 is also being studied as a single agent in hepatic cancer patients. The rationale to evaluate SGI-110 in these settings is supported by preclinical data demonstrating demethylation and re-expression of aberrantly silenced tumour suppressor genes.[62,63]

The pro-drug NPEOC-DAC contains a carbamate moiety designed to protect the N-4 from deamination; however the compound proved no more active than DAC *in vivo*.[109] CP-4200, an elaidate ester of AZA, showed enhanced efficacy compared with AZA in an orthotopic ALL mouse tumour model, possibly due to reduced dependency on nucleoside transporters for cellular uptake.[110]

An alternative approach to prolonging the half-life of DAC is to co-administer an inhibitor of the primary metabolising enzyme CDA. This has been demonstrated in non-human primates using tetrahydrouridine (THU).[111]

Several other nucleoside analogues are reported to inhibit DNMT activity after incorporation into DNA. The AZA analogue 5,6-dihydroazacytidine proved ~10-fold weaker than AZA at inducing DNA hypomethylation in cells, ascribed to inefficient conversion to the deoxyribonucleoside triphosphate required for DNA incorporation.[112,113] When incorporated into DNA *in vitro*, 5-fluoro-2′-deoxycytidine inhibits bacterial DNMT by formation of a similar covalent complex as for DAC and AZA,[114] but in cells the compound undergoes rapid metabolism to 5-fluoro-2′-deoxyuridylate which elicits cytotoxicity through inhibition of thymidylate synthase rather than DNMT.[115] Inhibition of DNMT by zebularine incorporated into DNA has been shown to be covalent, but unlike AZA and DAC, is reversible.[116] AML cells are significantly less sensitive to zebularine than to AZA or DAC, presumably again due to inefficient metabolic activation and incorporation into DNA.

4.7.2 Non-Nucleoside DNMT Inhibitors

The development of reversible DNMT inhibitors remains of current interest for treatment of cancers and other diseases characterised by aberrant DNA methylation. As described above, the activity of AZA and DAC is dependent on nucleoside transporter uptake into cells, metabolic activation by a cascade of enzymes, as well as metabolic degradation: each of these steps provides a mechanism by which cells might exhibit primary or acquired resistance to these agents. Non-nucleoside inhibitors that do not require incorporation into DNA could circumvent these dependencies, potentially leading to higher response rates. In addition, non-incorporated DNMT inhibitors could offer a wider window between hypomethylation and cytotoxicity, leading to enhanced therapeutic indexes.

Over the last decade a variety of non-nucleoside inhibitors has been reported to show inhibition of DNA methylation in biochemical or cellular assays (see Table 4.4).

Table 4.4 Non-nucleoside-based DNMT inhibitors.

Inhibitor	Structure	Phase	References
Hydralazine	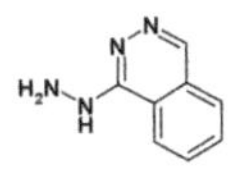	Phase II	117–119
Procainamide	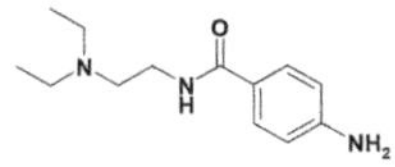	Preclinical	117 and 118
Procaine	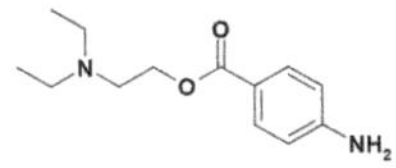	Preclinical	123 and 124
Compound **12**	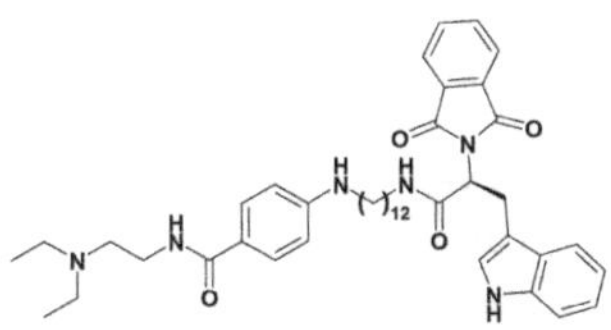	Preclinical	125
Isoxazoline 7b	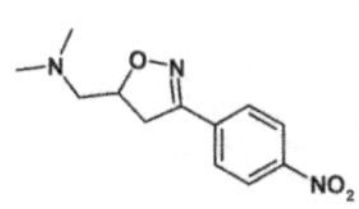	Preclinical	156
(−)-Epigallocatechin-3-gallate	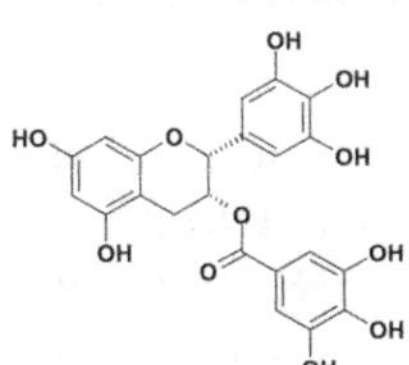	Preclinical	126
Mithramycin A	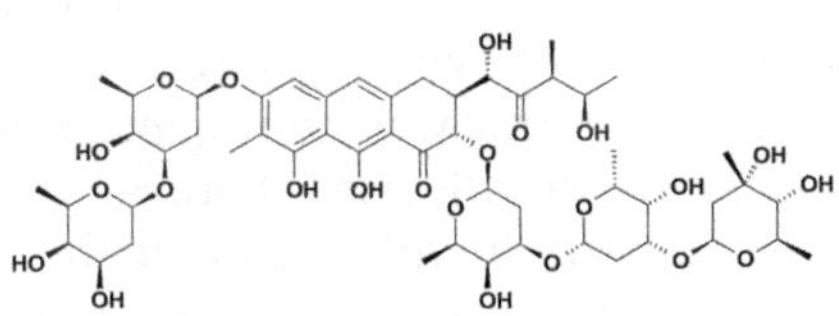	Preclinical	127
Nanaomycin	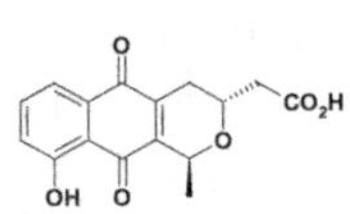	Preclinical	128

(*continued*)

Table 4.4 (*continued*)

Inhibitor	Structure	Phase	References
RG-108	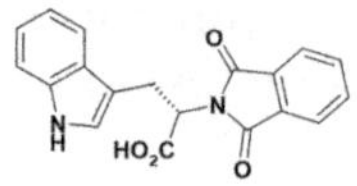	Preclinical	130 and 131
NSC-14778	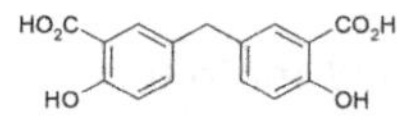	Preclinical	157
NSC-106084	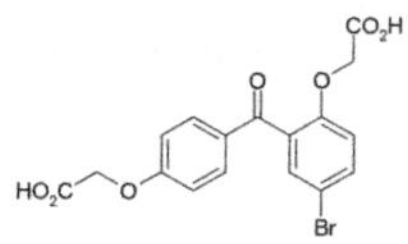	Preclinical	157
SGI-1027	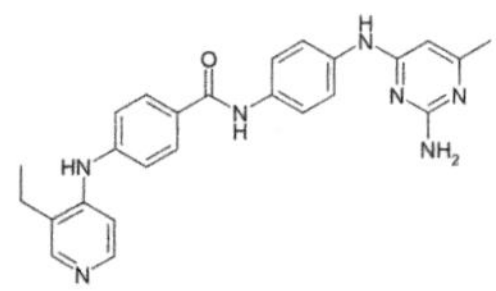	Preclinical	132
SW155246	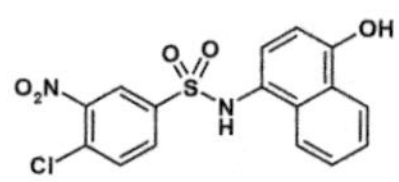	Preclinical	133
Dichlone	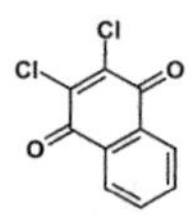	Preclinical	134
Laccaic acid	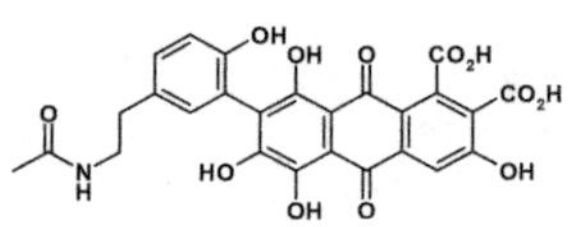	Preclinical	135

The antihypertensive drug hydralazine has been reported to inhibit DNA methylation in cellular assays but not *in vitro*,[117–119] although subsequent testing failed to reproduce the effects on DNA methylation in cells.[120] Despite this discrepancy, hydralazine is being studied in combination with magnesium valproate (a histone deacetylase inhibitor) in a phase II clinical trial in MDS.[121,122]

The antiarrhythmic drug procainamide was similarly reported to inhibit DNA methyltransferase activity in cells[117,118] that was subsequently not reproduced.[120] Reported cellular activity for the structurally related procaine[123] also failed to be reproduced independently.[124] The procainamide template has also been derivatised to make conjugates of RG-108 such as Compound **12**, with reportedly significant increases in biochemical and cellular potency,[125] and into constrained analogues exemplified by isoxazoline 7b, which showed weak biochemical and cellular anti-proliferative activity (IC_{50} 150 μM/GI_{50} 360 μM, respectively).

The green tea extract (−)-epigallocatechin-3-gallate (EGCG) was reported to show inhibition of DNA methylation in a nuclear extract biochemical assay, in addition to hypomethylating DNA and re-expressing tumour suppressor genes in cells;[126] like the other compounds described above, this activity was not reproduced in independent studies.[120,124]

The antibiotic mithramycin A has been reported to deplete levels of DNMT1 in lung cancer cells, and reduce the CpG island methylation of antimetastasis tumour suppressor genes, leading to re-expressed mRNA levels for these genes.[127] Consistent with this activity, mithramycin A inhibited invasiveness in cancer cell migration models. The mode of action has not been elucidated, but the authors hypothesise that the compound forms a covalent adduct with DNMTs in an analogous fashion to DAC, triggering proteasomal degradation of the enzyme.

Another antibiotic, nanaomycin, has been reported to inhibit DNMT3b selectively in cells, leading to DNA demethylation and re-expression of the tumour suppressor RASSF1A.[128]

The use of DNMT structures for drug discovery and design has been reviewed recently.[129] The phthalimide RG-108 was identified by virtual screening of NCI compounds against a three-dimensional model of human DNMT1, based on homology to bacterial DNMT X-ray crystal structures.[130,131] RG-108 inhibited DNMT activity in a bacterial DNMT biochemical assay, reduced DNA methylation in leukaemic cells with an IC_{50} of ~50–100 μM, and reactivated tumour suppressor genes.

Virtual screening was also used to identify the carboxylic acids NSC-14778 and NSC-106084, which in biochemical assays showed isoform selectivity for DNMT3B and DNMT1, respectively.

The quinolone SGI-1027 and related compounds have been reported to inhibit mammalian DNMT1, DNMT3a and DNMT3b in biochemical assays, with IC_{50} values in the range of 10 μM.[132] Although the precise mode of inhibition is unknown, the compound was shown to be non-competitive with respect to SAM, and did not interact with DNA or RNA. In RKO cells, SGI-1027 caused hypomethylation and re-expression of tumour suppressor genes, but unlike AZA and DAC, did not cause global hypomethylation. SGI-1027 caused selective degradation of DNMT1 without affecting DNMT3A/B levels. In this respect its behaviour partly resembles AZA and DAC, even though it is not incorporated into DNA, and there is no evidence of covalent modification of DNMTs.

HTS (high-throughput screening) using a ^{3}H-SAM scintillation proximity assay led to the identification of SW155246, which showed a 30-fold preference for inhibition of DNMT1 (IC_{50} ~ 1 μM) *versus* DNMT3A or 3B, inhibited global methylation in HeLa cells, and reactivated expression of the tumour suppressor gene RASSF1 in A549 cells.[133]

Medium-throughput screening using a non-radioactive biochemical assay identified several hits including the pesticide dichlone (IC_{50} 0.46 μM).[134] Dichlone reactivated YFP gene expression in a stable HEK293 cell line by promoter demethylation.

Finally, the insect-derived natural product laccaic acid was identified by HTS, using a fluorogenic assay determining production of methylated DNA by a highly active construct of DNMT1 in which the auto-inhibitory RFTS domain was deleted.[135] Laccaic acid was shown to inhibit DNMT directly by competition with the DNA substrate, with a K_i of 0.3 μM, and showed a ~2 °C thermostabilisation of DNMT1 using DSF. In MCF7 cells, high concentrations (~200 μM) elicited re-expression of a subset of silenced genes, and the compound showed synergistic effects on methylation when co-treated with DAC.

4.8 Outlook

The currently approved DNMT inhibitors have unfavourable pharmacological properties which limit exposure to tumour cells during S-phase. This limitation restricts their incorporation into DNA, which is required for these agents to covalently bind DNMTs and thereby reverse aberrant DNA hypermethylation. A major focus of research efforts to develop pro-drugs is to modulate the exposure profile of DAC or AZA in such a manner as to increase the exposure of malignant cells to these agents during S-phase of the cell cycle. The relatively rapid clearance of these agents from plasma may be one of the factors that has limited the effectiveness of DNMT inhibitors in solid malignancies. Inasmuch as solid tumours tend to have a much lower proliferative fraction relative to AML and high-risk myelodysplasia, it is reasonable to hypothesise that adequate incorporation of DAC or AZA into the DNA of solid tumours may be limiting. An effective pro-drug approach in which the exposure of either DAC or AZA can be extended could increase the anti-tumour activity of these agents in solid tumours. Aberrant hypermethylation is well known to be present in many solid malignancies and the therapeutic potential of DNMT inhibition in these diseases is significant.

New DNMT inhibitors could improve over DAC and AZA by targeting DNMTs without the requirement for incorporation into DNA for therapeutic activity. It is believed that the treatment-related myelosuppression associated with these agents is caused by the formation of DNA adducts and the subsequent dose-dependent saturation of excision repair pathways. Clearly, preventing treatment-emergent myelosuppression and febrile neutropenia would benefit patients as well as open up the possibility of chronic dosing and combinations which are currently not practical due to overlapping myelosuppression with other agents. One such area of drug development in which chronic administration of a DNMT inhibitor may be beneficial is in the area of sickle cell anaemia. Prior studies with DAC have demonstrated that the re-expression of epigenetically silenced foetal haemoglobin can reduce the frequency and severity of sickle-cell crisis in patients who do not respond to hydroxyurea.

However, based on the findings summarised in the preceding sections, the identification of reversible non-nucleoside inhibitors of DNMTs is challenging. Even in the context of biochemical assays, there are numerous opportunities for the identification of apparent inhibitors that do not directly inhibit DNMT, for example through intercalation of DNA leading to disruption of its ability to act as DNMT substrate. The majority of the compounds disclosed show relatively weak inhibitory activity in biochemical assays, and although many also show hypomethylation effects in cells, there is no evidence that this is mediated by direct DNMT inhibition in the cellular context. Clearly, these observations underscore the need for validation of newly discovered inhibitors in multiple orthogonal biochemical and biophysical assays, and ideally, reproduction in independent labs.[136]

Finally, although continued efforts to target DNMTs directly are well justified, it is also worth considering alternative approaches. In particular, because DNMTs function within multi-protein complexes, disruption of the function of other proteins within these complexes, or indeed of the architecture of the complexes, might elicit similar effects to direct DNMT inhibition. Targeting of UHRF1 may provide one such approach.[17] A variety of independent studies have implicated UHRF1 in breast,[137] NSCLC[138] and other cancers, suggesting that disruption of binding of UHRF1 to DNMT1,[139] to histones[140] or to hemi-methylated DNA[16] leads to similar hypomethylation as with direct DNMT inhibition. Small-molecule disruption of UHRF1 protein–protein or protein–DNA interactions would clearly present alternative challenges and opportunities to those experienced to date in seeking reversible DNMT inhibitors.

Acknowledgements

The authors wish to thank present and past colleagues at Astex Pharmaceuticals for many helpful discussions and critical reading of this chapter.

References

1. A. P. Bird, *Nature*, 1986, **321**, 209.
2. M. F. Fraga, E. Ballestar, M. F. Paz, S. Ropero, F. Setien, M. L. Ballestar, D. Heine-Suñer, J. C. Cigudosa, M. Urioste, J. Benitez, M. Boix-Chornet, A. Sanchez-Aguilera, C. Ling, E. Carlsson, P. Poulsen, A. Vaag, Z. Stephan, T. D. Spector, Y.-Z. Wu, C. Plass and M. Esteller, *Proc. Natl. Acad. Sci. U. S. A.*, 2005, **102**, 10604.
3. Z. A. Kaminsky, T. Tang, S.-C. Wang, C. Ptak, G. H. T. Oh, A. H. C. Wong, L. A. Feldcamp, C. Virtanen, J. Halfvarson, C. Tysk, A. F. McRae, P. M. Visscher, G. W. Montgomery, I. I. Gottesman, N. G. Martin and A. Petronis, *Nat. Genet.*, 2009, **41**, 240.
4. H. Heyn, F. J. Carmona, A. Gomez, H. J. Ferreira, J. T. Bell, S. Sayols, K. Ward, O. A. Stefansson, S. Moran, J. Sandoval, J. E. Eyfjord, T. D. Spector and M. Esteller, *Carcinogenesis*, 2012, **34**, 102.
5. B. M. Javierre, A. F. Fernandez, J. Richter, F. Al-Shahrour, J. I. Martin-Subero, J. Rodriguez-Ubreva, M. Berdasco, M. F. Fraga, T. P. O'Hanlon, L. G. Rider, F. V. Jacinto, F. J. Lopez-Longo, J. Dopazo, M. Forn, M. A. Peinado, L. Carreño, A. H. Sawalha, J. B. Harley, R. Siebert, M. Esteller, F. W. Miller and E. Ballestar, *Genome Res.*, 2010, **20**, 170.
6. J. Zhao, J. Goldberg, J. D. Bremner and V. Vaccarino, *Psychosom. Med.*, 2013, **75**, 523.
7. A. K. Maunakea, I. Chepelev and K. Zhao, *Circ. Res.*, 2010, **107**, 327.

8. L. Gordon, J. E. Joo, J. E. Powell, M. Ollikainen, B. Novakovic, X. Li, R. Andronikos, M. N. Cruickshank, K. N. Conneely, A. K. Smith, R. S. Alisch, R. Morley, P. M. Visscher, J. M. Craig and R. Saffery, *Genome Res.*, 2012, **22**, 1395.
9. W. Reik, W. Dean and J. Walter, *Science*, 2001, **293**, 1089.
10. S. Seisenberger, J. R. Peat, T. A. Hore, F. Santos, W. Dean and W. Reik, *Philos. Trans. R. Soc., B*, 2013, **368**, 20110330.
11. M. Kaneda, M. Okano, K. Hata, T. Sado, N. Tsujimoto, E. Li and H. Sasaki, *Nature*, 2004, **429**, 900.
12. S. Feng, S. E. Jacobsen and W. Reik, *Science*, 2010, **330**, 622.
13. K. Hata, M. Okano, H. Lei and E. Li, *Development*, 2002, **129**, 1983.
14. H. Gowher, K. Liebert, A. Hermann, G. Xu and A. Jeltsch, *J. Biol. Chem.*, 2005, **280**, 13341.
15. A. D. Riggs, Z. Xiong, L. Wang and J. M. LeBon, *Novartis Found. Symp.*, 1998, **214**, 214.
16. M. Bostick, J. K. Kim, P.-O. Estève, A. Clark, S. Pradhan and S. E. Jacobsen, *Science*, 2007, **317**, 1760.
17. M. Unoki, J. Brunet and M. Mousli, *Biochem. Pharmacol.*, 2009, **78**, 1279.
18. P. A. Jones, *Nat. Rev. Genet.*, 2012, **13**, 484.
19. S. Saxonov, P. Berg and D. L. Brutlag, *Proc. Natl. Acad. Sci. U. S. A.*, 2006, **103**, 1412.
20. R. S. Illingworth, U. Gruenewald-Schneider, S. Webb, A. R. W. Kerr, K. D. James, D. J. Turner, C. Smith, D. J. Harrison, R. Andrews and A. P. Bird, *PLoS Genet.*, 2010, **6**, e1001134.
21. A. K. Maunakea, R. P. Nagarajan, M. Bilenky, T. J. Ballinger, C. D'Souza, S. D. Fouse, B. E. Johnson, C. Hong, C. Nielsen, Y. Zhao, G. Turecki, A. Delaney, R. Varhol, N. Thiessen, K. Shchors, V. M. Heine, D. H. Rowitch, X. Xing, C. Fiore, M. Schillebeeckx, S. J. M. Jones, D. Haussler, M. A. Marra, M. Hirst, T. Wang and J. F. Costello, *Nature*, 2010, **466**, 253.
22. J. P. Thomson, P. J. Skene, J. Selfridge, T. Clouaire, J. Guy, S. Webb, A. R. W. Kerr, A. Deaton, R. Andrews, K. D. James, D. J. Turner, R. Illingworth and A. Bird, *Nature*, 2010, **464**, 1082.
23. X. Wu, J. V. Johansen and K. Helin, *Mol. Cell*, 2013, **49**, 1134.
24. N. P. Blackledge, J. C. Zhou, M. Y. Tolstorukov, A. M. Farcas, P. J. Park and R. J. Klose, *Mol. Cell*, 2010, **38**, 179.
25. R. Meehan, J. Lewis, S. Cross, X. Nan, P. Jeppesen and A. Bird, *J. Cell Sci.*, 1992, **1992**, 9.
26. A. P. Bird and A. P. Wolffe, *Cell*, 1999, **99**, 451.
27. A. Bird, *Genes Dev.*, 2002, **16**, 6.
28. H. Suzuki, E. Gabrielson, W. Chen, R. Anbazhagan, M. van Engeland, M. P. Weijenberg, J. G. Herman and S. B. Baylin, *Nat. Genet.*, 2002, **31**, 141.
29. K. Pruitt, R. L. Zinn, J. E. Ohm, K. M. McGarvey, S.-H. L. Kang, D. N. Watkins, J. G. Herman and S. B. Baylin, *PLoS Genet.*, 2006, **2**, e40.
30. Y. Kondo, L. Shen, A. S. Cheng, S. Ahmed, Y. Boumber, C. Charo, T. Yamochi, T. Urano, K. Furukawa, B. Kwabi-Addo, D. L. Gold, Y. Sekido, T. H.-M. Huang and J.-P. J. Issa, *Nat. Genet.*, 2008, **40**, 741.

31. H. Cedar and Y. Bergman, *Nat. Rev. Genet.*, 2009, **10**, 295.
32. N. J.-M. Raynal, J. Si, R. F. Taby, V. Gharibyan, S. Ahmed, J. Jelinek, M. R. H. Estécio and J.-P. J. Issa, *Cancer Res.*, 2012, **72**, 1170.
33. E. Vire, C. Brenner, R. Deplus, L. Blanchon, M. Fraga, C. Didelot, L. Morey, A. Van Eynde, D. Bernard, J.-M. Vanderwinden, M. Bollen, M. Esteller, L. Di Croce, Y. de Launoit and F. Fuks, *Nature*, 2006, **439**, 871.
34. S. Kriaucionis and N. Heintz, *Science*, 2009, **324**, 929.
35. M. Tahiliani, K. P. Koh, Y. Shen, W. A. Pastor, H. Bandukwala, Y. Brudno, S. Agarwal, L. M. Iyer, D. R. Liu, L. Aravind and A. Rao, *Science*, 2009, **324**, 930.
36. S. Ito, L. Shen, Q. Dai, S. C. Wu, L. B. Collins, J. A. Swenberg, C. He and Y. Zhang, *Science*, 2011, **333**, 1300.
37. Y.-F. He, B.-Z. Li, Z. Li, P. Liu, Y. Wang, Q. Tang, J. Ding, Y. Jia, Z. Chen, L. Li, Y. Sun, X. Li, Q. Dai, C.-X. Song, K. Zhang, C. He and G.-L. Xu, *Science*, 2011, **333**, 1303.
38. A. Inoue and Y. Zhang, *Science*, 2011, **334**, 194.
39. T. Nakamura, Y.-J. Liu, H. Nakashima, H. Umehara, K. Inoue, S. Matoba, M. Tachibana, A. Ogura, Y. Shinkai and T. Nakano, *Nature*, 2012, **486**, 415.
40. J. A. Hackett, R. Sengupta, J. J. Zylicz, K. Murakami, C. Lee, T. A. Down and M. A. Surani, *Science*, 2013, **339**, 448.
41. Y. Xu, F. Wu, L. Tan, L. Kong, L. Xiong, J. Deng, A. J. Barbera, L. Zheng, H. Zhang, S. Huang, J. Min, T. Nicholson, T. Chen, G. Xu, Y. Shi, K. Zhang and G. Shi Yujiang, *Mol. Cell*, 2011, **42**, 451.
42. K. Williams, J. Christensen, M. T. Pedersen, J. V. Johansen, P. A. C. Cloos, J. Rappsilber and K. Helin, *Nature*, 2011, **473**, 343.
43. C. G. Spruijt, F. Gnerlich, A. H. Smits, T. Pfaffeneder, P. W. T. C. Jansen, C. Bauer, M. Münzel, M. Wagner, M. Müller, F. Khan, H. C. Eberl, A. Mensinga, A. B. Brinkman, K. Lephikov, U. Müller, J. Walter, R. Boelens, H. van Ingen, H. Leonhardt, T. Carell and M. Vermeulen, *Cell*, 2013, **152**, 1146.
44. M. J. Booth, M. R. Branco, G. Ficz, D. Oxley, F. Krueger, W. Reik and S. Balasubramanian, *Science*, 2012, **336**, 934.
45. M. Esteller, P. G. Corn, S. B. Baylin and J. G. Herman, *Cancer Res.*, 2001, **61**, 3225.
46. M. Esteller, *Nat. Rev. Genet.*, 2007, **8**, 286.
47. P. C. Taberlay and P. A. Jones, *Prog. Drug Res.*, 2011, **67**, 1.
48. D. Sproul, R. Kitchen, C. Nestor, J. M. Dixon, A. Sims, D. Harrison, B. Ramsahoye and R. Meehan, *Genome Biol.*, 2012, **13**, R84.
49. C. N. Arnold, A. Goel, D. Niedzwiecki, J. M. Dowell, L. Wasserman, C. Compton, R. J. Mayer, M. M. Bertagnolli and C. R. Boland, *Cancer Biol. Ther.*, 2004, **3**, 960.
50. J. Chen, C. Rocken, C. Lofton-Day, H. U. Schulz, O. Muller, N. Kutzner, P. Malfertheiner and M. P. Ebert, *Carcinogenesis*, 2005, **26**, 37.
51. M. Toyota, N. Ahuja, M. Ohe-Toyota, J. G. Herman, S. B. Baylin and J.-P. J. Issa, *Proc. Natl. Acad. Sci.*, 1999, **96**, 8681.
52. W. S. Samowitz, H. Albertsen, J. Herrick, T. R. Levin, C. Sweeney, M. A. Murtaugh, R. K. Wolff and M. L. Slattery, *Gastroenterology*, 2005, **129**, 837.

53. L. A. E. Hughes, V. Melotte, J. de Schrijver, M. de Maat, V. T. H. B. M. Smit, J. V. M. G. Bovee, P. J. French, P. A. van den Brandt, L. J. Schouten, T. de Meyer, W. Van Criekinge, N. Ahuja, J. G. Herman, M. P. Weijenberg and M. van Engeland, *Cancer Res.*, 2013, **73**, 5858.
54. A. Dobrovic and D. Simpfendorfer, *Cancer Res.*, 1997, **57**, 3347.
55. M. Esteller, J. M. Silva, G. Dominguez, F. Bonilla, X. Matias-Guiu, E. Lerma, E. Bussaglia, J. Prat, I. C. Harkes, E. A. Repasky, E. Gabrielson, M. Schutte, S. B. Baylin and J. G. Herman, *J. Natl. Cancer Inst.*, 2000, **92**, 564.
56. R. G. Lapidus, A. T. Ferguson, Y. L. Ottaviano, F. F. Parl, H. S. Smith, S. A. Weitzman, S. B. Baylin, J. P. Issa and N. E. Davidson, *Clin. Cancer Res.*, 1996, **2**, 805.
57. L. C. Li, R. Chui, K. Nakajima, B. R. Oh, H. C. Au and R. Dahiya, *Cancer Res.*, 2000, **60**, 702.
58. J. G. Herman, A. Merlo, L. Mao, R. G. Lapidus, J. P. Issa, N. E. Davidson, D. Sidransky and S. B. Baylin, *Cancer Res.*, 1995, **55**, 4525.
59. C. Stirzaker, D. S. Millar, C. L. Paul, P. M. Warnecke, J. Harrison, P. C. Vincent, M. Frommer and S. J. Clark, *Cancer Res.*, 1997, **57**, 2229.
60. J. Roman-Gomez, X. Agirre, A. Jimenez-Velasco, V. Arqueros, A. Vilas-Zornoza, P. Rodriguez-Otero, I. Martin-Subero, L. Garate, L. Cordeu, E. San Jose-Eneriz, V. Martin, J. A. Castillejo, E. Bandres, M. J. Calasanz, R. Siebert, A. Heiniger, A. Torres and F. Prosper, *J. Clin. Oncol.*, 2009, **27**, 1316.
61. X. Agirre, A. Vilas-Zornoza, A. Jimenez-Velasco, J. I. Martin-Subero, L. Cordeu, L. Garate, E. San Jose-Eneriz, G. Abizanda, P. Rodriguez-Otero, P. Fortes, J. Rifon, E. Bandres, M. J. Calasanz, V. Martin, A. Heiniger, A. Torres, R. Siebert, J. Roman-Gomez and F. Prosper, *Cancer Res.*, 2009, **69**, 4443.
62. M. Li, C. Balch, J. Montgomery, M. Jeong, J. Chung, P. Yan, T. Huang, S. Kim and K. Nephew, *BMC Med. Genomics*, 2009, **2**, 34.
63. C. Zeller and R. Brown, *Ther. Adv. Med. Oncol.*, 2010, **2**, 319.
64. D. Hanahan and R. A. Weinberg, *Cell*, 2000, **100**, 57.
65. M. Esteller, M. F. Fraga, M. Guo, J. Garcia-Foncillas, I. Hedenfalk, A. K. Godwin, J. Trojan, C. Vaurs-Barriere, Y. J. Bignon, S. Ramus, J. Benitez, T. Caldes, Y. Akiyama, Y. Yuasa, V. Launonen, M. J. Canal, R. Rodriguez, G. Capella, M. A. Peinado, A. Borg, L. A. Aaltonen, B. A. Ponder, S. B. Baylin and J. G. Herman, *Hum. Mol. Genet.*, 2001, **10**, 3001.
66. D. D. De Carvalho, S. Sharma, J. S. You, S.-F. Su, P. C. Taberlay, T. K. Kelly, X. Yang, G. Liang and P. A. Jones, *Cancer Cell*, 2012, **21**, 655.
67. H. Shen and P. W. Laird, *Cell*, 2013, **153**, 38.
68. Y. Bergman and H. Cedar, *Nat. Struct. Mol. Biol.*, 2013, **20**, 274.
69. T. J. Ley, L. Ding, M. J. Walter, M. D. McLellan, T. Lamprecht, D. E. Larson, C. Kandoth, J. E. Payton, J. Baty, J. Welch, C. C. Harris, C. F. Lichti, R. R. Townsend, R. S. Fulton, D. J. Dooling, D. C. Koboldt, H. Schmidt, Q. Zhang, J. R. Osborne, L. Lin, M. O'Laughlin, J. F. McMichael, K. D. Delehaunty, S. D. McGrath, L. A. Fulton, V. J. Magrini, T. L. Vickery, J. Hundal, L. L. Cook, J. J. Conyers, G. W. Swift, J. P. Reed, P. A. Alldredge, T. Wylie, J. Walker, J. Kalicki, M. A. Watson, S. Heath, W. D. Shannon,

N. Varghese, R. Nagarajan, P. Westervelt, M. H. Tomasson, D. C. Link, T. A. Graubert, J. F. DiPersio, E. R. Mardis and R. K. Wilson, *N. Engl. J. Med.*, 2010, **363**, 2424.

70. M. Y. Shah and J. D. Licht, *Nat. Genet.*, 2011, **43**, 289.
71. M. J. Walter, L. Ding, D. Shen, J. Shao, M. Grillot, M. McLellan, R. Fulton, H. Schmidt, J. Kalicki-Veizer, M. O'Laughlin, C. Kandoth, J. Baty, P. Westervelt, J. F. Dipersio, E. R. Mardis, R. K. Wilson, T. J. Ley and T. A. Graubert, *Leukemia*, 2011, **25**, 1153.
72. X. J. Yan, J. Xu, Z. H. Gu, C. M. Pan, G. Lu, Y. Shen, J. Y. Shi, Y. M. Zhu, L. Tang, X. W. Zhang, W. X. Liang, J. Q. Mi, H. D. Song, K. Q. Li, Z. Chen and S. J. Chen, *Nat. Genet.*, 2011, **43**, 309.
73. E. A. Eklund, *Curr. Opin. Hematol.*, 2007, **14**, 85.
74. S. M. C. Langemeijer, R. P. Kuiper, M. Berends, R. Knops, M. G. Aslanyan, M. Massop, E. Stevens-Linders, P. van Hoogen, A. G. van Kessel, R. A. P. Raymakers, E. J. Kamping, G. E. Verhoef, E. Verburgh, A. Hagemeijer, P. Vandenberghe, T. de Witte, B. A. van der Reijden and J. H. Jansen, *Nat. Genet.*, 2009, **41**, 838.
75. S. Pradhan, H. G. Chin, P. O. Esteve and S. E. Jacobsen, *Epigenetics*, 2009, **4**, 383.
76. J. Wang, S. Hevi, J. K. Kurash, H. Lei, F. Gay, J. Bajko, H. Su, W. Sun, H. Chang, G. Xu, F. Gaudet, E. Li and T. Chen, *Nat. Genet.*, 2009, **41**, 125.
77. S. Beck, B. E. Bernstein, R. M. Campbell, J. F. Costello, D. Dhanak, J. R. Ecker, J. M. Greally, J.-P. Issa, P. W. Laird, K. Polyak, B. Tycko and P. A. Jones, *Cancer Res.*, 2012, **72**, 6319.
78. T. D. Heightman, *Expert Opin. Ther. Targets*, 2011, **15**, 729.
79. K. Nakano, D. L. Boyle and G. S. Firestein, *J. Immunol.*, 2013, **190**, 1297.
80. Y. Liu, M. Aryee, L. Padyukov, M. D. Fallin, E. Hesselberg, A. Runarsson, L. Reinius, N. Acevedo, M. Taub, M. Ronninger, A. Shchetynsky, J. Kere, L. Alfredsson, L. Klareskog, T. J. Ekstrom and A. P. Feinberg, *Nat. Biotechnol.*, 2013, **31**, 142.
81. D. Rastogi, M. Suzuki and J. M. Greally, *Sci. Rep.*, 2013, **3**, 2164.
82. S. Brand, D. A. Kesper, R. Teich, E. Kilic-Niebergall, O. Pinkenburg, E. Bothur, M. Lohoff, H. Garn, P. I. Pfefferle and H. Renz, *J. Allergy Clin. Immunol.*, 2012, **129**, 1602.
83. D. Stefanowicz, T.-L. Hackett, F. S. Garmaroudi, O. P. Günther, S. Neumann, E. N. Sutanto, K.-M. Ling, M. S. Kobor, A. Kicic, S. M. Stick, P. D. Paré and D. A. Knight, *PLoS One*, 2012, **7**, e44213.
84. B. Richardson, L. Scheinbart, J. Strahler, L. Gross, S. Hanash and M. Johnson, *Arthritis Rheum.*, 1990, **33**, 1665.
85. J. Song, O. Rechkoblit, T. H. Bestor and D. J. Patel, *Science*, 2011, **331**, 1036.
86. D. Jia, R. Z. Jurkowska, X. Zhang, A. Jeltsch and X. Cheng, *Nature*, 2007, **449**, 248.
87. G. V. Avvakumov, J. R. Walker, S. Xue, Y. Li, S. Duan, C. Bronner, C. H. Arrowsmith and S. Dhe-Paganon, *Nature*, 2008, **455**, 822.
88. C. Holz-Schietinger, D. M. Matje, M. F. Harrison and N. O. Reich, *J. Biol. Chem.*, 2011, **286**, 41479.

89. A. Quintas-Cardama, F. P. S. Santos and G. Garcia-Manero, *Nat. Rev. Clin. Oncol.*, 2010, **7**, 433.
90. D. V. Santi, A. Norment and C. E. Garrett, *Proc. Natl. Acad. Sci. U. S. A.*, 1984, **81**, 6993.
91. K. Ghoshal, J. Datta, S. Majumder, S. Bai, H. Kutay, T. Motiwala and S. T. Jacob, *Mol. Cell Biol.*, 2005, **25**, 4727.
92. H. Khan, C. Vale, T. Bhagat and A. Verma, *Semin. Hematol.*, 2013, **50**, 16.
93. T. Qin, E. M. Youssef, J. Jelinek, R. Chen, A. S. Yang, G. Garcia-Manero and J. P. Issa, *Clin. Cancer Res.*, 2007, **13**, 4225.
94. P. W. Hollenbach, A. N. Nguyen, H. Brady, M. Williams, Y. Ning, N. Richard, L. Krushel, S. L. Aukerman, C. Heise and K. J. MacBeth, *PLoS One*, 2010, **5**, e9001.
95. L. R. Silverman, E. P. Demakos, B. L. Peterson, A. B. Kornblith, J. C. Holland, R. Odchimar-Reissig, R. M. Stone, D. Nelson, B. L. Powell, C. M. DeCastro, J. Ellerton, R. A. Larson, C. A. Schiffer and J. F. Holland, *J. Clin. Oncol.*, 2002, **20**, 2429.
96. A. B. Kornblith, J. E. Herndon 2nd, L. R. Silverman, E. P. Demakos, R. Odchimar-Reissig, J. F. Holland, B. L. Powell, C. DeCastro, J. Ellerton, R. A. Larson, C. A. Schiffer and J. C. Holland, *J. Clin. Oncol.*, 2002, **20**, 2441.
97. L. R. Silverman, D. R. McKenzie, B. L. Peterson, J. F. Holland, J. T. Backstrom, C. L. Beach and R. A. Larson, *J. Clin. Oncol.*, 2006, **24**, 3895.
98. P. Fenaux, G. J. Mufti, E. Hellstrom-Lindberg, V. Santini, C. Finelli, A. Giagounidis, R. Schoch, N. Gattermann, G. Sanz, A. List, S. D. Gore, J. F. Seymour, J. M. Bennett, J. Byrd, J. Backstrom, L. Zimmerman, D. McKenzie, C. Beach and L. R. Silverman, *Lancet Oncol.*, 2009, **10**, 223.
99. H. Kantarjian, J. P. Issa, C. S. Rosenfeld, J. M. Bennett, M. Albitar, J. DiPersio, V. Klimek, J. Slack, C. de Castro, F. Ravandi, R. Helmer 3rd, L. Shen, S. D. Nimer, R. Leavitt, A. Raza and H. Saba, *Cancer*, 2006, **106**, 1794.
100. J. P. Issa, G. Garcia-Manero, F. J. Giles, R. Mannari, D. Thomas, S. Faderl, E. Bayar, J. Lyons, C. S. Rosenfeld, J. Cortes and H. M. Kantarjian, *Blood*, 2004, **103**, 1635.
101. H. Kantarjian, Y. Oki, G. Garcia-Manero, X. Huang, S. O'Brien, J. Cortes, S. Faderl, C. Bueso-Ramos, F. Ravandi, Z. Estrov, A. Ferrajoli, W. Wierda, J. Shan, J. Davis, F. Giles, H. I. Saba and J. P. Issa, *Blood*, 2007, **109**, 52.
102. D. P. Steensma, M. R. Baer, J. L. Slack, R. Buckstein, L. A. Godley, G. Garcia-Manero, M. Albitar, J. S. Larsen, S. Arora, M. T. Cullen and H. Kantarjian, *J. Clin. Oncol.*, 2009, **27**, 3842.
103. A. F. Cashen, G. J. Schiller, M. R. O'Donnell and J. F. DiPersio, *J. Clin. Oncol.*, 2010, **28**, 556.
104. L. Shen, H. Kantarjian, Y. Guo, E. Lin, J. Shan, X. Huang, D. Berry, S. Ahmed, W. Zhu, S. Pierce, Y. Kondo, Y. Oki, J. Jelinek, H. Saba, E. Estey and J. P. Issa, *J. Clin. Oncol.*, 2010, **28**, 605.
105. C. J. van Groeningen, A. Leyva, A. M. P. O'Brien, H. E. Gall and H. M. Pinedo, *Cancer Res.*, 1986, **46**, 4831.
106. B. A. Chabner, J. C. Drake and D. G. Johns, *Biochem. Pharmacol.*, 1973, **22**, 2763.

107. G. G. Chabot, J. Bouchard and R. L. Momparler, *Biochem. Pharmacol.*, 1983, **32**, 1327.
108. E. A. Griffiths, G. Choy, S. Redkar, P. Taverna, M. Azab and A. R. Karpf, *Drugs Future*, 2013, **38**, 535.
109. H. M. Byun, S. H. Choi, P. W. Laird, B. Trinh, M. A. Siddiqui, V. E. Marquez and A. S. Yang, *Cancer Lett.*, 2008, **266**, 238.
110. B. Brueckner, M. Rius, M. R. Markelova, I. Fichtner, P. A. Hals, M. L. Sandvold and F. Lyko, *Mol. Cancer Ther.*, 2010, **9**, 1256.
111. D. Lavelle, K. Vaitkus, Y. Ling, M. A. Ruiz, R. Mahfouz, K. P. Ng, S. Negrotto, N. Smith, P. Terse, K. J. Engelke, J. Covey, K. K. Chan, J. DeSimone and Y. Saunthararajah, *Blood*, 2012, **119**, 1240.
112. P. A. Jones and S. M. Taylor, *Cell*, 1980, **20**, 85.
113. D. B. McGregor, A. G. Brown, P. Cattanach, W. Shepherd, C. Riach, D. S. Daston and W. J. Caspary, *Carcinogenesis*, 1989, **10**, 2003.
114. D. G. Osterman, G. D. DePillis, J. C. Wu, A. Matsuda and D. V. Santi, *Biochemistry*, 1988, **27**, 5204.
115. E. M. Newman and D. V. Santi, *Proc. Natl. Acad. Sci. U. S. A.*, 1982, **79**, 6419.
116. C. Champion, D. Guianvarc'h, C. Senamaud-Beaufort, R. Z. Jurkowska, A. Jeltsch, L. Ponger, P. B. Arimondo and A. L. Guieysse-Peugeot, *PLoS One*, 2010, **5**, e12388.
117. E. Cornacchia, J. Golbus, J. Maybaum, J. Strahler, S. Hanash and B. Richardson, *J. Immunol.*, 1988, **140**, 2197.
118. B. Segura-Pacheco, C. Trejo-Becerril, E. Perez-Cardenas, L. Taja-Chayeb, I. Mariscal, A. Chavez, C. Acuna, A. M. Salazar, M. Lizano and A. Duenas-Gonzalez, *Clin. Cancer Res.*, 2003, **9**, 1596.
119. C. Deng, Q. Lu, Z. Zhang, T. Rao, J. Attwood, R. Yung and B. Richardson, *Arthritis Rheum.*, 2003, **48**, 746.
120. J. C. Chuang, C. B. Yoo, J. M. Kwan, T. W. Li, G. Liang, A. S. Yang and P. A. Jones, *Mol. Cancer Ther.*, 2005, **4**, 1515.
121. J. Coronel, L. Cetina, I. Pacheco, C. Trejo-Becerril, A. Gonzalez-Fierro, E. de la Cruz-Hernandez, E. Perez-Cardenas, L. Taja-Chayeb, D. Arias-Bofill, M. Candelaria, S. Vidal and A. Duenas-Gonzalez, *Med. Oncol.*, 2011, **28**(Suppl 1), S540–S546.
122. M. Candelaria, A. Herrera, J. Labardini, A. Gonzalez-Fierro, C. Trejo-Becerril, L. Taja-Chayeb, E. Perez-Cardenas, E. de la Cruz-Hernandez, D. Arias-Bofill, S. Vidal, E. Cervera and A. Duenas-Gonzalez, *Ann. Hematol.*, 2011, **90**, 379.
123. A. Villar-Garea, M. F. Fraga, J. Espada and M. Esteller, *Cancer Res.*, 2003, **63**, 4984.
124. C. Stresemann, B. Brueckner, T. Musch, H. Stopper and F. Lyko, *Cancer Res.*, 2006, **66**, 2794.
125. L. Halby, C. Champion, C. Sénamaud-Beaufort, S. Ajjan, T. Drujon, A. Rajavelu, A. Ceccaldi, R. Jurkowska, O. Lequin, W. G. Nelson, A. Guy, A. Jeltsch, D. Guianvarc'h, C. Ferroud and P. B. Arimondo, *ChemBioChem*, 2012, **13**, 157.

126. M. Z. Fang, Y. Wang, N. Ai, Z. Hou, Y. Sun, H. Lu, W. Welsh and C. S. Yang, *Cancer Res.*, 2003, **63**, 7563.
127. R. K. Lin, C. H. Hsu and Y. C. Wang, *Anticancer Drugs*, 2007, **18**, 1157.
128. D. Kuck, T. Caulfield, F. Lyko and J. L. Medina-Franco, *Mol. Cancer Ther.*, 2010, **9**, 3015.
129. J. Yoo and J. Medina-Franco, *J. Comput.-Aided Mol. Des.*, 2011, **25**, 555.
130. P. Siedlecki, R. Garcia Boy, T. Musch, B. Brueckner, S. Suhai, F. Lyko and P. Zielenkiewicz, *J. Med. Chem.*, 2006, **49**, 678.
131. B. Brueckner, R. Garcia Boy, P. Siedlecki, T. Musch, H. C. Kliem, P. Zielenkiewicz, S. Suhai, M. Wiessler and F. Lyko, *Cancer Res.*, 2005, **65**, 6305.
132. J. Datta, K. Ghoshal, W. A. Denny, S. A. Gamage, D. G. Brooke, P. Phiasivongsa, S. Redkar and S. T. Jacob, *Cancer Res.*, 2009, **69**, 4277.
133. J. A. Kilgore, X. Du, L. Melito, S. Wei, C. Wang, H. G. Chin, B. Posner, S. Pradhan, J. M. Ready and N. S. Williams, *J. Biol. Chem.*, 2013, **288**, 19673.
134. A. Ceccaldi, A. Rajavelu, S. Ragozin, C. Sénamaud-Beaufort, P. Bashtrykov, N. Testa, H. Dali-Ali, C. Maulay-Bailly, S. Amand, D. Guianvarc'h, A. Jeltsch and P. B. Arimondo, *ACS Chem. Biol.*, 2013, **8**, 543.
135. R. L. Fagan, D. E. Cryderman, L. Kopelovich, L. L. Wallrath and C. Brenner, *J. Biol. Chem.*, 2013, **288**, 23858.
136. S. V. Frye, *Nat. Chem. Biol.*, 2010, **6**, 159.
137. W. Jin, L. Chen, Y. Chen, S.-g. Xu, G.-h. Di, W.-j. Yin, J. Wu and Z.-m. Shao, *Breast Cancer Res. Treat.*, 2010, **123**, 359.
138. A. Daskalos, U. Oleksiewicz, A. Filia, G. Nikolaidis, G. Xinarianos, J. R. Gosney, A. Malliri, J. K. Field and T. Liloglou, *Cancer*, 2011, **117**, 1027.
139. J. Sharif, M. Muto, S. Takebayashi, I. Suetake, A. Iwamatsu, T. A. Endo, J. Shinga, Y. Mizutani-Koseki, T. Toyoda, K. Okamura, S. Tajima, K. Mitsuya, M. Okano and H. Koseki, *Nature*, 2007, **450**, 908.
140. S. B. Rothbart, K. Krajewski, N. Nady, W. Tempel, S. Xue, A. I. Badeaux, D. Barsyte-Lovejoy, J. Y. Martinez, M. T. Bedford, S. M. Fuchs, C. H. Arrowsmith and B. D. Strahl, *Nat. Struct. Mol. Biol.*, 2012, **19**, 1155.
141. J. G. Herman, J. Jen, A. Merlo and S. B. Baylin, *Cancer Res.*, 1996, **56**, 722.
142. A. Batova, M. B. Diccianni, J. C. Yu, T. Nobori, M. P. Link, J. Pullen and A. L. Yu, *Cancer Res.*, 1997, **57**, 832.
143. J. R. Graff, J. G. Herman, R. G. Lapidus, H. Chopra, R. Xu, D. F. Jarrard, W. B. Isaacs, P. M. Pitha, N. E. Davidson and S. B. Baylin, *Cancer Res.*, 1995, **55**, 5195.
144. J. R. Graff, V. E. Greenberg, J. G. Herman, W. H. Westra, E. R. Boghaert, K. B. Ain, M. Saji, M. A. Zeiger, S. G. Zimmer and S. B. Baylin, *Cancer Res.*, 1998, **58**, 2063.
145. M. F. Kane, M. Loda, G. M. Gaida, J. Lipman, R. Mishra, H. Goldman, J. M. Jessup and R. Kolodner, *Cancer Res.*, 1997, **57**, 808.
146. J. G. Herman, A. Umar, K. Polyak, J. R. Graff, N. Ahuja, J. P. Issa, S. Markowitz, J. K. Willson, S. R. Hamilton, K. W. Kinzler, M. F. Kane, R. D. Kolodner, B. Vogelstein, T. A. Kunkel and S. B. Baylin, *Proc. Natl. Acad. Sci. U. S. A.*, 1998, **95**, 6870.

147. M. Esteller, R. Levine, S. B. Baylin, L. H. Ellenson and J. G. Herman, *Oncogene*, 1998, **17**, 2413.
148. S. Y. Leung, S. T. Yuen, L. P. Chung, K. M. Chu, A. S. Chan and J. C. Ho, *Cancer Res.*, 1999, **59**, 159.
149. A. Agathanggelou, S. Honorio, D. P. Macartney, A. Martinez, A. Dallol, J. Rader, P. Fullwood, A. Chauhan, R. Walker, J. A. Shaw, S. Hosoe, M. I. Lerman, J. D. Minna, E. R. Maher and F. Latif, *Oncogene*, 2001, **20**, 1509.
150. Y. L. Choi, S. Y. Kang, Y. K. Shin, J. S. Choi, S. H. Kim, S. J. Lee, D. S. Bae and G. Ahn, *Virchows Arch.*, 2006, **448**, 331.
151. P. Gonzalez-Gomez, M. J. Bello, M. E. Alonso, D. Arjona, J. Lomas, J. M. de Campos, A. Isla and J. A. Rey, *Br. J. Cancer*, 2003, **88**, 109.
152. J. G. Herman, F. Latif, Y. Weng, M. I. Lerman, B. Zbar, S. Liu, D. Samid, D. S. Duan, J. R. Gnarra, W. M. Linehan, *et al.*, *Proc. Natl. Acad. Sci. U. S. A.*, 1994, **91**, 9700.
153. S. C. Clifford, A. H. Prowse, N. A. Affara, C. H. Buys and E. R. Maher, *Genes, Chromosomes Cancer*, 1998, **22**, 200.
154. C. B. Yoo, S. Jeong, G. Egger, G. Liang, P. Phiasivongsa, C. Tang, S. Redkar and P. A. Jones, *Cancer Res.*, 2007, **67**, 6400.
155. J. C. Chuang, S. L. Warner, D. Vollmer, H. Vankayalapati, S. Redkar, D. J. Bearss, X. Qiu, C. B. Yoo and P. A. Jones, *Mol. Cancer Ther.*, 2010, **9**, 1443.
156. S. Castellano, D. Kuck, M. Viviano, J. Yoo, F. López-Vallejo, P. Conti, L. Tamborini, A. Pinto, J. L. Medina-Franco and G. Sbardella, *J. Med. Chem.*, 2011, **54**, 7663.
157. D. Kuck, N. Singh, F. Lyko and J. L. Medina-Franco, *Bioorg. Med. Chem.*, 2010, **18**, 822.

CHAPTER 5

Targeting Histone Acetylation

DANIEL VITT*[a], ROLAND BAUMGARTNER[a], AND HELLA KOHLHOF[a]

[a]4SC AG, Am Klopferspitz 19a, 82152 Planegg-Martinsried, Germany
*E-mail: Daniel.Vitt@4sc.com

5.1 Introduction

The pioneering work of Philips[1] and Pogo *et al.*[2] on histone acetylation in the 1960s laid out the ground plan of the entirely new field of epigenetics. Subsequently, a whole set of proteins which are able to enzymatically attach or remove acetyl groups to and from lysine side chains of histones was identified. The former group of enzymes was named histone acetylases (HATs); the latter has been dubbed histone deacetylases (HDACs). Very soon researchers established a link between histone acetylation and the control of gene expression. The contribution of epigenetically controlled dysregulation of gene expression in certain pathophysiological conditions, including cancer, was recognised and HDAC enzymes were established as a promising and innovative target class. Starting from there, it took 40 years until the approval of vorinostat **1** (Zolinza™, SAHA, suberoylanilide hydroxamic acid) for treatment of cutaneous T-cell lymphoma as the first HDAC inhibitor (HDI).[3–7] Despite the fact that vorinostat achieved only partial commercial success, this approval has significantly boosted the entire field of epigenetic drug development (Figure 5.1).

Today we are faced with not only an ever-growing field of basic research but also clinical drug development efforts. Our mechanistic understanding

RSC Drug Discovery Series No. 48
Epigenetics for Drug Discovery
Edited by Nessa Carey

Published by the Royal Society of Chemistry, www.rsc.org

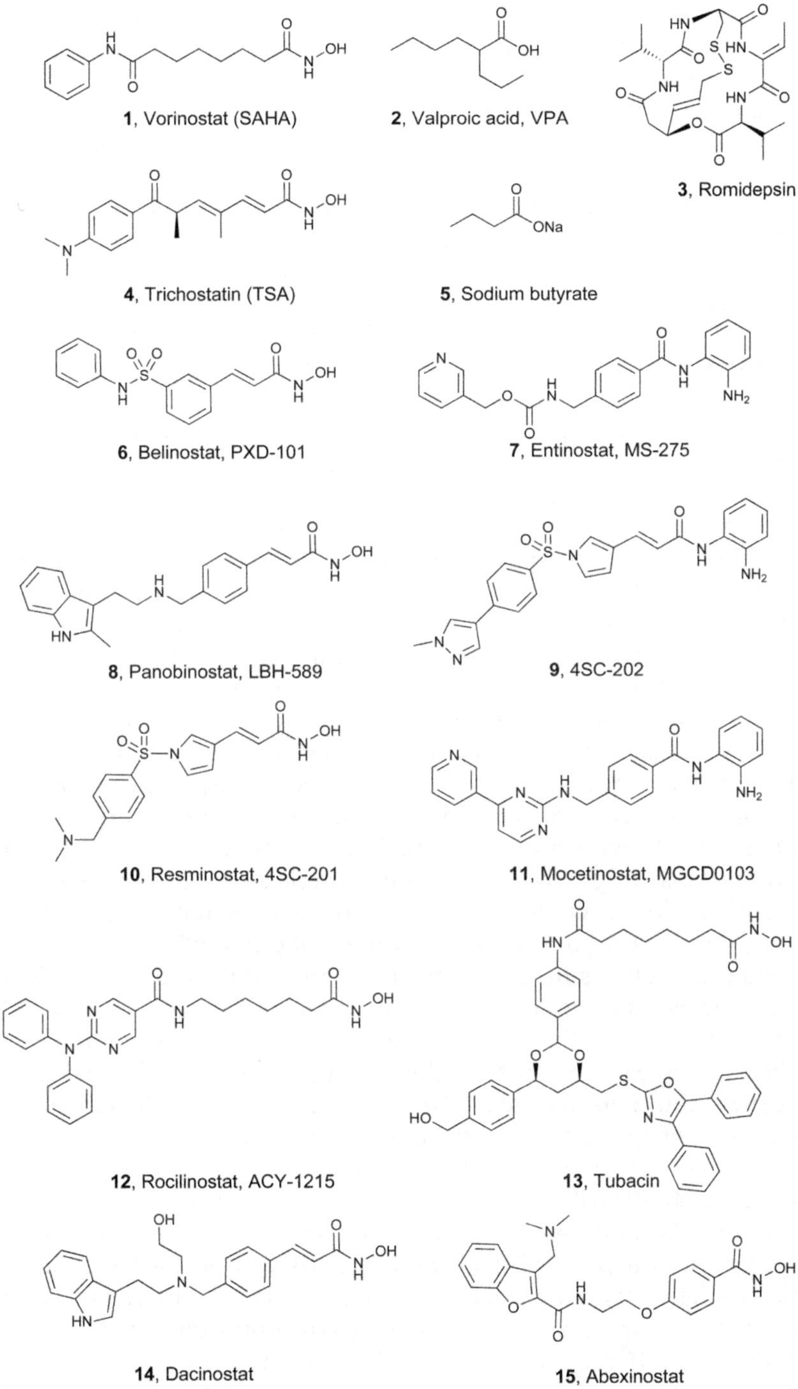

Figure 5.1 Selected list of HDAC inhibitors which are in development or which have been important milestones in the progress of HDAC inhibitor development.

of pharmacological inhibition of HDAC enzymes has evolved from the simple picture of transcriptional regulation by open or condensed chromatin states,[8,9] towards a complex and dynamical signature of histone marks that eventually give rise to diverse phenotypic outcomes such as induction of apoptosis or cell differentiation induced by epigenetic reprogramming. However, a complete understanding of how the so-called histone code[10] might be linked to specific disease states and how it can be modulated by pharmacological intervention at the level of epigenetic targets is still in its infancy. Recent work in the field has emphasised the need to adopt a holistic view in order to fully appreciate the complex flow of epigenetic information and its influence on the transcriptome, starting from the histone code up to the resulting phenotypes observed at a cellular level and even in living organisms.[11,12]

An important initiative to improve our understanding of epigenetic control is the ENCODE project,[13] in which researchers are systematically analysing the interplay between, for example, epigenetic modifications, transcription factors, repressor complexes or even the role of open chromatin states of non-coding regions of the genome.[12,13] As our knowledge of the histone code and further post-translational modifications increases, it is evident that acetylation and deacetylation are neither the sole epigenetic modification nor are restricted to histone proteins. The spectrum of protein targets modified by HDAC enzymes has extended to transcription factors and beyond. Acetylation and deacetylation should not be discussed in isolation but in the context of a variety of other post-translational histone modifications, such as methylation or demethylation of arginine and lysine residues.

HDAC enzymes are organised into highly specialised protein complexes consisting of transcription factors, enhancer molecules and additional epigenetic modifiers, like writers and erasers, but also reader proteins.[14] Chromatin structures consisting of huge enhancer complexes, called super-enhancers, add another level of context specificity to the epigenetic regulation of the transcriptome in the context of embryonic cell differentiation.[15,16]

Overall, the authors firmly believe that our understanding of epigenetics will constantly increase in the years to come and epigenetic research will continue to reveal exciting insights and results to us. Translational researchers should take care to translate these results into clinical development of next-generation HDIs. As such, HDAC inhibitors will establish themselves as the mainstay of epigenetic therapy in the near future.

In this chapter, we will review the current status and thinking in the field of HDAC research. We will provide an overview of the HDAC family of enzymes and discuss the underlying molecular pharmacology and signalling pathways. Important aspects of isoform selectivity will be addressed as well as medicinal chemistry efforts to further improve on existing HDAC inhibitors. Furthermore, we will provide some clues to the important capability of HDIs to induce cell differentiation. This unique feature of HDIs might allow us to intervene with the so-called cancer stem cell phenotype and to potentially change clinical treatment regimens in the field of oncology. Finally, we summarise the lessons learned from ongoing clinical trials with regard to clinical pharmacology, issues of pharmacokinetics and pharmacodynamics, biomarkers as well as potential toxicities and drug combination strategies.

5.2 HDAC Proteins

The family of HDACs comprises four classes, which are grouped by cellular localisation, interacting metal ion or NAD dependency. Class I isoforms HDAC1, HDAC2, HDAC3 and HDAC8 are all localised in the nucleus, although HDAC3 is additionally found in cytosol. Class IIA isoforms HDAC4, HDAC5, HDAC7 and HDAC9 are localised in nucleus and cytoplasm whereas class IIB enzymes HDAC6 and HDAC10 are preferentially found in the cytoplasm.[8,17] Class III HDACs are also called sirtuins and consist of SIRT1–7.[18] HDAC11 is the only representative of Class IV and is localised in nucleus and cytoplasm. Whereas HDAC6 and HDAC11 contain two catalytic sites, all other HDACs only have one.

The Zn-binding motive in the active site is the common feature of Class I, Class II and Class IV enzymes. In contrast to this, Class III HDACs use either a NAD^+-dependent deacetylase mechanism or in case of Sirt4 act as an ADP-ribosyltransferase. Sirtuins (Class III) have even more biological functions than the Zn-dependent HDACs and are described in detail in a number of reviews.[18,19]

The landscape of protein interactions clearly differentiates Class II HDACs from the Class I forms. Class IIa HDACs (HDAC4, HDAC5, HDAC7 and HDAC9) share a MEF2 (myocyte enhancer factor 2) binding site, a NLS and a 14-3-3 binding motif.[20] These features enable regulation of intracellular distribution of Class IIa HDACs, *e.g.* by phosphorylation and thus modulation of 14-3-3/HDAC binding and transport. These N-terminal binding motifs are not present in Class IIb HDACs (HDAC6 and HDAC10).

5.2.1 Function of Histone Deacetylases

The function of HDACS can be two-fold: enzymatic activity and protein interactions. As enzymes, HDACs are able to remove acetyl moieties from proteins. More precisely, the enzymes of this class should be named protein deacetylases or even more specifically, lysine deacetylases, abbreviated as KDACs rather than HDACs. However, for the sake of clarity we will adopt the term HDAC, which is commonly used in the literature. HDACs are also able to remove acetyl groups from non-histone proteins, *e.g.* transcription factors (*e.g.* GLI1/2, NFkB, HIF1a, p53), phosphatases, kinases, and many more (Figure 5.2).[21–24]

Because the acetylation state is such an important aspect of transcriptional regulation, the specificity of HDAC functions is tightly controlled by a complex network of protein–protein interactions of HDACs and other relevant proteins. In order to activate the enzymatic activity of Class I HDACs, the recruitment to multi-subunit co-repressor complexes is required.[25,26]

The multiple interactions of the 11 HDACs are not yet completely characterised. For Class I HDACs, it has been shown that HDAC1 and HDAC2 together with RBBP4 (RBAP48) and RBBP7 (RBAP46) are part of the core deacetylase complex. This core complex is an essential part of the Sin3,[27] NuRD[28] and CoREST[29] complexes. By interacting with the CoREST complex, HDAC1 and HDAC2 are also bound to lysine demethylase KDM1A, which is an example of coupling histone demethylation and histone deacetylation.[30] A cartoon of the structure of the NuRD complex is shown in Figure 5.3.

A crystal structure of a complex between HDAC3 and SMRT (NCOR2) co-repressor highlights that in addition to the direct binding properties of the involved proteins, inositol phosphate is required for the complex assembly and activation of Class I HDACs.[25] This principle has also been confirmed recently by resolving the structure of HDAC1 in complex with MTA1 which is part of the NuRD complex.[26] This is the first structure of HDAC1 to be published. The success of this work underlines the importance of co-crystallising HDAC enzymes with relevant binding partners in order to stabilise the protein structure (Figure 5.4).

In addition to increased rigidity due to formation of multi-protein complexes, post-translational modifications can alter HDAC activities. For example, phosphorylation of HDAC1 by CK2 leads to modified binding with NuRD complex and altered enzymatic activity.[31] Another kinase, Mirk1 (Dyrk1B),

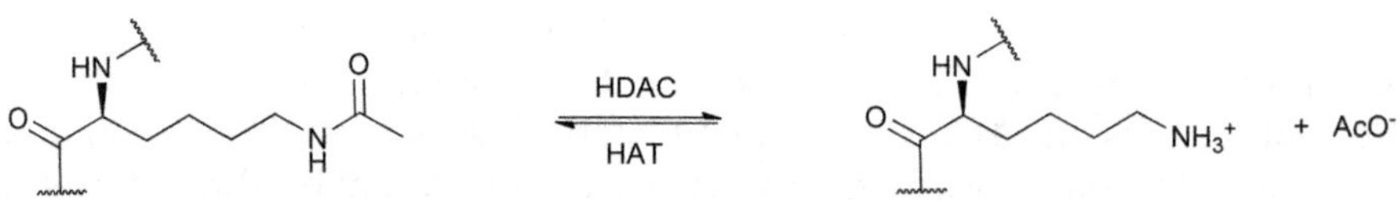

Figure 5.2 HAT enzymes catalyse the transfer of an acetyl moiety to lysines of proteins. HDACs are catalysing the backward reaction of deacetylation.

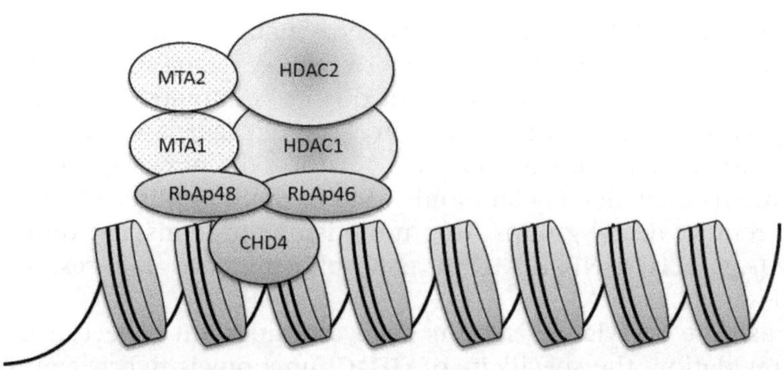

Figure 5.3 Representation of the assembly of the NuRD repressor complex with two HDAC isoforms HDAC1 and HDAC2 being an essential part of the complex.

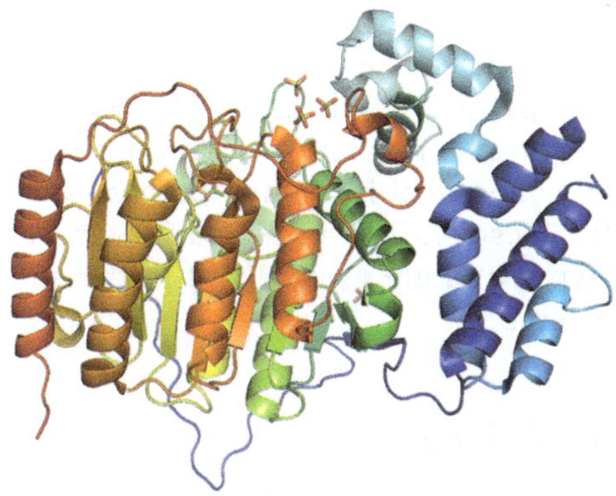

Figure 5.4 Representation of HDAC1 in complex with MTA1 and the involvement of phosphoinositol in stabilisation of the complex (PDB code 4BKX[26]).

can phosphorylate Class IIa HDACs at the 14-3-3 binding site and thus regulate intracellular distribution of these enzymes.[32]

Generally, HDACs have a high number of interacting and binding partners for context-specific regulation of transcription. Recently, Joshi *et al.*[14] investigated the functional interactome landscape for the 11 human Class I, II and IV HDACs in human T-cells. Interactions clustered nicely according to functional classification and revealed a high number of known and previously unknown interactions of various HDACs. These findings may also lead to a broader understanding of why kinetics of histone acetylation do not directly correlate with effects on gene transcription.[33] In a recent work, Lauffer *et al.*[33] have shown that kinetics of acetylation changes correlate well with enzyme inhibition kinetics of respective HDAC proteins, but the impact on regulation of target gene expression turned out to be more complex. It can be speculated that the broad variety of interactions of HDACs is causal in the fine tuning of transcription regulation.

5.2.2 Regulation of Signalling Pathways and Phenotypes

Inhibition of HDAC proteins with small-molecule compounds can result in regulation of therapeutically relevant signalling pathways. Therefore, a careful analysis of the impact of HDAC inhibitors on target genes of relevant pathways may lead to the identification of the links between the complex acetylome and relevant phenotypes. In order to understand the level of intracellular changes in acetylation, several groups have analysed the cell-wide modulation of acetylation and investigated the influence of HDIs on the acetylome with advanced mass spectrometry and bioinformatics analysis.[34–37] On the other end of this investigation, modulation of complete pathways by change of acetylation patterns have been investigated. For example, NFκB signalling can be directly linked to HDAC activity through a direct association of HDAC1 with the Rel homology domain of p65 and an indirect effect of HDAC2.[24,38] In the field of oncology, a very important group of molecular pathways is heavily influenced by acetylation status of involved proteins. This group consists of pathways linked to embryonic stem cell features and growth biology. For example, regulation of GLI proteins, which are the core of canonical and non-canonical hedgehog pathways, are particularly regulated by some HDIs. These clusters of regulation link biochemical changes and cellular phenotypes and are exemplified in the context of stemness in Section 5.5 in more detail.

5.3 HDAC Inhibitors

5.3.1 First- and Second-Generation HDAC Inhibitors

Since identification of the first HDAC enzyme[39] and the first HDAC inhibitors such as valproate **2** and trichostatin **4**, lots of molecules with different chemotypes have been isolated or synthesised, which have been summarised in a number of reviews.[3,7,8,10,19,40–42] Identification and development of second-generation HDAC inhibitors was dedicated towards improvements based on the insights gained from first-generation molecules. These include HDAC subtype selectivity, specific regulation of signalling pathways, and optimised physicochemical properties which allow for long-term safe and efficacious treatment.

However, there are still challenges associated with development of third-generation HDAC inhibitors: What biochemical isoform selectivity is beneficial for specific indications? Can we leverage the findings from the HDAC interactome? How is enzymatic selectivity translating into specific histone code patterns, if at all? Are some of these patterns correlated with relevant phenotypes in cells and organisms? Can we design the right third-generation HDAC inhibitor based on structural information? Should we combine HDAC inhibition and secondary pharmacological target activities in one molecule?

A core feature of all HDACs from Class I, II and IV is a Zn atom at the catalytic site that can be addressed by small-molecule inhibitors. Chemical space in HDIs can be divided into four main classes of inhibitors: hydroxamates, carboxylates, aminophenylanilides and others, as shown in Figure 5.3. Most hydroxamates are claimed to be pan-HDAC inhibitors, whereas aminophenylanilides have a preference for Class I HDAC inhibition. Selectivity of compounds, however, was determined with different biochemical test systems and recent papers showed that the isoform-specific inhibition depends on the substrate used and that addition of proteins which interact with the HDACs or co-expression with, for example, SMRT in the case of HDAC3 have a significant impact on biochemical activity values.[43] Bradner *et al.*[44] described some newer synthetic substrates which allow for testing of the HDACs with lower enzymatic activity including Class IIa HDACs 4, 5, 7 and HDAC 9. Jones *et al.*[45] reported that in addition, impurities in preparations of Class IIa HDACs can lead to incorrect assessments of inhibition and selectivities.

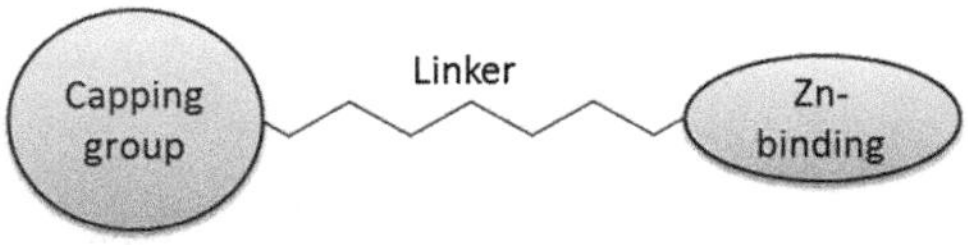

Figure 5.5 Basic concept of most HDAC inhibitors consisting of a Zn-binding group, a linker and a capping unit.

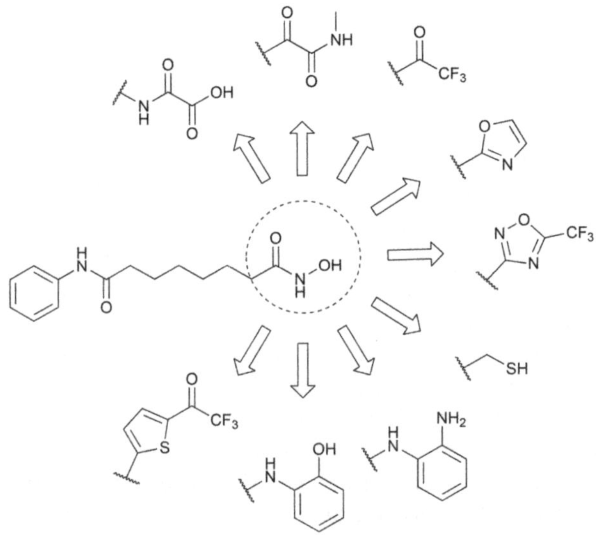

Figure 5.6 Starting from hydroxamates, as exemplified by vorinostat structure shown, chemists identified a diverse set of alternative Zn-binding moieties.

Newer data suggests that vorinostat is not a pan-HDAC inhibitor but is selective for Class I and IIb.[44] A recent paper by Lauffer *et al.*[33] added yet another complicating aspect of the kinetics of binding. It was shown that the two hydroxamates vorinostat **1** and TSA **4** have fast binding kinetics, whereas aminophenylanilides like entinostat **7** are slow binders and have a longer residence time on HDACs. Because residence time turned out to vary among different isoforms, this concept can thus be used to develop molecules with high kinetic selectivity. In this study, entinostat (**7**) was found to be ten-fold selective for HDAC2 over HDAC1.

Most HDAC inhibitors share a common design consisting of a Zn-binding moiety, a linker which is a mimetic of a substrate lysine and a capping group, as shown in Figure 5.5.

Vorinostat is probably the best investigated prototype of this concept. Zinc binding is performed by a hydroxamate and an aliphatic linker connects the hydroxamate with a small aromatic system, the capping group, as shown in Figure 5.6.

Figure 5.7 shows a high-resolution X-ray structure of vorinostat **1** (left) and compound **16** (right), a selective Class IIa HDI bound to HDAC7.[46] Researchers

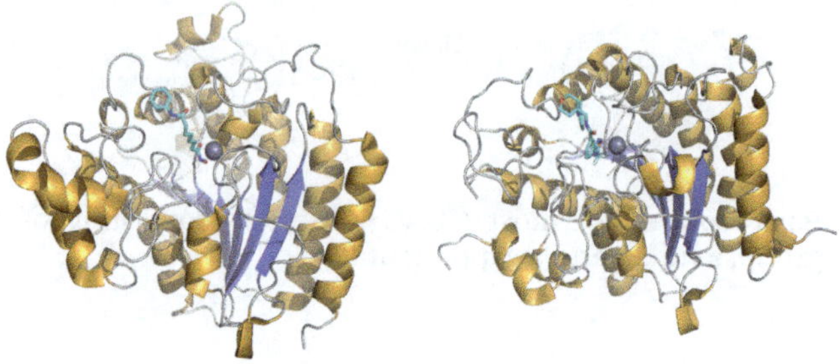

Figure 5.7 High-resolution crystal structures of vorinostat (left, PDB code 3ZNR) and compound 16 (right, PDB code 4LX) co-crystallised with HDAC7.[46]

put much effort into trying to vary the Zn-binding motif with the aim of improving selectivity, and the pharmacokinetic and physicochemical properties of HDIs. However, most published HDIs share the hydroxamate moiety as a Zn-binding motif. Optimisation of the linker group has been successful in order to optimise the pharmaceutical properties of HDIs. Panobinostat, resminostat and belinostat represent molecules which are in advanced clinical testing and have significantly improved properties over vorinostat.[47–50]

The search for alternative binding motifs which are able to interact with the important Zn in the HDAC proteins has led to a number of interesting new approaches. A selection of examples is shown in Figure 5.6. Groups which are capable of Zn-binding and which are of different chemistry generate new subtype selectivity and kinetics. Some examples are ketoamides,[51] ketoisoxazoles like compound **18**,[52] trifluoromethyloxadiazoles,[46,53] sulfamides,[54] trithiocarbonates,[55] or even primary amides.[56]

5.3.2 Selectivity within HDAC Subtypes

A number of academic and industrial programmes are aiming for biochemically selective HDAC inhibitors. In this area a lot of interest has been attributed to HDAC1, HDAC2, HDAC4, HDAC5 and HDAC6 selective molecules. More recently there is a growing interest in the use of, for example, HDAC9[57–60] and HDAC11 selective inhibitors for treatment of diseases outside oncology.[61]

Because HDAC6 has some unique properties in its interaction with HSP90 and its cellular distribution is highest in cytosol, it has been the subject of increasing interest. Implications of selective HDAC6 inhibition have been discussed extensively.[62] Starting from tubacin (**13**), a number of HDAC6 selective molecules have been discovered.[62] In particular rocilinostat (ACY1215, **12**) has entered clinical phase I development in patients with multiple myeloma. A couple of interesting concepts have been developed to obtain higher selectivity towards HDAC isoforms. For example, as early as 2003, heterocyclic ketones like compound **18** and ketoacylamides were identified as potent HDIs.[63]

Lobera *et al.*[46] developed trifluoromethyloxadiazole (TFMO) containing inhibitors which show extraordinary Class IIa selectivity and low nanomolar

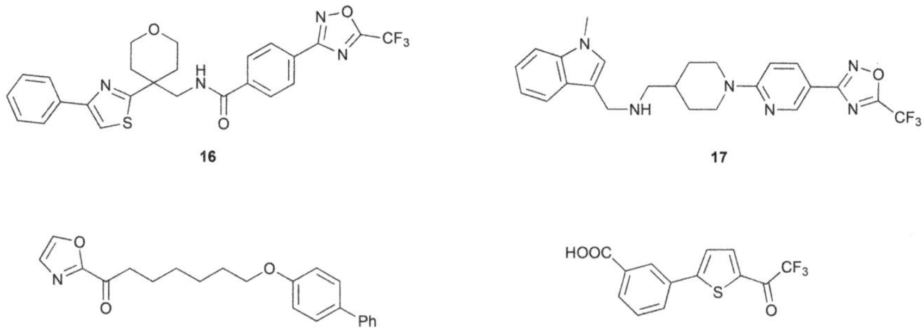

16 17 18 19

Figure 5.8 HDI compounds with HDAC isoform selectivity.

inhibition. The solution of an HDAC7 co-crystal structure of molecule **16** shows that the TFMO head group is binding to the Zn in the active site but is not chelating the ion (see Figure 5.7). Compound **16** and the *N*-methylpyrimidine analogue are selective inhibitors of HDAC9 with IC_{50}s of 19 nM and 8 nM, respectively. Another example of an HDAC9 inhibitor of the same group is compound **17** (Figure 5.8).[53]

Ontoria *et al.*[64] identified lead compound **19**, a trifluoromethylketone with ten-fold selectivity on HDAC4 (100 nM) and HDAC6 (70 nM) with respect to HDAC1 (13 μM).

5.4 Histone Acetyl Transferases (HATs)

Another class of emerging targets are histone acetyl transferases (HATs), the natural antagonists of HDACs.[65] This field of research is focusing mostly on the two enzymes and transcription factors CBP and p300.

The search for selective HAT inhibitors has turned out to be challenging. A number of relatively weak small-molecule modulators were found, *e.g.* hydroxyquinolines,[66] and long chain alkylidenemalonates.[67] After publication of a high-resolution crystal structure of p300/CBP by Liu *et al.* in 2008,[68] Cole *et al.*[69] identified interesting compounds using a virtual screening approach. Compound **20** (C646, Figure 5.9) is an inhibitor of p300 with an IC_{50} of 1.6 μM.

An alternative approach to p300/CBP inhibition is an indirect one, by blocking the chromatin binding of p300/CBP. A chemical probe molecule, SGC-CBP30, was co-crystallised with CREBBP and shown to be a potent inhibitor of CREBBP bromodomain with a K_d of 21 nM.[70]

5.5 Stemness, Chromatin and HDACs

The role of HDACs for the phenotype of cells has been investigated by inhibition and depletion of the enzymes. It is important to note that all phenotypes described might be mediated by specific inhibition patterns of individual

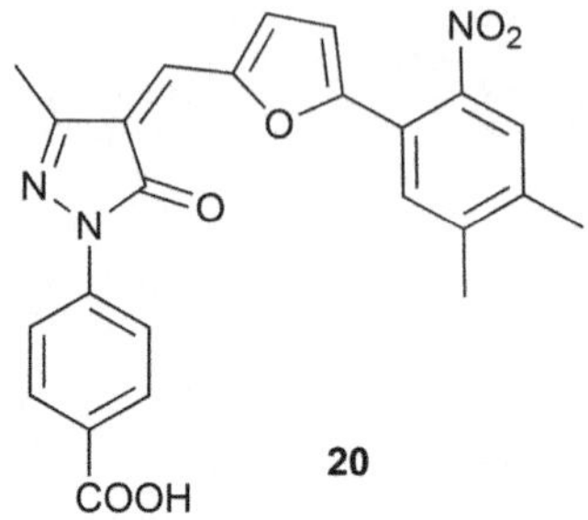

20

Figure 5.9 Compound 20 inhibits p300 HAT with an IC_{50} of 1.6 μM.

HDAC inhibitors. Some HDAC inhibitors may mediate quite similar phenotypic changes while others may not, even though those inhibitors have a similar biochemical HDAC isoform selectivity profile. Additionally, the inhibition of HDACs and the depletion of HDAC enzymes may or may not have the same outcome for a cell.

Because the DNA is basically the same in every cell of an organism (despite mutations in single cells), the differentiation of an oocyte towards a whole organism with functional organs is orchestrated by epigenetic changes. So, the chromatin structure and chromatin status are very important for the phenotype and differentiation status of a cell.

5.5.1 Stem Cells and Chromatin

For example, the pluripotency of embryonic stem cells (ESC) appears to be due to an active chromatin remodelling orchestrated by the transcription factor Oct-4.[71] The balance between acetylation and deacetylation by HATs and HDACs seems to be the key for this chromatin status. The initiation of differentiation can be pushed by disturbing this balance, by either repressing Oct-4 expression or inhibiting HDAC1 activity, thereby enhancing the acetylation of histones.[71] Those effects have been investigated by siRNA mediated depletion of Oct-4 and HDAC inhibition by TSA (**4**) inhibiting HDACs 1, 3, 4, 6 and 10.

Important for ESC maintenance and key cell identity, Whyte *et al.*[15] described the abundance of so-called super-enhancers, a special chromatin structure consisting of huge enhancer regions, high transcription factor (TF) density and content and enhanced ability to activate gene transcription.[15] H3K27ac and H3K4me1 are described as histone marks for active transcription sites and could be detected in their study with much higher levels than in normal enhancer regions. Genes expressed from these super-enhancer regions seem to be the master regulators of ESC maintenance. On the other hand, perturbation of this feedback loop by depletion of Oct-4 or the general transcription factor Mediator leads to preferential loss of super-enhancer associated genes compared to normal enhancer associated genes. Therefore, if the cell status is controlled or maintained by specific super-enhancer structures, this status is much more sensitive to imbalance of certain components. Lovén *et al.*[16] described the role of super-enhancers in oncogenesis and shed new light on oncogene addiction and epigenetic cancer targets. They showed that super-enhancer structures were present at key oncogenic drivers like the MYC oncogene and others in multiple myeloma cells and that the inhibition of BRD4 as a binding element for H3K27ac leads to gene-specific inhibition in those cells. They concluded that key oncogenes are regulated by super-enhancers and are therefore more sensitive to disturbance of the epigenetic environment.

5.5.2 HDACs, Chromatin and Transcription

The general view that HDACs are associated with transcriptional repression and HATs with transcriptional activation must be revised because the genome-wide mapping of HATs and HDACs uncovered distinct functions in active and inactive genes.[72] Actively transcribed genes are associated with high levels of both HATs and HDACs. HATs seem to be important for the acetylation and subsequent recruitment of RNA polymerase II for transcriptional initiation and elongation, whereas HDACs seem to be important in those actively transcribed genes to reset the chromatin structure for the next round of transcription. Inactive genes that are primed for transcription by H3K4 methylation are associated with low levels of transiently bound HATs and HDACs. Deacetylation within the promoter region by HDACs prevents the binding of RNA polymerase II and subsequently transcription. Disturbing the balance between low levels of HATs and low levels of HDACs in H3K4me primed genes by, for example, inhibition of HDACs can result in a remarkable increase of RNA polymerase II binding in the relevant promotor region.

5.5.3 HDACs and Differentiation

Because HDACs are important in the differentiation of organs during embryonic development, conditional knockout models have been developed. The deletion of HDAC1, 2 or 3 is embryonically lethal, whereas in cell-type specific deletions of specific HDACs their function appears to be redundant and can be replaced by other HDACs. For example, HDACs are associated with the differentiation towards liver, cardiovascular, neuronal and haematopoietic lineages.[73]

Literature analysis leads to an inconsistent picture of the role of HDACs in differentiation and de-differentiation, when investigated pharmacologically. Summarising those publications, results seem to be highly dependent on the HDI and concentration used, the duration of intervention and the origin and developmental stage of the cell used if differentiation or de-differentiation results. Experiments of Karantzali *et al.*[74] shed more light on this apparently conflicting topic by investigating TSA-treated ESC for changes in chromatin structure and gene expression pattern. They showed that treatment with TSA on a global scale leads to increased H3 acetylation and H3K4 trimethylation and a concomitant loss of the repressive H3K27 trimethylation. They conclude that before full commitment to a differentiated phenotype, there might be an indifferent and highly permissive state for the cell. Therefore, it is possible that HDAC inhibitors in general lead to this kind of permissive state by opening the chromatin and facilitating the establishment of active chromatin marks. Depending on the origin of the treated cell and its developmental status, those cells can then differentiate or de-differentiate dependent on further signals. An example of the permissive status of the chromatin induced by HDAC inhibitors is the creation of induced pluripotent stem (iPS) cells as described by Huangfu and colleagues.[75] Reprogramming of murine embryonic fibroblasts (MEFs) towards induced pluripotent stem (iPS) cells, artificially induced by expression of the master transcription factors c-Myc, Klf4, Sox2 and Oct4, can be accelerated and reinforced by treatment with chromatin-modifying chemicals like HDAC and DNMT inhibitors. By testing vorinostat, TSA, valproic acid (VPA) and 5′-azacytidine it was shown that these HDIs enhance reprogramming more potently than the DNMTi used. VPA was the most potent HDI. On the other hand, treatment of untransfected MEFs with VPA alone does not lead to induction of iPS cells. Clearly, additional signals are necessary for the subsequent de-differentiation but modifying the chromatin status seems to act like a gatekeeper.

An example where the outcome of shifting the balance between acetylation/deacetylation of histones in stem cells leads to differentiation induction was published by Dong and colleagues.[76] They describe the induction of differentiation of bone marrow derived mesenchymal stem cells into liver cells by treatment with VPA. They showed that the inhibition of Class I HDACs by VPA leads to an expression pattern of hepatocyte-specific genes and liver-specific functions like glycogen storage, cytochrome P450 activity and synthesis of AFP and ALB.

5.5.4 Cancer Stem Cells and Signalling Pathways

Cancer stem cells are described as a subpopulation or a transient status of cancer cells with high plasticity and tumour-initiating capacity, presumably responsible for the initiation of the tumour, metastasis formation and recurrence. So far there are no reliable surface markers to clearly define a cancer stem cell population. Therefore a more phenotypic description and the abundance of active signalling pathways should be the basis for further investigation. Cancer stem cells share multiple properties with ESCs, especially their dependency on developmentally relevant pathways like Wnt, Hedgehog, Notch and TGF signalling. Because there is crosstalk and interaction between these pathways, and HDACs have several ways to modify each single pathway, the contribution of HDACs to this network is quite complex and additionally cell status dependent (Figure 5.10).

Apart from their epigenetic role in chromatin modification, HDACs can also act by direct interaction with other regulatory proteins. It has been shown that HDACs can be recruited by the co-repressor Groucho, which is

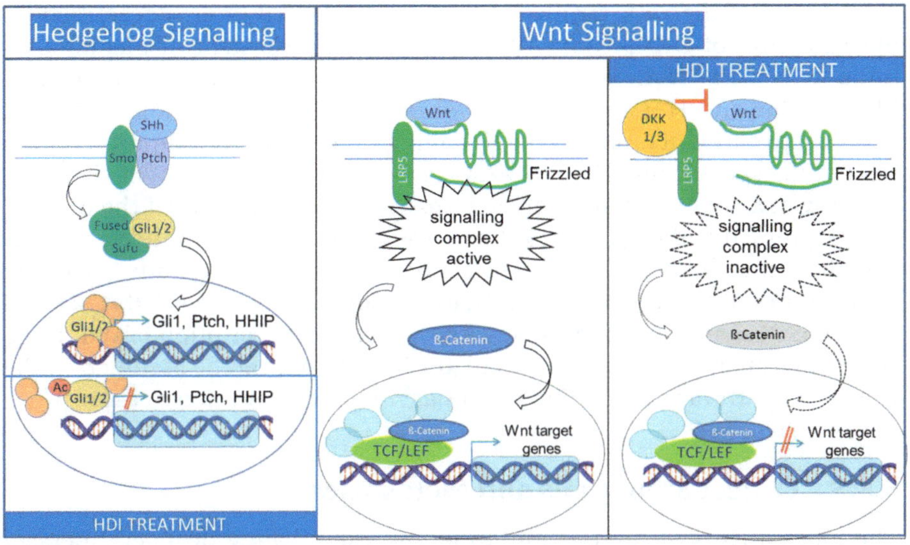

Figure 5.10 Schematic representation of Wnt and Hh pathways.

involved in the Wnt and Notch signalling pathway.[77,78] Additionally, enzymatic activity of HDACs leads to deacetylation of non-histone proteins, thereby modifying their activation status, their stability or their potential for interaction with other partners. To investigate the influence of HDACs and HDIs on these pathways, reporter assays are often used. But this model is overly simplistic because the normal heterogeneity in promotor composition is not captured. This disregards the enormous modulation capacity of specific HDIs on very complex promotor-specific complexes. Therefore, an analysis of multiple and meaningful primary target genes and phenotypic outcomes is absolutely necessary.

The key transcription factors of the Hedgehog (Hh) signalling pathway are Gli1, Gli2 and Gli3. Gli1 and Gli2 have activating capacity for transcription and transforming activity. Post-translational modifications like phosphorylation, ubiquitination and acetylation are key regulators of the transcriptional activity of Gli1 and Gli2. Gli2 acetylation by the HAT p300 appears to prevent Gli2 recruitment to the chromatin, possibly by precluding the binding to the chromatin remodelling factor Brg1.[79] Brg1 has been reported to interact with both HDACs and Gli, and plays a crucial role in Hh-induced chromatin recruitment and target gene.[80]

The impact of HDACs on pathway regulation might be even more complicated, as described for the Notch signalling pathway, in smooth muscle cell differentiation. In this example CBF-1-mediated Notch signalling is not suppressed by HDAC inhibition, but the HDAC activity is required for Notch differentiation signals through mitogen-activated protein kinase and PI3K pathways.[81]

The Wnt signalling pathway can be regulated on multiple levels and is aberrantly activated in medulloblastoma by epigenetic repression of the antagonists of the DKK family DKK1 and DKK3. Two groups described the up-regulation of DKK1[82] and DKK3[83] following treatment of medulloblastoma cells with TSA. Activating promotor-specific histone modifications like H3K9ac and H3K4me2 were involved in this up-regulation. Exceeding this negative modulation of Wnt signalling on the ligand/receptor level, the HDAC1,2,3 specific inhibitor 4SC-202 negatively regulates the Wnt signalling pathway in a concerted manner such that positive regulators of the pathway were repressed and negative regulators were induced. This inhibition of Wnt signals is mediated at multiple levels of the signalling cascade, including transcriptional repression of chromatin modifiers such as CARM1, DOT1L and WHSC1 which are associated with the active transcription of Wnt target genes. Modulation of a master regulator by 4SC-202 can be assumed in this concerted pathway modulation but it has not been identified to date.[84]

5.5.5 Cancer Stem Cells and Metastases

The mechanisms of tumour initiation and establishing of macro metastases share multiple features, namely: (i) the fruitful interaction with a new environment, (ii) maintenance of self-renewal capacity, and (iii) potency to establish a heterogeneous tumour.[85] Because those processes must be well coordinated and highly flexibly controlled, epigenetic mechanisms seem to be strong candidates for this dynamic interplay. Bernardo and colleagues[86] described how the endogenous HDAC inhibitor maspin regulates extracellular matrix (ECM) remodelling and inhibits TGFβ signalling and its downstream networks. In tumour bone metastasis, TGFβ release from the stroma seems to be the driver for tumour/bone interactions with subsequent tumour bone remodelling and establishment of macro metastasis.[87] Therefore, one impact of HDAC inhibition on the establishment of macro metastasis might be the prevention of proper stroma/tumour interaction by ECM proteins and paracrine signalling cascades. The specific HDAC1,2,3 inhibitor 4SC-202 was shown to prevent the formation of metastasis in an orthotopic RENCA mouse model.[84] Even though it has not been investigated in this specific animal model, the negative influence of 4SC-202 on the interface of tumour cells could be assumed.

Considering this knowledge about the importance of epigenetics in the cancer stem cell theory we should rethink clinical trial design for further development of epigenetic modifying drugs.

5.6 HDIs in Clinical Development

Without any doubt HDIs have established themselves as promising treatment modalities in the field of epigenetic therapy and have made their mark on our therapeutic armoury available to fight cancer. When vorinostat (**1**) and romidepsin (**3**) were approved by the FDA and EMA for the treatment of haematological malignancies, HDIs obtained their validation. To date, this excitement has triggered approximately 500 clinical trials in oncology alone[41] to further evaluate this therapeutic concept. Currently, there are approximately 300 clinical trials listed as 'ongoing' and 'recruiting' at clinicaltrial.gov. Many of them are investigating the use of HDIs in combination with other chemotherapeutics. As our clinical experience with epigenetic strategies has increased, HDIs have also been shown to have some shortcomings. The most prominent observation consists of a strong bias towards haematological malignancies, especially against cancers derived from T-cells, with respect to clinical effectiveness. What is missing is a robust manifestation of clinical effects in solid tumour entities.

Many excellent reviews are available in the current literature that summarise not only the clinical experience gathered up to now,[88–90] but also focus on observed toxicities,[91] molecular pharmacology,[9,92] combination therapy[93,94] and strategies to position HDIs in the field of solid tumours and to expand their clinical utility to other disease areas.[95] Therefore, it is not our intention to provide another comprehensive review of available data. In this chapter we will present a bird's eye overview of the 'lessons learned' from the ongoing clinical development. We will also restrict ourselves to oncology, although the application of HDIs in other therapeutic areas like inflammatory diseases, diabetes, heart failure, virology or neurological disorders like multiple sclerosis is gaining considerable importance.[95–97] With regard to oncology, a pertinent and still open question concerns the appropriateness of doses and dosing schemes currently used in the clinic along with the limited sensitivity of solid tumours towards HDIs. Another important aspect is the identification of suitable biomarkers. Whereas pharmacodynamic markers are almost abundantly available, predictive markers are urgently required. Overall, the infamous five 'R's[98] of personalised medicine (right drug, right dose, right time, right indication, right patient) need to be assessed with the same scrutiny in the field of epigenetic drugs as has been accomplished in other areas of targeted therapy.

5.6.1 Clinical Pharmacology of HDIs

Research and development efforts in the field of HDIs culminated in the FDA approval of vorinostat and romidepsin, both in cutaneous T-cell lymphoma in 2009. Later, romidepsin was approved for relapsed peripheral T-cell lymphoma in the USA. Since then, researchers all over the world have tried to advance the concept of HDAC inhibition to tumour indications not limited to haematological malignancies. Although inhibition of HDAC enzymes is usually considered a targeted therapy, researchers and clinicians are facing a remarkable spectrum of possible pharmacological effects.[88,92] On the one hand, this might be the result of inhibiting several HDAC isoforms to different degrees by a given HDI. On the other hand, particular isoforms are involved in several hallmarks of cancer biology, like control of proliferation, migration and angiogenesis or inhibition of apoptosis and cell differentiation.[88] Understanding the mechanism by which any given HDI is exhibiting its particular pharmacological effect will determine its clinical use. Based on *in vitro* results no less than 10 molecular mechanisms have been postulated and we believe there are yet more to emerge.[88,92] These different effects are clearly dose dependent and it will be important to evaluate which clinical drug exposures are necessary to elicit the desired pharmacology. A still open and intensively discussed topic constitutes the rather limited clinical success of HDIs in monotherapy in solid tumour entities. A recent review of 30 clinical trials in solid tumour entities concluded there was an obvious lack of therapeutic effect in HDI monotherapy (based on 20 trials) but increased therapeutic effects can be stated for combination therapy (based on 10 trials).[99] To put the question differently, why is HDI monotherapy best suited for haematological malignancies, particularly in T-cell lymphomas and Hodgkin's lymphoma? Currently, we cannot provide a final answer and we must rely on rather speculative assumptions. One reason could be the lower number of mutations found in haematological tumours compared to solid ones. Therefore, haematological tumours might be furnished with a more intact apoptotic pathway signature that can be reactivated upon HDI treatment.[92]

One feature of the wide-ranging pharmacology of HDIs is their immune-modulatory effect, which might contribute to clinical effectiveness in CTCL.[100] It has been demonstrated that vorinostat is able to shift the prevailing TH2 milieu that characterises CTCL to a more favourable TH1 milieu. In particular, HDAC inhibition can down-regulate the expression of TH2 cytokines IL-10 and IL-4, whereas the expression of TH1 cytokine IFN-γ is up-regulated. These effects are most likely mediated by HDI-induced attenuation of STAT3-dependent transcription.[101] A recent investigation has pinpointed these effects to the involvement of isoform HDAC3, at least for panobinostat in STAT3-positive diffuse large B-cell lymphoma cells.[102] Replication stress was also found to be a result of selective HDAC3 inhibition in a CTCL cell line using the HDAC3 selective inhibitor RGFP966.[103] Both romidepsin and vorinostat are potent HDAC3 inhibitors[41] and this would be consistent with a hypothesis postulating HDAC3 inhibition as a possible rationale for the clinical effects observed in lymphomas.

Another possible explanation for the difference in sensitivity towards HDIs between solid and haematological tumours might be that optimal dosing and dosing schemes for solid tumours are not yet established. Recently, the concept of treating patients at the maximum tolerated dose with epigenetic drugs has been seriously challenged.[90] The effects of HDIs might be classified as cytotoxic or epigenetic in nature, depending on the prevailing drug concentration. Generally speaking, cytotoxic effects are triggered rapidly at higher drug concentrations whereas epigenetic reprogramming might need time to develop and can be achieved at sub-cytotoxic concentrations. This concept has been demonstrated for DNMTi, however a clinical confirmation for HDIs is still missing. For a more in-depth discussion on optimal dosing and dosing schedules for epigenetic drugs the reader is referred to Chapter 10 by Steve Baylin.

The ability of HDIs to induce cell differentiation by epigenetic reprogramming is of particular interest. The introduction and experimental verification of the concept of the so-called cancer stem cells or tumour-initiating cells has catalysed this interest. As already discussed above, cancer stem cells are believed to comprise a poorly differentiated subpopulation within the bulk tumour. These cells are remarkably resistant to conventional chemotherapy and are believed to be responsible for relapse after apparently successful eradication of the bulk tumour. Accordingly, induction of a differentiation programme by epigenetic drugs to 're-transform' cancer stem cells into chemosensitive tumour cells comprises an intriguing alternative therapeutic concept.

5.6.2 Combination Therapy

Confronted with the moderate success of HDAC inhibition in haematological and solid tumours in monotherapy, attention has shifted towards combining HDIs with existing chemotherapeutic regimens in a rational manner. HDIs possess an acceptable toxicity profile and can, at least in principle, address a broad spectrum of pharmacological mechanisms which can be matched to an optimal combination partner. Judged by their prime mode of action the most obvious combination consists of HDIs plus DNA-intercalating agents. Because inhibition of HDAC enzymes results in an open and de-condensed nucleosome assembly, intercalating agents like topoisomerase inhibitors are granted access to DNA more easily and synergise with HDAC inhibition.[104–106] Currently, the combination of doxorubicin, an intercalating topoisomerase II inhibitor, and vorinostat is being investigated in two phase I/II trials in patients with advanced solid tumours or relapsed/refractory lymphomas, respectively. The ability of class I/II HDIs to affect crucial components of the DNA repair response provides the molecular basis for the combination with DNA-damaging therapies, including radiation.[107] A plethora of *in vitro* data demonstrate that HDIs can increase p53 activity, reduce DNA repair kinetics of Ku70, an important component of the non-homologous end joining, and might influence the physiological role of BRAC1, ATM and ATR.[104] This concept is exemplified by several clinical phase I trials including the evaluation of CUDC-101 together with radiation and cisplatin in head and neck cancer, or the combination of radiation therapy and panobinostat in patients with brain tumours. Other approaches involving DNA-damaging agents include the clinical evaluation of HDIs in combination with topoisomerase I inhibitors, like irinotecan. Preclinical data demonstrate that HDIs potentiate the effect of topoisomerase inhibitors, however a clear schedule dependency was observed. Greater efficacy is gained upon pretreating cells with HDIs and as such sensitising the cells to DNA-damaging agents or radiation. This aspect needs to be carefully considered in the design of combination studies.

In the context of combination therapy, isoform HDAC6 recently took centre stage. First of all, this HDAC isoform is thought to be responsible for deacetylation of microtubules and promoting microtubule depolymerisation.[108]

Taxanes, including paclitaxel and docetaxel, preferentially bind to polymerised microtubules and act as stabilising agents by themselves. *In vitro* studies indicated that taxanes exhibit an even higher affinity towards HDI-induced, pre-stabilised microtubules. Recent evidence demonstrates that microtubules show increased acetylation by treating MCF-7 cells either with TSA or HDAC6 directed siRNA. However, only pharmacological inhibition of HDAC6 by TSA resulted in increased microtubule stability. Additionally, upon inhibitor treatment, HDAC6 proteins displayed an increased binding to microtubules. It is speculated that HDI-induced HDAC6 binding contributes to microtubule stabilisation rather than its acetylation activity.[108] Currently, the combination of HDIs with paclitaxel alone, or in combination with paclitaxel and other chemotherapeutics, is under intensive clinical investigation.

Inhibition of isoform HDAC6 also provides a mechanistic rationale for combining HDIs with hormonally-targeted therapy. In oestrogen or androgen receptor positive breast and prostate cancers tumorigenesis is driven by hormone receptor-mediated transactivation of oncogene expression with concurrent inhibition of tumour suppressor genes. HDIs in general can abolish oestrogen and androgen receptor-mediated signalling by down-regulating their gene expression. Isoform HDAC6 in particular positively regulates heat-shock protein Hsp90, which is an essential constituent of chaperone complexes. Chaperone complexes stabilise physiologically important signalling proteins and guarantee their functional integrity. Hsp90 is known to interact with hormone receptors and by such means maintains the active ligand-binding receptor conformation. Upon acetylation the hormone receptors are released from the complex and diverted to proteasomal degradation. Pharmacological intervention by means of inhibiting HDAC6 will shift the distribution of Hsp90/hormone receptor complex to a degradation prone state.

The concept of combining HDIs with hormone-targeted therapy has received attention in several clinical trials, albeit with a different mechanistic rational. Vorinostat was administered together with tamoxifen, an oestrogen receptor antagonist, to patients with ER-positive metastatic breast cancer already progressing on endocrine therapy.[109] In preclinical models tamoxifen is able to stabilise ER, whereas vorinostat down-regulates ER transcription and reverses tamoxifen-induced stabilisation. These effects are attributed mainly to the inhibition of isoform HDAC2. An objective response of 19% and a clinical benefit rate of 40% were reported for this trial. Other clinical strategies include the combination of HDIs with aromatase inhibitors, like anastrozole or letrozole, which prevent the enzymatic conversion of androgen to oestrogen. Clinical trials with vorinostat in combination with aromatase inhibitors are currently ongoing or awaiting results.

Due to their increased metabolic turnover, cancer cells are sensitive to proteasome inhibitors such as bortezomib (Velcade). Proteasome inhibition increases cellular stress and eventually results in cell death. Many HDIs are also known to increase reactive oxygen species (ROS) and synergise in proteasome inhibitor-induced cellular stress.[94] Again, isoform HDAC6 might play a pivotal role because its activity is required to maintain aggresome integrity.

Currently, clinicaltrials.gov lists 16 recruiting clinical trials where bortezomib is combined with HDIs, either alone or in combination with other chemotherapeutic drugs.

Combining HDI with existing therapy modalities seems a promising approach to expand the clinical utility of HDIs and might pave the way into the field of solid tumours. Most studies that strive for a mechanistic understanding of combination strategies rely on *in vitro* or preclinical *in vivo* data. Certainly, it is a difficult task to perform mechanistic studies on patient material and we have to rely on preclinical findings to some extent. Therefore, what one should at least take care of is a pharmacologically meaningful translation of preclinical drug exposure into the clinical setting to ensure safety and benefit for patients.

5.6.3 PK/PD Considerations

The definition of a clinically useful dose and dosing scheme represents the ultimate goal of clinical development. Especially in oncology there seems to be a marked discordance between the derived recommended phase 2 dose (RP2D) and the dose used during late stage phase III development. A recent survey of 84 non-paediatric oncology trials, conducted between 1990 and 2012, revealed that only in 60% of the cases did the dose in late stage development exactly match the RP2D.[110] The determination of a clinically useful dose usually results from the joint effort of molecular pharmacology, pharmacokinetics (PK) and pharmacodynamics (PD) to deliver drugs with an optimal and balanced PK/PD profile given the prevailing mode of action. Understanding the PK and PD properties of current HDIs might provide useful insights for further development of next-generation compounds.

Most pan-HDIs belonging to the hydroxamic acid class of compounds display rather short plasma half-life in humans after oral single dosing. The average terminal $t_{1/2}$ of vorinostat ranges between 1.5 and 2 hours,[100,111] similar to belinostat (1.5 hours).[112] Prolonged half-lives are reported for resminostat (3 hours) and givinostat (5–7 hours),[113] whereas the half-life of panobinostat is around 16 hours.[114] Generally, dose proportional pharmacokinetics within the investigated dose range have been reported. The situation is different for Class I specific aminophenylanilide-like HDIs. The plasma half-life of mocetinostat amounts to 7–11 hours,[115] 4SC-202 is in the range of 20 hours and entinostat produces a terminal half-life of 50–150 hour.[116] These rather long half-life times were not expected from preclinical data.

There has been some debate in the HDI field regarding the pharmacokinetic properties that need to be met in order to elicit a pharmacodynamic effect.[111] Most HDIs are administered in a discontinuous manner in order to prevent or allow resolution of unwanted side effects. Given the rather short half-life times of some of the pan-HDIs, resulting in only transient target coverage, the question remains as to how a sufficient and clinically relevant pharmacological effect can be elicited. During the first wave of clinical trials most inhibitors were administered intravenously. As soon as pharmacokinetic data from the oral route became available it was evident that higher C_{max} values could be achieved under i.v. administration compared to the oral route. In contrast, oral AUCs exceeded the ones from i.v. administration, as was demonstrated for vorinostat.[117] This finding triggered the discussion about whether HDAC inhibition and consequent modulation of the epigenome represents either a C_{max} or an AUC driven effect.

A comparison of the time dependence of *in vitro* IC_{50} values on human tumour cell lines revealed a rapid onset of hydroxamate compounds like vorinostat or dacinostat (LAQ-824), whereas members of the benzamide class required several hours to develop their full pharmacological response.[118] In line with the above findings, wash-out experiments demonstrated a sustained duration of HDAC inhibition in the case of mocetinostat.[118] In contrast, prolonged target inhibition after drug removal was not observed for vorinostat. These findings suggest a direct relation of plasma concentration with target modulation for pan-HDIs belonging to the hydroxamic acid class of compounds and might suggest a C_{max} contribution. For benzamide compounds using AUC as a suitable measure of exposure might be the right pharmacokinetic strategy.[119] However, Novotny-Diermayr and colleagues reported the pharmacodynamic evaluation of SB939,[120] an oral pan-HDI. Histone H3 acetylation was measured in PBMCs (peripheral blood mononuclear cells) from patients with advanced solid malignancies as PD-marker. They found a good correlation (r = 0.97) for both C_{max} as well as AUC of SB939 plasma concentration when compared to pharmacodynamic biomarker response at 3 hours post-dosing.

The above findings have been strengthened by experimental data on inhibition kinetics of a diverse panel of HDIs.[33] The k_{on} and k_{off} rates of TSA, vorinostat and entinostat, along with two experimental Class I selective aminophenylanilide compounds, have been investigated on purified HDAC1 and HDAC2 isoforms. Aminophenylanilide compounds were characterised as slow tight-binders. Their drug residence times were in the range of 2 to 38 hours, depending on the respective compound within the benzamide class and on the respective HDAC isoform. Entinostat exhibited a drug residence time of 2.2 hours on HDAC1 and 12.3 hours on HDAC2. Drug residence times were relatively short for TSA (0.3 hours on HDAC2 and 0.5 hours on HDAC2) and not quantifiable in the case of vorinostat (below 1.4 minutes). This marked difference in the kinetic binding properties can be rationalised by comparing the X-ray structures of HDAC2 in complex with either vorinostat or benzamide compounds. The accommodation of the lipophilic biaryl moiety of the experimental benzamide compound requires the disruption of a hydrogen bond and the conformational rearrangement in parts of the binding site. On the cellular level the inhibitor binding kinetics reflect the onset and duration of HDI-induced histone acetylation. Wash-out experiments confirmed the rather direct concentration–response relationship of hydroxamate inhibitors, as well as the prolonged drug action seen for benzamide-based inhibitors. Surprisingly, transcriptional regulation of a set of cell cycle genes seems to be governed solely by association kinetics and is independent of drug residence times. During drug wash-out, normalisation of gene expression was similar for all compounds. It is also noteworthy that constant exposure to vorinostat resulted in a comparable number of up- and down-regulated genes, whereas after pulsed compound application a five-fold gene up-regulation compared to gene down-regulation could be observed. Recent work by Wilson and colleagues reports that vorinostat and panobinsotat require permanent exposure in order to inhibit growth of several colon-derived tumour cell lines.[121] In summary, these observations might indicate that constant exposure is necessary to elicit cytotoxic effects. It would be of utmost interest to investigate how binding kinetics of HDI might influence other gene clusters, *e.g.* genes responsible for cell differentiation. One could speculate on the utility of short-time exposure to selectively modulate the up- or down-regulation of certain clusters of genes.

Elucidating relevant exposure schemes might be crucial not only for a successful clinical implementation of future HDIs but for epigenetic modulators in general. However, monitoring PK/PD relationships on the level of HDAC target inhibition or histone acetylation might not be the relevant correlate to the underlying pharmacology on the transcriptional level and even less so for the overall clinical response.

5.6.4 Biomarkers for HDIs

Most common pharmacodynamic biomarkers monitored in clinical trials comprise the measurement of HDAC inhibition, acetylation of histone 3 or 4 and regulation of gene expression. These markers are employed during early development in order to demonstrate target engagement. These PD-markers are usually measured in PBMCs or whole blood cells because routine use of tumour biopsies is considered too invasive and demanding for the patients. It would be desirable to correlate these measures of target inhibition with clinical outcome. To the best of our knowledge so far, markers of target inhibition have not correlated with clinical end points.

These PD-markers have their limitations. There are issues with regard to substrate specificity to measure overall or isoform-specific HDAC inhibition directly. Other critical points are availability of validated antibodies to measure H3 acetylation, or the issue of overlapping epitopes in the case of H4 acetylation. The biggest dispute remains to what extent the epigenetic make-up of PBMCs reflects the situation in the tumour. Studies of Class I HDAC isoenzyme distribution in PBMCs compared to CLL samples revealed similar expression levels in both cell populations and provide a positive hint.[122] Another interesting finding originates from the study of epigenetic changes during the development of schizophrenia.[123] According to these studies pathological influences imposed from the environment on the patients are reflected in epigenetic changes in both target tissue and PBMCs.

However, the most valuable biomarker would be a biomarker predictive for response to therapy with HDIs. Ideally, a 'the target is the marker' scenario, well established in the field of kinase inhibitors, would be the most favourable situation in order to move epigenetic therapy towards the personalised medicine paradigm. To identify an epigenetic signature in terms of certain combinations of histone marks seems to be a reasonable approach for epigenetic drugs. However, we believe that methodological limitations and the complexity of the underlying system biology might prevent such an endeavour, at least for the present.

So far there have been no clinical trials performed that utilise a biomarker to predict the response towards HDIs or to define a responsive subpopulation. However, the protein HR23B (RAD23B) has been described as a predictive marker for HDI sensitivity in haematological tumours.[124] Originally, the protein was discovered in the course of a genome-wide loss-of-function screen to identify genes that render tumour cells sensitive towards HDIs.[125] HR23B actively delivers ubiquitinated proteins to proteasomal degradation. Under normal conditions HDACs inhibit HR23B mediated transport and regulate proteasome activity. Inhibition of the responsible HDAC isoforms leads to an overloaded proteasome complex and eventually results in ER stress and apoptosis. Very recently, New *et al.*[126] provided a more detailed insight into the mechanistic basis of this marker. Their work demonstrated that isoform HDAC6 regulates HR23B activity, albeit independent of HDAC6 enzymatic activity. This concept has been demonstrated for vorinostat, which is able to sensitise CTCL cells for proteasome inhibitors.

In a recent review article Timp and Feinberg[127] suggest a unified model for cancer wherein both the genetic and epigenetic levels of tumorigenesis and tumour progression are considered. Most importantly, the authors demonstrate that mutations in epigenetic enzymes are rather frequently observed in many solid and haematological tumour entities and as such might present 'epigenetic driver mutations'. Unfortunately, although the list of mutations in epigenetic targets is quite impressive, there are only two examples for HDAC enzymes. Isoform HDAC4 was found mutated at a frequency of 4% ($N = 24$) in breast adenocarcinomas and HDAC9 was mutated at a frequency of 42.9% ($N = 7$) in prostate adenocarcinomas. Targeting mutated epigenetic enzymes could create a situation similar to the kinase inhibitor arena. This might open up the possibility for prognostic as well as for predictive biomarkers or, as the authors suggest, even for tailored chemoprevention.

5.6.5 Clinical Toxicities

From a structural point of view HDIs currently in clinical development comprise a varied collection of chemical entities. A representative selection of HDIs discussed in this chapter is presented in Figure 5.1. Chemical structures include members of short-chain fatty acid compounds (*e.g.* valproic acid), hydroxamic acids (*e.g.* vorinostat), depsipeptides (*e.g.* romidepsin) and the class of benzamide compounds (*e.g.* entinostat). Despite this diverse panel of chemical functionalities, observed toxicities during clinical trials are relatively similar. The most prominent toxicities consist of fatigue, nausea, vomiting and anorexia, which were dose-limiting in phase I clinical trials in monotherapy. Frequently, neutropenia and thrombocytopenia have been observed as the most prominent haematological toxicities. Interestingly, the duration and intensity of neutropenia and thrombocytopenia were less pronounced when compared to conventional cytotoxic agents and resolved upon discontinuation of treatment within a short period of time. *In vitro* studies with romidepsin revealed an almost 1000-fold reduced sensitivity on bone marrow progenitor cells when compared to B-cell CLL cell lines. These findings might provide, at least in the case of romidepsin, an experimental explanation of the less pronounced myelosuppressive effects observed in the clinic.

However, cardiac safety was one of the most intensively disputed side effects observed during clinical trials. This discussion was mainly fuelled by a clinical trial of romidepsin in patients with neuroendocrine tumours,[128] where asymptomatic ventricular tachycardia and QT prolongation was observed in some patients. The study was terminated after one sudden death. Consequently, intensive monitoring of possible cardiac events has been implemented in HDAC development programmes. Changes in ST- and T-wave, as well as QT-prolongation, were reported for vorinostat and entinostat.[129] Over time more routine clinical use of HDIs has suggested that cardiac toxicities could be observed with structurally differing HDIs and might constitute a common finding for HDIs. However, extensive clinical experience with a range of HDIs including dacinostat, belinostat and romidepsin, revealed that cardiac toxicities are manageable and sometimes due to a predisposition resulting from an uncorrected deficiency in potassium and magnesium levels. Very experienced clinicians reported that non-specific asymptomatic T-wave flattening, which has been observed for several HDIs, can explain the apparent detection of QT-prolongation.[130] In conclusion, the occurrence of cardiac events during clinical use of HDIs does not justify a delay in further clinical development of this inhibitor class. Low dosing of HDIs might remove the risk of cardiac events.[130]

5.7 Conclusions

HDACs and DNMTs were the first targets identified in the epigenetic field. Today, an impressive portfolio of different HDI chemotypes is available and the number is still growing. These compounds offer to us a broad spectrum of selectivity towards HDAC classes or even particular HDAC isoforms. This diversity needs to be fully exploited in order to investigate the pleiotropic effects of HDAC inhibition. This might lead to a better understanding of how HDAC inhibition interferes with complex cellular processes and how their mode of action can be converted into useful treatment modalities for oncological and other diseases.

One should be aware that studying HDIs on the level of target inhibition alone might provide a rather limited picture of its biological consequences. Future HDAC research should focus on profound understanding of the interplay between modulating the histone code signature, its consequences on relevant molecular pathways and the resulting expression of phenotypes. Probably the greatest potential of HDIs might be their ability to epigenetically reprogramme certain cellular programmes. Driving cancer stem cells into differentiation appears to be the most attractive clinical application of HDAC inhibition. The large number of ongoing clinical trials underlines the enormous importance of the concept of epigenetic intervention. However, a critical rethinking of clinical trial designs as well as of clinical response criteria is of utmost importance to guarantee successful clinical development of epigenetic drugs in the near future.

Acknowledgements

We would like to thank Dr Christine Wenzkowski for the graphics of protein structures and Dr Stefan Strobl for proofreading the chapter.

References

1. D. M. P. Phillips, *Biochem. J.*, 1963, **87**, 258–263.
2. B. G. Pogo, V. G. Allfrey and A. E. Mirsky, *Proc. Natl. Acad. Sci. U. S. A.*, 1966, **55**, 805–812.
3. V. M. Richon, Y. Webb, R. Merger, T. Sheppard, B. Jursic, L. Ngo, F. Civoli, R. Breslow, R. A. Rifkind and P. A. Marks, *Proc. Natl. Acad. Sci. U. S. A.*, 1996, **93**, 5705–5708.
4. P. A. Marks, V. M. Richon and R. A. Rifkind, *JNCI, J. Natl. Cancer Inst.*, 2000, **92**, 1210–1216.
5. S. Grant, C. Easley and P. Kirkpatrick, *Nat. Rev. Drug Discovery*, 2007, **6**, 21–22.
6. M. Duvic, R. Talpur, X. Ni, C. Zhang, P. Hazarika, C. Kelly, J. H. Chiao, J. F. Reilly, J. L. Ricker, V. M. Richon and S. R. Frankel, *Blood*, 2007, **109**, 31–39.
7. P. A. Marks, R. A. Rifkind, V. M. Richon and R. Breslow, *Clin. Cancer Res.*, 2001, **7**, 759–760.
8. A. J. M. de Ruijter, A. H. van Gennip, H. N. Caron, S. Kemp and A. B. P. van Kuilenburg, *Biochem. J.*, 2003, **370**, 737.
9. C. H. Arrowsmith, C. Bountra, P. V. Fish, K. Lee and M. Schapira, *Nat. Rev. Drug Discovery*, 2012, **11**, 384–400.
10. B. D. Strahl and C. D. Allis, *Nature*, 2000, **403**, 41–45.
11. L. Shen, I. Choi, E. J. Nestler and K.-J. Won, *Genomics & Informatics*, 2013, **11**, 60.
12. M. Yavartanoo and J. K. Choi, *Genomics Inform*, 2013, **11**, 2–6.
13. T. E. P. Consortium, *Science*, 2004, **306**, 636–640.
14. P. Joshi, T. M. Greco, A. J. Guise, Y. Luo, F. Yu, A. I. Nesvizhskii and I. M. Cristea, *Mol. Syst. Biol.*, 2013, **9**, 1–21.
15. W. A. Whyte, D. A. Orlando, D. Hnisz, B. J. Abraham, C. Y. Lin, M. H. Kagey, P. B. Rahl, T. I. Lee and R. A. Young, *Cell*, 2013, **153**, 307–319.
16. J. Lovén, H. A. Hoke, C. Y. Lin, A. Lau, D. A. Orlando, C. R. Vakoc, J. E. Bradner, T. I. Lee and R. A. Young, *Cell*, 2013, **153**, 320–334.

17. E. Verdin, F. Dequiedt and H. G. Kasler, *Trends Genet.*, 2003, **19**, 286–293.
18. J. David Adams and L. K. Klaidman, *Lett. Drug Des. Discovery*, 2007, **4**, 44–48.
19. L. Guarente, *N. Engl. J. Med.*, 2011, **364**, 2235–2244.
20. X.-J. Yang and S. Grégoire, *Mol. Cell. Biol.*, 2005, **25**, 2873–2884.
21. S. Spange, T. Wagner, T. Heinzel and O. H. Krämer, *Int. J. Biochem. Cell Biol.*, 2009, **41**, 185–198.
22. N. Carey and N. B. La Thangue, *Curr. Opin. Pharmacol.*, 2006, **6**, 369–375.
23. M. A. Glozak, N. Sengupta, X. Zhang and E. Seto, *Gene*, 2005, **363**, 15–23.
24. B. N. Singh, G. Zhang, Y. L. Hwa, J. Li, S. C. Dowdy and S.-W. Jiang, *Expert Rev. Anticancer Ther.*, 2010, **10**, 935–954.
25. P. J. Watson, L. Fairall, G. M. Santos and J. W. R. Schwabe, *Nature*, 2012, **481**, 335–340.
26. C. J. Millard, P. J. Watson, I. Celardo, Y. Gordiyenko, S. M. Cowley, C. V. Robinson, L. Fairall and J. W. R. Schwabe, *Mol. Cell*, 2013, **51**, 57–67.
27. C. A. Hassig, T. C. Fleischer, A. N. Billin, S. L. Schreiber and D. E. Ayer, *Cell*, 1997, **89**, 341–347.
28. J. K. Tong, C. A. Hassig, G. R. Schnitzler, R. E. Kingston and S. L. Schreiber, *Nature*, 1998, **395**, 917–921.
29. A. You, J. K. Tong, C. M. Grozinger and S. L. Schreiber, *Proc. Natl. Acad. Sci. U. S. A.*, 2001, **98**, 1454–1458.
30. Y.-J. Shi, C. Matson, F. Lan, S. Iwase, T. Baba and Y. Shi, *Mol. Cell*, 2005, **19**, 857–864.
31. M. K. H. Pflum, J. K. Tong, W. S. Lane and S. L. Schreiber, *J. Biol. Chem.*, 2001, **276**, 47733–47741.
32. X. Deng, D. Z. Ewton, S. E. Mercer and E. Friedman, *J. Biol. Chem.*, 2005, **280**, 4894–4905.
33. B. E. L. Lauffer, R. Mintzer, R. Fong, S. Mukund, C. Tam, I. Zilberleyb, B. Flicke, A. Ritscher, G. Fedorowicz, R. Vallero, D. F. Ortwine, J. Gunzner, Z. Modrusan, L. Neumann, C. M. Koth, P. J. Lupardus, J. S. Kaminker, C. E. Heise and P. Steiner, *J. Biol. Chem.*, 2013, 26926–26943.
34. M. Klammer, K. Godl, A. Tebbe and C. Schaab, *BMC Bioinf.*, 2010, **11**, 351.
35. M. Bantscheff, C. Hopf, M. M. Savitski, A. Dittmann, P. Grandi, A.-M. Michon, J. Schlegl, Y. Abraham, I. Becher, G. Bergamini, M. Boesche, M. Delling, B. Dümpelfeld, D. Eberhard, C. Huthmacher, T. Mathieson, D. Poeckel, V. Reader, K. Strunk, G. Sweetman, U. Kruse, G. Neubauer, N. G. Ramsden and G. Drewes, *Nat. Biotechnol.*, 2011, **29**, 255–265.
36. J. Li, R. Liu, Y. Lei, K. Wang, Q. C. Lau, N. Xie, S. Zhou, C. Nie, L. Chen, Y. Wei and C. Huang, *Autophagy*, 2010, **6**, 711–724.
37. Q. Zhou, R. Chaerkady, P. G. Shaw, T. W. Kensler, A. Pandey and N. E. Davidson, *Proteomics*, 2010, **10**, 1029–1039.
38. B. P. Ashburner, S. D. Westerheide and A. S. Baldwin, *Mol. Cell. Biol.*, 2001, **21**, 7065–7077.
39. J. Taunton, C. A. Hassig and S. L. Schreiber, *Science*, 1996, **272**, 408–411.
40. M. Jung, *Curr. Med. Chem.*, 2001, **8**, 1505–1511.

41. B. E. Gryder, Q. H. Sodji and A. K. Oyelere, *Future Med. Chem.*, 2012, **4**, 505–524.
42. P. Jones, in *Epigenetic Targets in Drug Discovery*, ed. W. Sippl and M. Jung, Wiley-VCH Verlag GmbH & Co. KGaA, 2010, pp. 185–223.
43. N. Khan, M. Jeffers, S. Kumar, C. Hackett, F. Boldog, N. Khramtsov, X. Qian, E. Mills, S. C. Berghs, N. Carey, P. W. Finn, L. S. Collins, A. Tumber, J. W. Ritchie, P. B. Jensen, H. S. Lichenstein and M. Sehested, *Biochem. J.*, 2008, **409**, 581.
44. J. E. Bradner, N. West, M. L. Grachan, E. F. Greenberg, S. J. Haggarty, T. Warnow and R. Mazitschek, *Nat. Chem. Biol.*, 2010, **6**, 238–243.
45. P. Jones, S. Altamura, R. De Francesco, P. Gallinari, A. Lahm, P. Neddermann, M. Rowley, S. Serafini and C. Steinkühler, *Bioorg. Med. Chem. Lett.*, 2008, **18**, 1814–1819.
46. M. Lobera, K. P. Madauss, D. T. Pohlhaus, Q. G. Wright, M. Trocha, D. R. Schmidt, E. Baloglu, R. P. Trump, M. S. Head, G. A. Hofmann, M. Murray-Thompson, B. Schwartz, S. Chakravorty, Z. Wu, P. K. Mander, L. Kruidenier, R. A. Reid, W. Burkhart, B. J. Turunen, J. X. Rong, C. Wagner, M. B. Moyer, C. Wells, X. Hong, J. T. Moore, J. D. Williams, D. Soler, S. Ghosh and M. A. Nolan, *Nat. Chem. Biol.*, 2013, **9**, 319–325.
47. G. Giaccone, A. Rajan, A. Berman, R. J. Kelly, E. Szabo, A. Lopez-Chavez, J. Trepel, M.-J. Lee, L. Cao, I. Espinoza-Delgado, J. Spittler and P. J. Loehrer, *J. Clin. Oncol.*, 2011, **29**, 2052–2059.
48. S. Mandl-Weber, F. G. Meinel, R. Jankowsky, F. Oduncu, R. Schmidmaier and P. Baumann, *Br. J. Haematol.*, 2010, **149**, 518–528.
49. P. Revill, N. Mealy, N. Serradell, J. Bolós and E. Rosa, *Drugs Future*, 2007, **32**, 315.
50. J. A. Plumb, N. Steele, P. W. Finn and R. Brown, *Biochem. Soc. Trans.*, 2004, **32**, 1095–1097.
51. C. K. Wada, R. R. Frey, Z. Ji, M. L. Curtin, R. B. Garland, J. H. Holms, J. Li, L. J. Pease, J. Guo, K. B. Glaser, P. A. Marcotte, P. L. Richardson, S. S. Murphy, J. J. Bouska, P. Tapang, T. J. Magoc, D. H. Albert, S. K. Davidsen and M. R. Michaelides, *Bioorg. Med. Chem. Lett.*, 2003, **13**, 3331–3335.
52. S. Tapadar, R. He, D. N. Luchini, D. D. Billadeau and A. P. Kozikowski, *Bioorg. Med. Chem. Lett.*, 2009, **19**, 3023–3026.
53. E. Baloglu, S. Ghosh, M. Lobera and D. Schmidt, Patent application WO2011088192, 2011.
54. A. Wahhab, D. Smil, A. Ajamian, M. Allan, Y. Chantigny, E. Therrien, N. Nguyen, S. Manku, S. Leit, J. Rahil, A. J. Petschner, A.-H. Lu, A. Nicolescu, S. Lefebvre, S. Montcalm, M. Fournel, T. P. Yan, Z. Li, J. M. Besterman and R. Déziel, *Bioorg. Med. Chem. Lett.*, 2009, **19**, 336–340.
55. F. Dehmel, T. Ciossek, T. Maier, S. Weinbrenner, B. Schmidt, M. Zoche and T. Beckers, *Bioorg. Med. Chem. Lett.*, 2007, **17**, 4746–4752.
56. B. Attenni, J. M. Ontoria, J. C. Cruz, M. Rowley, C. Schultz-Fademrecht, C. Steinkühler and P. Jones, *Bioorg. Med. Chem. Lett.*, 2009, **19**, 3081–3084.
57. T. I. S. G. C. (isgc), the W. T. C. C. C. 2 (wtccc2), C. Bellenguez, S. Bevan, A. Gschwendtner, C. C. A. Spencer, A. I. Burgess, M. Pirinen, C. A. Jackson, M. Traylor, A. Strange, Z. Su, G. Band, P. D. Syme, R. Malik, J. Pera,

B. Norrving, R. Lemmens, C. Freeman, R. Schanz, T. James, D. Poole, L. Murphy, H. Segal, L. Cortellini, Y.-C. Cheng, D. Woo, M. A. Nalls, B. Müller-Myhsok, C. Meisinger, U. Seedorf, H. Ross-Adams, S. Boonen, D. Wloch-Kopec, V. Valant, J. Slark, K. Furie, H. Delavaran, C. Langford, P. Deloukas, S. Edkins, S. Hunt, E. Gray, S. Dronov, L. Peltonen, S. Gretarsdottir, G. Thorleifsson, U. Thorsteinsdottir, K. Stefansson, G. B. Boncoraglio, E. A. Parati, J. Attia, E. Holliday, C. Levi, M.-G. Franzosi, A. Goel, A. Helgadottir, J. M. Blackwell, E. Bramon, M. A. Brown, J. P. Casas, A. Corvin, A. Duncanson, J. Jankowski, C. G. Mathew, C. N. A. Palmer, R. Plomin, A. Rautanen, S. J. Sawcer, R. C. Trembath, A. C. Viswanathan, N. W. Wood, B. B. Worrall, S. J. Kittner, B. D. Mitchell, B. Kissela, J. F. Meschia, V. Thijs, A. Lindgren, M. J. Macleod, A. Slowik, M. Walters, J. Rosand, P. Sharma, M. Farrall, C. L. M. Sudlow, P. M. Rothwell, M. Dichgans, P. Donnelly and H. S. Markus, *Nat. Genet.*, 2012, **44**, 328–333.

58. R. Glauben, E. Sonnenberg, T. Stroh, A. Batra, P. Mascagni, M. Zeitz and B. Siegmund, *Z. Gastroenterol.*, 2009, 47.
59. Z. Chen, F. Lin, Y. Gao, Z. Li, J. Zhang, Y. Xing, Z. Deng, Z. Yao, A. Tsun and B. Li, *Int. Immunopharmacol.*, 2011, **11**, 536–542.
60. X. Hou, G. Shi and H. Wan, *Am. Thorac. Soc.*, 2012, A5655.
61. A. Villagra, F. Cheng, H.-W. Wang, I. Suarez, M. Glozak, M. Maurin, D. Nguyen, K. L. Wright, P. W. Atadja, K. Bhalla, J. Pinilla-Ibarz, E. Seto and E. M. Sotomayor, *Nat. Immunol.*, 2009, **10**, 92–100.
62. J. H. Kalin and J. A. Bergman, *J. Med. Chem.*, 2013, 6297–6313.
63. A. Vasudevan, Z. Ji, R. R. Frey, C. K. Wada, D. Steinman, H. R. Heyman, Y. Guo, M. L. Curtin, J. Guo, J. Li, L. Pease, K. B. Glaser, P. A. Marcotte, J. J. Bouska, S. K. Davidsen and M. R. Michaelides, *Bioorg. Med. Chem. Lett.*, 2003, **13**, 3909–3913.
64. J. M. Ontoria, S. Altamura, A. Di Marco, F. Ferrigno, R. Laufer, E. Muraglia, M. C. Palumbi, M. Rowley, R. Scarpelli, C. Schultz-Fademrecht, S. Serafini, C. Steinkühler and P. Jones, *J. Med. Chem.*, 2009, **52**, 6782–6789.
65. F. J. Dekker and H. J. Haisma, *Drug Discovery Today*, 2009, **14**, 942–948.
66. A. Mai, D. Rotili, D. Tarantino, A. Nebbioso, S. Castellano, G. Sbardella, M. Tini and L. Altucci, *Bioorg. Med. Chem. Lett.*, 2009, **19**, 1132–1135.
67. G. Sbardella, S. Castellano, C. Vicidomini, D. Rotili, A. Nebbioso, M. Miceli, L. Altucci and A. Mai, *Bioorg. Med. Chem. Lett.*, 2008, **18**, 2788–2792.
68. X. Liu, L. Wang, K. Zhao, P. R. Thompson, Y. Hwang, R. Marmorstein and P. A. Cole, *Nature*, 2008, **451**, 846–850.
69. E. M. Bowers, G. Yan, C. Mukherjee, A. Orry, L. Wang, M. A. Holbert, N. T. Crump, C. A. Hazzalin, G. Liszczak, H. Yuan, C. Larocca, S. A. Saldanha, R. Abagyan, Y. Sun, D. J. Meyers, R. Marmorstein, L. C. Mahadevan, R. M. Alani and P. A. Cole, *Chem. Biol.*, 2010, **17**, 471–482.
70. P. Filippakopoulos, S. Picaud, I. Felletar, D. Hay, O. Fedorov, S. Martin, T. Krojer, R. Nowak, F. Von Delft, P. Brennan, C. H. Arrowsmith, A. M. Edwards, C. Bountra and S. Knapp, 2013, http://www.ncbi.nlm.nih.gov/Structure/mmdb/mmdbsrv.cgi?uid=4NR7, to be published.

71. X. H. Parsons, *Anat. Physiol.*, 2012, DOI: 10.4172/2161-0940.S3-002.
72. Z. Wang, C. Zang, K. Cui, D. E. Schones, A. Barski, W. Peng and K. Zhao, *Cell*, 2009, **138**, 1019–1031.
73. A. Kretsovali, C. Hadjimichael and N. Charmpilas, *Stem Cells Int.*, 2012, **2012**, 184154.
74. E. Karantzali, H. Schulz, O. Hummel, N. Hubner, A. Hatzopoulos and A. Kretsovali, *Genome Biol.*, 2008, **9**, R65.
75. D. Huangfu, K. Osafune, R. Maehr, W. Guo, A. Eijkelenboom, S. Chen, W. Muhlestein and D. A. Melton, *Nat. Biotechnol.*, 2008, **26**, 1269–1275.
76. X. Dong, R. Pan, H. Zhang, C. Yang, J. Shao and L. Xiang, *PLoS One*, 2013, **8**, e63405.
77. H. Brantjes, J. Roose, M. van de Wetering and H. Clevers, *Nucl. Acids Res.*, 2001, **29**, 1410–1419.
78. T. Sekiya and K. S. Zaret, *Mol. Cell*, 2007, **28**, 291–303.
79. S. Coni, L. Antonucci, D. D'Amico, L. D. Magno, P. Infante, E. D. Smaele, G. Giannini, L. D. Marcotullio, I. Screpanti, A. Gulino and G. Canettieri, *PLoS One*, 2013, **8**, e65718.
80. X. Zhan, X. Shi, Z. Zhang, Y. Chen and J. I. Wu, *Proc. Natl. Acad. Sci. U. S. A.*, 2011, **108**, 12758–12763.
81. Y. Tang, J. M. Boucher and L. Liaw, *J. Am. Heart Assoc.*, 2012, 1.
82. R. Vibhakar, G. Foltz, J. Yoon, L. Field, H. Lee, G. Ryu, J. Pierson, B. Davidson and A. Madan, *Neuro-Oncol.*, 2007, **9**, 135–144.
83. F. Valdora, B. Banelli, S. Stigliani, S. Pfister, S. Moretti, M. Kool, M. Remke, A. Bai, C. Brigati, T. Hielscher, M. Romani, T. Servidei, M. Zollo, G. Cinalli, A. Oberthuer, G. Tonini and S. Coco, *Int. J. Mol. Sci.*, 2013, **14**, 7492–7505.
84. H. Kohlhof, *2nd Cancer Epigenetics Conference*, Oral Presentation, Boston, 2012.
85. C. Scheel and R. A. Weinberg, *Semin. Cancer Biol.*, 2012, **22**, 396–403.
86. M. M. Bernardo, Y. Meng, J. Lockett, G. Dyson, A. Dombkowski, A. Kaplun, X. Li, S. Yin, S. Dzinic, M. Olive, I. Dean, D. Krass, K. Moin, R. D. Bonfil, M. Cher, W. Sakr and S. Sheng, *Genes Cancer*, 2012, 1009–1022.
87. Y. Drabsch and P. ten Dijke, *J. Mammary Gland Biol. Neoplasia*, 2011, **16**, 97–108.
88. G. Giannini, W. Cabri, C. Fattorusso and M. Rodriquez, *Future Med. Chem.*, 2012, **4**, 1439–1460.
89. J. M. Wagner, B. Hackanson, M. Lübbert and M. Jung, *Clin. Epigenet.*, 2010, **1**, 117–136.
90. N. Azad, C. A. Zahnow, C. M. Rudin and S. B. Baylin, *Nat. Rev. Clin. Oncol.*, 2013, **10**, 256–266.
91. S. Subramanian, S. E. Bates, J. J. Wright, I. Espinoza-Delgado and R. L. Piekarz, *Pharmaceuticals*, 2010, **3**, 2751–2767.
92. R. L. Piekarz and S. E. Bates, *Clin. Cancer Res.*, 2009, **15**, 3918–3926.
93. K. T. Thurn, S. Thomas, A. Moore and P. N. Munster, *Future Oncol.*, 2011, **7**, 263–283.
94. C. P. Miller, M. M. Singh, N. Rivera-Del Valle, C. A. Manton and J. Chandra, *BioMed Res. Int.*, 2011, **2011**, 514261.

95. C. A. Dinarello, G. Fossati and P. Mascagni, *Mol. Med.*, 2011, **17**, 333–352.
96. V. A. DeWoskin and R. P. Million, *Nat. Rev. Drug Discovery*, 2013, **12**, 661–662.
97. D. Ritchie, R. L. Piekarz, P. Blombery, L. J. Karai, S. Pittaluga, E. S. Jaffe, M. Raffeld, J. E. Janik, H. M. Prince and S. E. Bates, *Haematologica*, 2009, **94**, 1618–1622.
98. F. W. Frueh, *Personalized Medicine: What Is It ? How Will It Affect Health Care?, 11th Annual FDA Science Forum*, Washington, D. C., 2005.
99. T. Qiu, L. Zhou, W. Zhu, T. Wang, J. Wang, Y. Shu and P. Liu, *Future Oncol.*, 2013, **9**, 255–269.
100. C. Zhang, S. Rangwala and M. Duvic, *Blood Lymphatic Cancer: Targets Ther.*, 2012, 17.
101. C. Tiffon, J. Adams, L. van der Fits, S. Wen, P. Townsend, A. Ganesan, E. Hodges, M. Vermeer and G. Packham, *Br. J. Pharmacol.*, 2011, **162**, 1590–1602.
102. M. Gupta, J. Han, M. Stenson, L. Wellik and T. Witzig, *Leukemia*, 2012, **26**, 1356–1364.
103. C. E. Wells, S. Bhaskara, K. R. Stengel, Y. Zhao, B. Sirbu, B. Chagot, D. Cortez, D. Khabele, W. J. Chazin, A. Cooper, V. Jacques, J. Rusche, C. M. Eischen, L. Y. McGirt and S. W. Hiebert, *PLoS One*, 2013, **8**, e68915.
104. L. Ellis and R. Pili, *Pharmaceuticals*, 2010, **3**, 2441–2469.
105. D. C. Marchion, E. Bicaku, A. I. Daud, D. M. Sullivan and P. N. Munster, *Mol. Cancer Ther.*, 2005, **4**, 1993–2000.
106. M. Namdar, G. Perez, L. Ngo and P. A. Marks, *Proc. Natl. Acad. Sci. U. S. A.*, 2010, **107**, 20003–20008.
107. B. Groselj, N. L. Sharma, F. C. Hamdy, M. Kerr and A. E. Kiltie, *Br. J. Cancer*, 2013, **108**, 748–754.
108. J. Asthana, S. Kapoor, R. Mohan and D. Panda, *J. Biol. Chem.*, 2013, **288**, 22516–22526.
109. P. N. Munster, K. T. Thurn, S. Thomas, P. Raha, M. Lacevic, A. Miller, M. Melisko, R. Ismail-Khan, H. Rugo, M. Moasser and S. E. Minton, *Br. J. Cancer*, 2011, **104**, 1828–1835.
110. D. L. Fontes Jardim, *J. Clin. Oncol.*, 2013, **31**(suppl; abstr 2509).
111. J. Zain and O. A. O'Connor, *Invest. New Drugs*, 2010, **28**, 58–78.
112. N. L. Steele, J. A. Plumb, L. Vidal, J. Tjørnelund, P. Knoblauch, P. Buhl-Jensen, R. Molife, R. Brown, J. S. de Bono and T. R. J. Evans, *Cancer Chemother. Pharmacol.*, 2011, **67**, 1273–1279.
113. A. Furlan, V. Monzani, L. L. Reznikov, F. Leoni, G. Fossati, D. Modena, P. Mascagni and C. A. Dinarello, *Mol. Med.*, 2011, **17**, 353–362.
114. P. Hamberg, M. M. Woo, L.-C. Chen, J. Verweij, M. G. Porro, L. Zhao, W. Li, D. van der Biessen, S. Sharma, T. Hengelage and M. de Jonge, *Cancer Chemother. Pharmacol.*, 2011, **68**, 805–813.
115. G. Garcia-Manero, S. Assouline, J. Cortes, Z. Estrov, H. Kantarjian, H. Yang, W. M. Newsome, W. H. Miller, C. Rousseau, A. Kalita, C. Bonfils, M. Dubay, T.-A. Patterson, Z. Li, J. M. Besterman, G. Reid, E. Laille, R. E. Martell and M. Minden, *Blood*, 2008, **112**, 981–989.
116. S. Kummar, M. Gutierrez, E. R. Gardner, E. Donovan, K. Hwang, E. J. Chung, M.-J. Lee, K. Maynard, M. Kalnitskiy, A. Chen, G. Melillo, Q. C.

Ryan, B. Conley, W. D. Figg, J. B. Trepel, J. Zwiebel, J. H. Doroshow and A. J. Murgo, *Clin. Cancer Res.*, 2007, **13**, 5411–5417.
117. W. K. Kelly, V. M. Richon, O. O'Connor, T. Curley, B. MacGregor-Curtelli, W. Tong, M. Klang, L. Schwartz, S. Richardson, E. Rosa, M. Drobnjak, C. Cordon-Cordo, J. H. Chiao, R. Rifkind, P. A. Marks and H. Scher, *Clin. Cancer Res.*, 2003, **9**, 3578–3588.
118. C. Bonfils, A. Kalita, M. Dubay, L. L. Siu, M. A. Carducci, G. Reid, R. E. Martell, J. M. Besterman and Z. Li, *Clin. Cancer Res.*, 2008, **14**, 3441–3449.
119. R. C. Jackson, *ISRN Pharmacol.*, 2012, DOI: 10.5402/2012/590626.
120. V. Novotny-Diermayr, N. Sausgruber, Y. K. Loh, M. K. Pasha, R. Jayaraman, H. Hentze, W.-P. Yong, B.-C. Goh, H.-C. Toh, K. Ethirajulu, J. Zhu and J. M. Wood, *Mol. Cancer Ther.*, 2011, **10**, 1207–1217.
121. P. M. Wilson, M. J. Labonte, S. C. Martin, S. T. Kuwahara, A. El-Khoueiry, H.-J. Lenz and R. D. Ladner, *Invest. New Drugs*, 2013, **31**, 845–857.
122. M. V. Damme, E. Crompot, N. Meuleman, P. Mineur, D. Bron, L. Lagneaux and B. Stamatopoulos, *Epigenetics*, 2012, **7**, 1403–1412.
123. D. P. Gavin and R. P. Sharma, *Cardiovasc. Psychiatry Neurol.*, 2009, DOI: 10.1155/2009/409562.
124. O. Khan, S. Fotheringham, V. Wood, L. Stimson, C. Zhang, F. Pezzella, M. Duvic, D. J. Kerr and N. B. L. Thangue, *Proc. Natl. Acad. Sci. U. S. A.*, 2010, **107**, 6532–6537.
125. S. Fotheringham, M. T. Epping, L. Stimson, O. Khan, V. Wood, F. Pezzella, R. Bernards and N. B. La Thangue, *Cancer Cell*, 2009, **15**, 57–66.
126. M. New, H. Olzscha, G. Liu, O. Khan, L. Stimson, J. McGouran, D. Kerr, A. Coutts, B. Kessler, M. Middleton and N. B. La Thangue, *Cell Death Differ.*, 2013, 1–11.
127. W. Timp and A. P. Feinberg, *Nat. Rev. Cancer*, 2013, **13**, 497–510.
128. M. H. Shah, P. Binkley, K. Chan, J. Xiao, D. Arbogast, M. Collamore, Y. Farra, D. Young and M. Grever, *Clin. Cancer Res.*, 2006, **12**, 3997–4003.
129. R. Kristeleit, P. Fong, G. W. Aherne and J. de Bono, *Clin. Lung Cancer*, 2005, 7(1), S19–S30.
130. R. Molife, P. Fong, M. Scurr, I. Judson, S. Kaye and J. de Bono, *Clin. Cancer Res.*, 2007, **13**, 1068.

CHAPTER 6

Targeting Histone Lysine Methyltransferases in Cancer

PATRICK TROJER*[a]

[a]Constellation Pharmaceuticals, 215 1st Street, Suite 200, Cambridge, MA 02142, USA
*E-mail: patrick.trojer@constellationpharma.com

6.1 Histone Lysine Methylation and Cancer

More than a decade ago the initial discovery of an enzyme catalysing the methylation of a specific histone lysine residue[1] heralded the beginning of a new era of chromatin research. Since then numerous studies have attempted to elucidate the intricacies of histone lysine methylation metabolism and to identify the functional consequences of individual and combined histone methylation states.[2–4] Searching the human genome for genes with homology to the catalytic domain of the initially identified methyltransferases led to the identification of 51 SET (which stands for Suppressor of variegation 3–9 [Su(var)3–9], Enhancer of Zeste [E(z)] and Trithorax [Trx]) domain-encoding gene products. Many of these SET domains have now been experimentally confirmed as active lysine methyltransferases (KMTs; for a recent review and detailed domain organisation of KMTs see Herz *et al.*[3]).

Several additional human proteins with SET domain similarity have been identified using computational methods,[5] potentially expanding the number of SET domain-containing KMTs to 55. Based on their SET domain sequence, KMT family members can be grouped into three subfamilies comprised of

RSC Drug Discovery Series No. 48
Epigenetics for Drug Discovery
Edited by Nessa Carey

Published by the Royal Society of Chemistry, www.rsc.org

'classic SET domain' (which is a heterogeneous group that can be further subdivided), 'divergent SET domain' and 'PR domain' (Figure 6.1). Only a single enzyme, with a catalytic domain that is structurally related to the protein arginine methyltransferase family, DOT1L, exhibits histone KMT activity.[6] All these enzymes utilise *S*-adenosyl-L-methionine (SAM) as the methyl group donor to transfer one, two or three methyl groups to histone lysine residues. Individual KMTs are exquisitely specific both with regard to the lysine residue upon which they operate as well as to the degree of methylation.

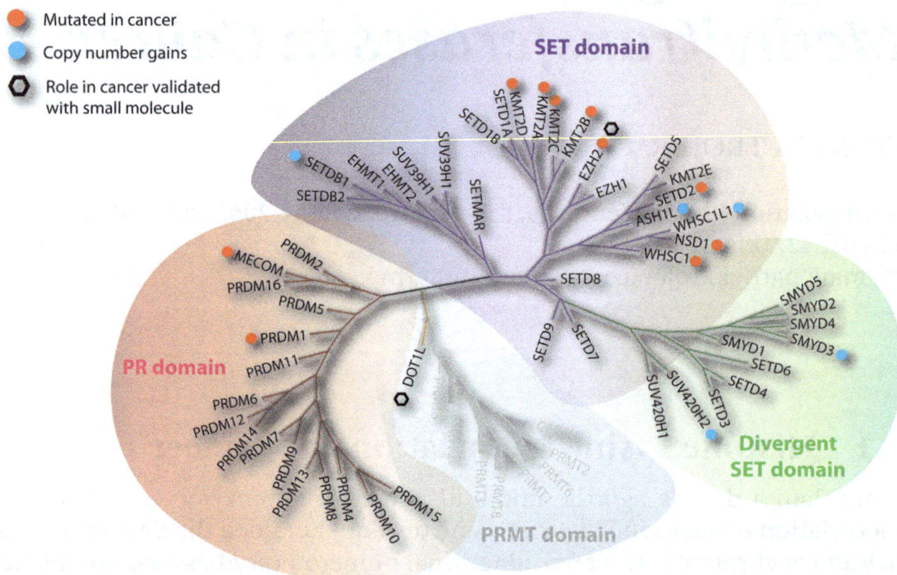

Figure 6.1 Phylogenetic tree of human lysine methyltransferases (KMTs). The SET domains of 51 SET domain-containing proteins were aligned using ClustalW and the tree was originally generated using ChromoHub.[238] DOT1L and protein arginine methyltransferases (PRMTs) were aligned separately. The composite tree was assembled from the original output files and individual branches represent a relative phylogenetic distance based on amino acid sequence similarity of the KMT catalytic domains. Branches are not drawn to scale and thus do not represent absolute phylogenetic distances. The displayed KMT names correspond to the official gene bank entries (as accessed in February 2014) and are categorised into four groups: SET domain (indicated in purple), divergent SET domain (indicated in green), PR domain (indicated in red) and PRMT domain (indicated in light grey and semi-transparent). PRMTs are not discussed in this review but shown here to illustrate their similarity with DOT1L. PRMTs are part of a large, heterogeneous methyltransferase superfamily that is omitted. Recurrent somatic mutations and translocations are indicated with red dots and copy number gains with blue dots. Black hexagons indicate KMTs for which high-quality small molecule inhibitors are available and have been used to credential the respective target for cancer therapy applications. See text for more detail.

The number of enzymes, the number of targetable histone lysines and the degree of methylation on individual lysine residues give rise to a highly complex repertoire of potential functional outputs with further layers of complexity introduced by the fact that histone methylation marks do not function in isolation but rather cooperatively with other types of histone modifications.[7] Moreover, individual methylated lysine residues within this histone methylation landscape are recognised by various families of proteins harbouring chromatin-binding modules.[8] Despite the subtle physicochemical nature of lysine methylation (methylated lysines retain their fundamentally basic character), the regulatory potential of this 'histone lysine methylation network' is tremendous. Although we have only begun to unravel the mechanistic complexity of this network, its critical impact on governing important cellular processes including transcription, cell identity and genomic stability is undisputed. It is now also appreciated that KMTs target non-histone as well as histone substrates (for review see Clarke[9] and Moore and Gozani[10]) but our understanding of the functional impact of non-histone methylation lags behind that of histone methylation events, which have proven roles in the regulation of processes that require access to DNA.

With the advent of techniques such as chromatin immunoprecipitation and histone mass spectrometry, and through the generation of highly specific histone methylation state-specific antibodies, it has been unambiguously confirmed that the methylation of lysines is a reversible process. It is now appreciated that the interplay of KMTs and histone lysine demethylases (KDMs) in addition to other chromatin-related processes such as histone exchange[11] and histone clipping[12,13] impact chromatin dynamics[14] and define histone lysine methylation turnover. In addition, the global turnover rates of individual histone lysine methylation marks are variable and may be either coupled or uncoupled from cellular replication and cell cycle progression.[15,16] The dynamics of a given methylation mark may also be influenced by its genomic location. Environmental stimuli can rapidly alter local chromatin structure (for instance at transcriptional start sites) and result in significant changes in spatially restricted histone methylation levels that coincide with rapid changes in the transcriptional output of a given gene.[17–19] Thus, even histone methylation marks that appear to be slowly turned over on a global scale can show remarkable dynamics in a spatially and temporally defined context. Manipulation of a given KMT (*e.g.* by RNAi, genome editing) or its inhibition by small molecules may affect global and local levels of the corresponding histone lysine methylation mark with distinct kinetics (Figure 6.2). Theoretically, critical downstream phenotypic events may thus surface even prior to changes in the global histone methylation landscape. Once the target is effectively engaged and the corresponding global/local histone lysine methylation mark is turned over, the kinetics of the downstream functional consequences of mark loss will dictate the timing of a relevant phenotypic response. With respect to transcription, it is not yet completely understood if loss of a mark that supports the transcription cycle causes faster transcriptional responses compared to a mark

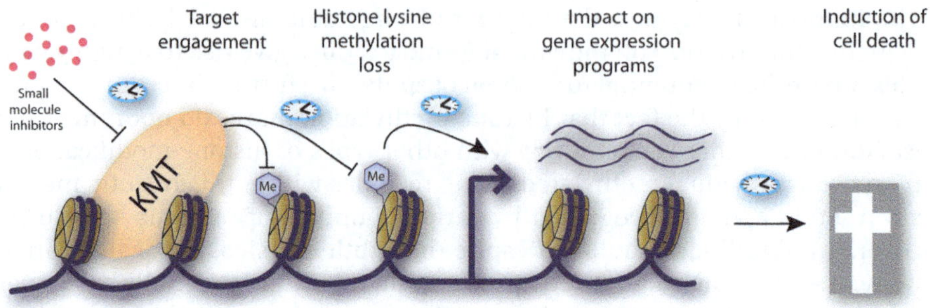

Figure 6.2 Potential phenotypic consequences of inhibiting a KMT in cancer cells are dictated by the kinetics of the molecular events in response to target engagement. Step 1 – comprehensive target coverage. Step 2 – global and/or local histone methylation loss. Step 3 – molecular consequences of mark loss, for instance impact on gene expression. Loss of histone methylation marks that support the transcription cycle may have different kinetics compared to loss of marks that maintain transcriptional repression; histone methylation marks can have different kinetics dependent on genomic location and cellular context. Step 4 – induction of a phenotypic response such as cell cycle arrest, differentiation, apoptosis, necrosis and senescence. The molecular determinants that define the response timing for each step are often not completely understood.

that functions in transcriptional repression. In instances where 'repression' is a roadblock for already initiated transcription rather than maintenance of stable gene silencing the transcriptional response timing upon removal of such a repressive mark might be relatively fast (Figure 6.2). Alternatively, phenotypic responses may not at all depend on histone lysine methylation but rather on non-histone methylation turnover. Thus it is expected that inhibitors of different KMTs will display not only different phenotypes but different kinetics in the relationship between the methylation mark change and the associated phenotype.

Large genomic sequencing campaigns of primary human tumour samples have provided evidence for recurrent mutations, translocations and somatic copy number gains or losses in KMT, KDM and other chromatin-associated protein-encoding genes (Figure 6.3; for instance see Gonzalez-Perez *et al.*,[20] Kandoth *et al.*,[21] Lawrence *et al.*,[22] Zack *et al.*[23]). Moreover, several KMTs are widely over-expressed in cancer and recent studies document aberrant global histone lysine methylation levels in cancer cell lines.[24] However, a direct causal relationship between alterations in histone lysine methylation and malignant transformation has not been demonstrated.[25] Depending on the nature of the genomic aberration, both increases and decreases in the levels or function of these enzymes have been identified. For instance, in follicular and diffuse large B-cell lymphomas mono-allelic gain-of-function mutations were identified in the histone H3 lysine 27 (H3K27)-specific KMT EZH2 (Figure 6.4). These mutations in EZH2 occur as single amino acid substitutions in the SET domain[26–28] that alter the substrate specificity of the enzyme

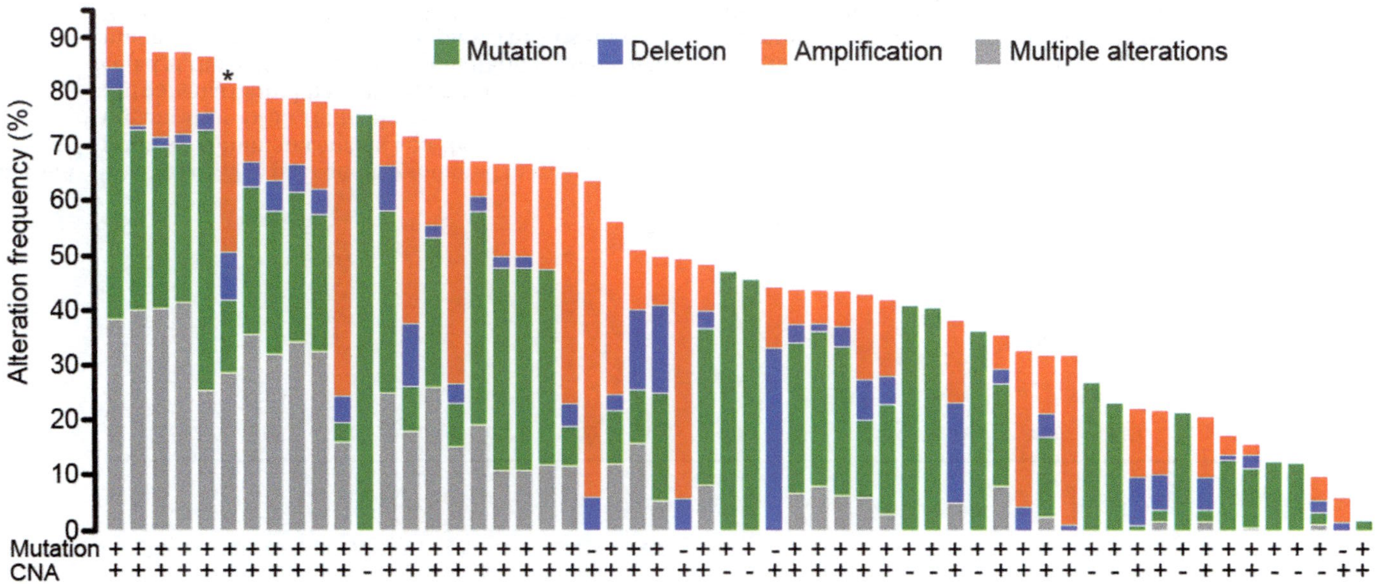

Figure 6.3 KMT-encoding genes are frequently subject to genomic aberrations in cancer. Fifty-six cancer genomics studies comprising primary human tumour samples as well as cancer cell lines (bar indicated by an asterisk) were queried for the frequency of mutations (shown in green) and gene amplifications (shown in red) in any of the 52 human KMTs. Each bar represents a separate study. A '+' below each bar indicates the availability of mutation data (top row) and copy number alteration (CNA) data (bottom row). The graph was generated using the cBioPortal for cancer genomics (http://www.cbioportal.org/public-portal/index.do).[239,240]

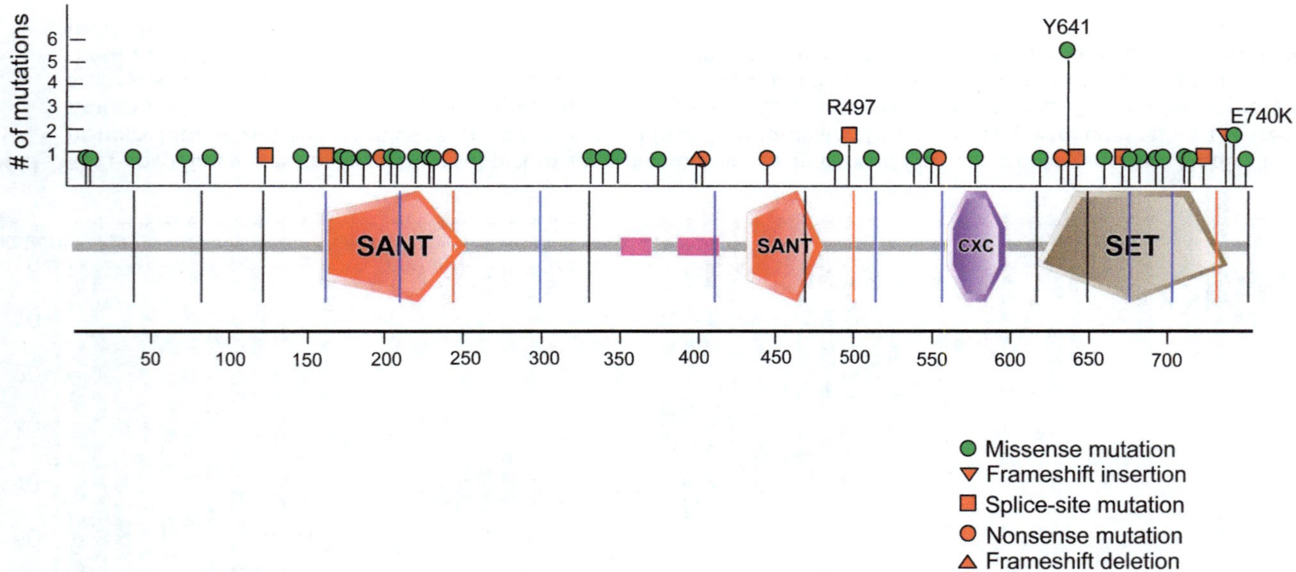

Figure 6.4 The KMT EZH2 is frequently mutated in cancer. Somatic, non-silent mutations in the EZH2 protein encoding gene from querying 4742 human cancers and their matched normal-tissue samples across 21 cancer types are represented.[22] The domain organisation of EZH2 (UniprotKB entry Q15910) was generated using the SMART domain architecture representation tool (http://smart.embl-heidelberg.de/). The amino acid positions from the N-terminus (left) to the C-terminus (right) are indicated at the bottom. The type of mutation is indicated in the figure legend; the *y*-axis captures the frequency of each observed mutation across all cancer samples. The position of the recurrently mutated tyrosine 641 (Y641) residue located within the EZH2 SET domain is indicated. Y641 mutations were identified in diffuse large B-cell lymphoma and melanoma but not in other cancer types.

and result in elevated levels of trimethylated H3K27.[29–33] Similarly, the H3K36-specific KMT WHSC1 (NSD2, MMSET) is translocated in multiple myeloma.[34,35] In this genomic rearrangement the immunoglobulin heavy chain enhancer is juxtaposed to the WHSC1 5′-coding region found in chromosome 4, resulting in substantially elevated WHSC1 transcription, globally elevated H3K36me2 levels and oncogenic transformation.[36,37] Loss-of-function mutations in KMT and KDM-encoding genes are in certain contexts heterozygous,[38,39] suggesting that haploinsufficiency may promote cancer growth but that total absence of function is detrimental to growth and viability. Homo- and heterozygous mutations in KDMs such as those observed in KDM6A (UTX),[39] result in aberrantly increased histone lysine methylation states that are controlled by corresponding KMTs, to which the cancer cell may thereby become addicted. Moreover, KMTs may become absolutely essential for cellular viability in the context of inactivating mutations in other chromatin modifiers, as was shown in the case of malignant rhabdoid tumour. SMARCB1, a subunit of the ATP-dependent chromatin remodelling complex SWI/SNF, is the primary driver in this aggressive paediatric malignancy.[40] The oncogenic effect of SMARCB1 deletions can be completely suppressed *in vivo* by genetic elimination of EZH2.[41]

Importantly, when the various factors that compromise histone lysine methylation metabolism in cancer cells are considered as one oncogenic network module it becomes evident how frequently such alterations in chromatin occur.[42] Initially, genetic studies were instrumental in revealing changes in the epigenome of malignant cells. And it is now becoming clear that both genomic and epigenomic changes like histone lysine methylation cooperate to drive the selection of cells with the greatest clonal advantage.[43] With this perspective, individual KMTs appear attractive as potential therapeutic targets for cancer. Gain-of-function aberrations are associated with oncogenic behaviour, while mono- and bi-allelic loss-of-function changes may render tumour cells perhaps sensitive to further inhibition of histone lysine methylation or create novel dependencies in chromatin regulatory pathways.

Academic and industrial efforts are under way to identify relevant targets for cancer treatment within these histone methylation networks. Consequently, small-molecule drug discovery efforts have been launched to identify KMT inhibitors. These inhibitors may prove valuable tools for investigating the contribution of the catalytic activity to the overall KMT function in normal tissues as well as in tumours. Importantly, high-quality, drug-like KMT inhibitors should provide a novel therapeutic approach to cancer. From early crystallographic studies it became evident that the SET domain architecture lends itself to targeting by small molecules but there were initial concerns as to how the complex organisation and conformational flexibility of the active site involving pre- and post-SET domains[44] would affect druggability. Also, early KMT drug discovery efforts were handicapped by the lack of adequate hit validation because artifactual inhibition of KMTs by small molecules in enzymatic assays was underestimated and biophysical methods for this target space still needed to be developed. The past 6 years have seen successful HTS campaigns and medicinal chemistry efforts that have led to the identification of high-quality small-molecule KMT inhibitors. These inhibitors have proven to be efficacious in various preclinical oncology models and several have already entered clinical trials. The rest of this chapter will review inhibitors of specific KMTs, their mechanism of action, the nature and kinetics of their molecular and cellular responses and their activity in disease-relevant models.

6.2 Pharmacological Inhibition of DOT1L

DOT1L is the only known histone H3 lysine 79-specific methyltransferase. It is conserved from yeast to humans.[45–47] Unlike other KMTs, DOT1L does not have a SET domain but has a catalytic domain that structurally resembles that of classic non-histone methyltransferases,[48] with the catechol *O*-methyltransferase COMPT[49] being the most closely related. DOT1L has a strong preference for nucleosomal substrates compared to histone octamers or free histone H3. Although K79 is located in the globular domain of histone H3, the structure of the nucleosome[50] clearly shows that the residue is exposed, located in the L1-loop between the α1 and α2 helices of the H3 histone fold motif. Moreover, the histone H4 N-terminal and the H2B C-terminal regions are in close proximity to K79, suggesting that DOT1L makes extensive contacts with the nucleosome. In fact, it was known from previous studies in yeast that H2BK123 mono-ubiquitination was required for H3K79 methylation.[51,52] This was corroborated by biochemical studies in which the presence of a ubiquitin moiety on human H2BK120 significantly stimulated DOT1L KMT activity for nucleosomal H3K79.[53] Collectively, these early data provide a strong basis for a chromatin-centric role of DOT1L. To date there is no evidence for a demethylase that removes methyl groups from H3K79. The turnover rate of H3K79me2 was found to be very slow,[16] which is consistent with the lack of a H3K79-specific demethylase.

Dot1L function is essential during embryonic development[54,55] and H3K79 methylation has been linked to subtelomeric gene silencing,[56,57] heterochromatin maintenance, meiotic checkpoint control,[58,59] DNA repair,[60] DNA replication,[61] cell cycle progression[62] and transcriptional regulation.[63–65] The following observations strongly implicate DOT1L and H3K79 methylation in transcriptional elongation: (i) H3K79 di- and trimethylation is found at the 5′-coding region of active genes; (ii) the PAF complex with its well-established role in elongation mediates H2BK120 mono-ubiquitination, which in turn stimulates H3K79 methylation; (iii) DOT1L is a component of several protein complexes implicated in elongation by association with the positive transcription elongation factor (P-TEFb; reviewed by Nguyen and Zhang[6]); and (iv) DOT1L physically interacts with the phosphorylated C-terminal domain of RNA polymerase II.[66] Interestingly, DOT1L is also required for transcriptional initiation after UV-induced DNA damage, so that RNA synthesis can resume after DNA lesions have been repaired.[67] Recently, DOT1L was identified in a chromatin modifier-specific RNAi screen as a suppressor of somatic cell reprogramming.[68] Reduction of DOT1L levels by RNAi increased the number of induced pluripotent stem cells, with the impact depending on the enzyme's catalytic activity. Reprogramming of fibroblasts usually requires the transcription factors OCT4, SOX2, KLF4 and MYC. Interestingly, suppression of DOT1L catalytic activity bypassed the requirement for KLF4 and MYC expression by down-regulation of mesenchymal genes and indirect up-regulation of the well-known stem cell factors NANOG and LIN28.

No recurrent mutations, translocations or somatic copy number gains have been discovered in the genomic locus encoding DOT1L.[22] However, DOT1L is functionally required for the malignant transformation of haematopoietic progenitors involving the oncogenic gene-fusion products MLL-AF10, CALM-AF10, MLL-AF4 and MLL-AF9.[69–72] These fusion proteins recruit DOT1L to target genes where it contributes to their misexpression (reviewed by Barry *et al.*,[73] Bernt and Armstrong[74]). The MLL-translocation partner proteins AF10, AF4, AF9 and ENL were found in various protein complexes with DOT1L,[75–79] which provided a plausible explanation for how DOT1L is physically recruited to MLL fusion protein-controlled target genes. These and other mechanistic studies[80] suggest that the oncogenic potential of the MLL fusion proteins is based on aberrant transcript elongation. However, DOT1L's oncogenic mechanism in MLL-translocated leukaemia is still not fully understood because MLL-GAS7 and MLL-AFX also seem to depend on DOT1L.[81] At this point there is no physical interaction described that places DOT1L in the proximity of these particular MLL fusion proteins. Moreover, DOT1L was shown recently to be required for the MLL-AF6 mediated transformation of bone marrow cells,[82] although AF6 is a predominantly cytoplasmic protein and no mechanistic basis exists for how AF6 recruits DOT1L to MLL-AF6 target genes. Another variation of MLL rearrangements in AML is found in MLL partial tandem duplications (MLL-PTDs), where the underlying molecular mechanism of its oncogenic role is still largely unknown. It has recently been suggested that various chromatin modifiers, including DOT1L, could be dependency factors in this context.[83]

Apart from its involvement in cases of acute leukaemias with MLL rearrangements, DOT1L was identified as a transcriptional co-activator of the key Wnt pathway transcriptional regulators TCF4 and beta-catenin.[75,84,85] However, the putative function of Dot1l in Wnt-controlled intestinal crypt cells was recently challenged by a study in which Dot1l was knocked out in intestinal stem cells or the entire intestinal epithelium. Dot1l loss did not impair intestinal homeostasis and only had modest impact on Wnt-controlled gene expression.[86] Despite these findings it is still possible that DOT1L plays a more prominent role in the context of a constitutively activated Wnt pathway such as that observed in colorectal cancer.

The discovery in 2011 of small molecules targeting DOT1L catalytic activity[87] was a breakthrough in this area of research. Aminonucleoside inhibitors were identified through mechanism-guided design principles using *S*-adenosylhomocysteine as a starting point and then optimised through analyses of structure–activity relationships and crystallographic approaches.[88] These compounds function by competing with SAM for binding to the enzyme's active site. The published molecule, EPZ004777, is potent, highly selective and causes dose-dependent decrease of global H3K79 methylation levels (Table 6.1) and bears testimony to the idea that it is possible to engineer selectivity into SAM analogues.[89]

This concept is supported by a number of published SAM analogues that show varying degrees of potency and selectivity against DOT1L (Table 6.1). Changes to the nucleoside portion of the molecules were generally not well tolerated. However, N6 substituents were found to increase selectivity[90] and the N7 nitrogen could be replaced with a carbon.[88,91] Bromination of this carbon increased potency, solubility and permeability while retaining exquisite selectivity (Table 6.1).[92,93] Other compounds have been described in which the ribose moiety is replaced by cyclopentane or cyclopentene groups.[94] These compounds retained potency and selectivity (Table 6.1) and yielded better microsomal stability.

The DOT1L-inhibitor dissociation kinetics inversely correlated with compound potency (potent inhibitor = slow off-rate) while the association kinetics remained invariant across the compound series.[88] Interestingly, DOT1L inhibitors seemed to bind the target with higher affinity in the presence of nucleosomes,[91] suggesting that DOT1L conformational changes upon nucleosome binding favour inhibitor binding. The binding mode of these inhibitors was assessed by crystallisation with DOT1L,[88,92] which addressed open questions regarding how the hydrophilic SAM-binding pocket could accommodate a hydrophobic and relatively large molecule such as EPZ004777. Indeed, the amino-nucleoside inhibitor occupied the cofactor binding site. However, significant conformational changes were observed in the activation and substrate binding loops and led to the formation of a new internal cavity complementary to the *para-tert*-butylphenyl appending group. Both the urea linker as well as the *t*-butylphenyl group were shown to make extensive contacts to the protein and are largely responsible for the compound's high affinity. Further structurally guided design led to the synthesis of EPZ-5676 (Table 6.1), the most potent DOT1L inhibitor described to date. Despite having a different appendix, the compound also induced conformational changes in the active site, retained a SAM-competitive mode of inhibition, and displayed exquisite selectivity and a remarkably slow off-rate.[95]

DOT1L aminonucleoside inhibitors demonstrated effective cell killing of MLL-AF9, MLL-AF10 and CALM-AF10-transformed murine myeloid progenitors and human AML cells harbouring MLL-AF4, MLL-AF9 and MLL-ENL translocations at sub-μM to low μM concentrations while non-rearranged leukaemia cell lines were affected to a lesser extent.[82,87,92,95,96] Pharmacological inhibition of DOT1L reversed the oncogenic transcriptional signature caused by the MLL fusion protein and eventually resulted in apoptosis and differentiation. Importantly, the functional downstream consequences of DOT1L inhibition were time-dependent, and no phenotypes were observed even at high doses after short treatment periods. These findings support the

Table 6.1 DOT1L inhibitors.

	Structure	Name	Mechanism of action	*In vitro* IC_{50}[a] (nM)	Fold selectivity[a]	Cellular potency[a,b] (nM)	References
1	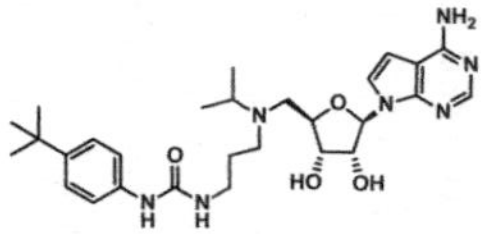	EPZ004777	SAM competitive	0.4 ± 0.1	1280 (PRMT5) >100 000 (other KMTs)	12–48 (MV4–11)	Daigle *et al.*[87]
2	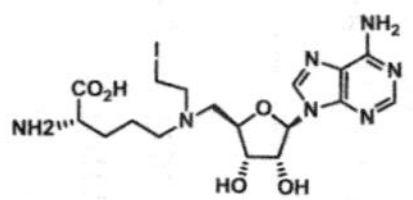	Example 4	SAM competitive	38	~30 (CARM1)~70 (PRMT1)~50 (EHMT2)>2600 (SUV39H1)	ND[d]	Yao *et al.*[90]
3	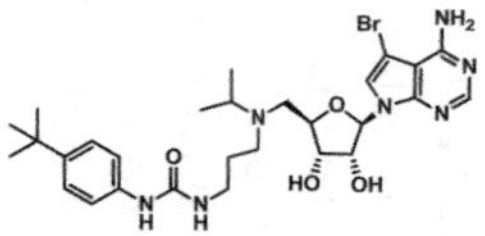	SGC9046	SAM competitive	0.3 ± 0.1	>100 000	8.8 ± 1.6 (MCF10A)	Yu *et al.*[93]
4	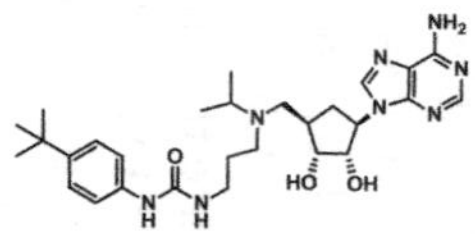	Example 6	SAM competitive	~2	>4500 (PRMT1, SUV39H1, CARM1)	~200 (MV4–11)	Deng *et al.*[94]
5	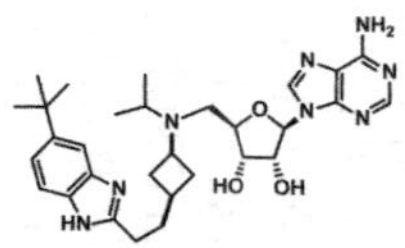	EPZ-5676	SAM competitive	<0.08[c]	>37 000	2.7 ± 0.9 (MV4–11)	Daigle *et al.*[95]

[a]As indicated in the primary reference.
[b]Global H3K79me2 levels were monitored as a surrogate of cellular target engagement; the cell line used is indicated in brackets.
[c]Refers to K_i not IC_{50}; probably underestimates potency.
[d]ND, not determined.

concept that any functional consequence of DOT1L target engagement is dictated in part by the turnover of global or local H3K79 methylation. Therefore, the transcriptional and potential phenotypic responses should be considerably delayed given the slow kinetics of H3K79 methylation exchange. The gene expression changes observed upon DOT1L inhibitor treatment were similar to the ones detected after conditional knockout of Dot1l in a MLL-AF9 leukaemia model.[97] EPZ004777 administered by continuous infusion in an MV4-11 subcutaneous xenograft model achieved target engagement, and in a disseminated MV4-11 model it achieved gains in overall survival. The relatively poor pharmacokinetic properties of the compound were not surprising for an early stage compound. However, these data perhaps hint at pharmacokinetic limitations for nucleoside-containing SAM analogues. Recently, the more potent and drug-like DOT1L inhibitor EPZ-5676 (Table 6.1) was shown to induce apoptosis in MLL-translocated leukaemia cell lines in a time- and dose-dependent manner.[95] Surprisingly, this compound was also efficacious in AML models harbouring MLL-PTD.[98] EPZ-5676 had improved properties but similar to its predecessors exhibited no oral bioavailability.[99] Continuous infusion of EPZ-5676 for 21 days resulted in complete regression of MV4-11 subcutaneous xenograft tumours in rats. The impact on tumour growth correlated well with the EPZ-5676 dose and reduction of H3K79me2 levels in the tumour.[95] The anti-tumour activity was significantly diminished when the infusion period was shortened or intermittent dosing schedules were employed, suggesting that continued target coverage is necessary to obtain maximal efficacy. There is now also evidence that EPZ-5676 shows synergy with the AML standard of care agents cytarabine and daunorubicin,[100] which could aid the clinical development of DOT1L inhibitors. EPZ-5676 has recently entered a phase I clinical trial (http://clinicaltrials.gov/) in which the response of patients with MLL-rearranged AML is being explored.

6.3 Pharmacological Inhibition of EZH2

The enhancer of Zeste homologue 2 (EZH2) and its closely related homologue EZH1 (Figure 6.1) are the only enzymes known to catalyse H3K27 methylation. Owing to decades of *Drosophila* genetics, EZH2 was recognised to be a chromatin-bound transcriptional repressor long before its H3K27-specific, SET domain-dependent histone KMT activity was discovered in 2002.[101–103] EZH2 is essential for embryonic development[104] and plays an important role in cell fate determination. EZH2 is part of the multi-subunit protein complex Polycomb repressive complex 2 (PRC2) and the complex components EED and SUZ12 are the minimal requirements for EZH2 to methylate chromatin at H3K27.[105] Interactions with other proteins further modulate EZH2 activity and its interactions with RNA and DNA.[106] PRC2 functions are tightly linked to the repression of transcription, and different genes are repressed by PRC2 in different cell types.[107] Chromatin-bound PRC2 mediates transcriptional repression at least in part by H3K27 methylation, although resultant chromatin architectural and functional consequences are not completely understood. H3K27me3 and H3K27me2 turnover rates are slow,[16] which one would expect for a modification that is implicated in the maintenance of gene silencing. Interestingly, H3K27me1 is turned over more rapidly, perhaps indicating that the kinetics of PRC2 catalysis is different for different degrees of H3K27 methylation. Moreover, non-catalytic PRC2 functions that contribute to transcriptional repression have also been described,[108–110] and the context in which the PRC2 catalytic activity is absolutely required to exert gene silencing remains to be determined. Removal of PRC2 core subunits[111,112] but not accessory factors[113–115] by genetic or RNAi-based approaches destabilises EZH2 and results in the abrogation of all PRC2 functions.

Pharmacological inhibition of EZH2 by the *S*-adenosylhomocysteine hydrolase inhibitor 3-deazaneplanocin A (DZNep) has been described[116] but it is likely that the observed effects are due to the indirect degradation of PRC2 and thus do not address questions regarding the contribution of EZH2 catalytic activity to overall PRC2 function. Moreover, certain questions regarding EZH1 and its importance for PRC2 function remain unresolved. EZH1, similar to EZH2, interacts with PRC2 components and requires these interactions to be catalytically active, suggesting an overlapping functional repertoire between these two homologous proteins. This is corroborated by genetic studies exploring PRC2 function in embryonic, skin and haematopoietic stem cells where loss of Ezh2 had a considerably milder phenotype compared to Ezh1 and Ezh2 double or Eed deletions.[117–119] However, it has been proposed that the EZH1-containing PRC2 functions by mechanisms different from EZH2-containing PRC2.[109] Genetic data also support the notion that the two enzymes have non-redundant functions including cell fate decisions for tissue-specific stem cells, cortical progenitor cells and B-cells; expansion of pancreatic islet cells; cardiac homeostasis; T-cell activation; and embryonic stem cell reprogramming capacity.[120–125] EZH1 cannot compensate for the loss of EZH2 during embryonic development[104,126] and cannot substitute for EZH2 to maintain H3K27 methylation profiles on a global scale. It is still possible that EZH1 cooperates with EZH2 in certain contexts, or even maintains H3K27 methylation on certain gene subsets in the absence of EZH2. Alternatively, EZH1 may function at certain genomic locations to support the transcription cycle rather than to suppress it.[127] In general, EZH1 and EZH2 are not similarly expressed across tissues. While EZH2 expression is high in stem and progenitor cells, expression is usually low in terminally differentiated cells.

EZH2 (but not EZH1) is frequently over-expressed in human cancer. High levels of expression correlate with increased H3K27me3 levels, late stage disease and poor outcome.[128–130] Moreover, a number of functional studies implicate PRC2 and specifically EZH2 in tumorigenesis, cancer progression and metastasis.[41,131–134] However, in these studies the contribution of EZH2 KMT activity to the observed cancer biology was not directly addressed. The identification of recurrent mutations within the EZH2 catalytic domain (Figure 6.4) provided a strong argument that certain cancers might be dependent on EZH2's catalytic activity. The mutated residues Y641, A677 and A687 alter EZH2 substrate specificity, facilitating the conversion of H3K27 from a dimethylated to a trimethylated state,[29–33] and thus mutant EZH2-containing cancer cells exhibit higher global H3K27me3 levels compared to those with wild-type EZH2. Gain-of-function EZH2 mutations have been found in germinal centre B-cell-like diffuse large B-cell lymphoma (GCB-DLBCL) and follicular lymphoma, as well as melanoma (Figure 6.3).[26–28,135] In all these cases, no homozygous EZH2 mutations were discovered, suggesting that these cancer cells are dependent on both wild-type and mutant EZH2. However, when all recurrently mutated genes in DLBCL were interrogated for potential imbalance in allelic expression only 17% of mutated genes, including EZH2, showed preferential expression of the mutant allele.[28] Collectively, genomic and transcriptomic data in lymphoma strongly argue that EZH2 is an oncogenic driver. In contrast, inactivating homo- and heterozygous somatic mutations in EZH2 (Figure 6.4) and other PRC2 components have been discovered in AML, myelodysplastic syndrome, myeloproliferative neoplasms, chronic myelomonocytic leukaemia and early T-cell precursor acute lymphoblastic leukaemia,[136–140] suggesting that EZH2 may function in these malignancies as a tumour suppressor.

Small molecules have been reported that inhibit the catalytic activity of EZH2 and thereby suppress global H3K27 methylation.[141–149] All inhibitors published to date are exquisitely selective over other KMTs (Table 6.2) and function by the same mechanism of action, namely competition with SAM. Most of the compounds are very potent (low nM to pM range) and elicit cellular responses such as modulation of global H3K27me3 levels at submicromolar concentrations. Despite the fact that these inhibitors were initially derived from independent high-throughput screening efforts they share a common structural feature, a pyridone group, which is absolutely required for target binding. Due to converging SAR, these molecules occupy similar chemical space suggesting at least at present a narrow window of opportunity one has to operate in to identify EZH2 inhibitors. The exception is a molecule containing a tetramethylpiperidine head group (example 6 in Table 6.2) that shares no structural features with any of the pyridone-containing inhibitors. This compound is also SAM-competitive, is completely selective over other KMTs and inhibits EZH1 and EZH2 to a similar extent.[142] Despite comprehensive enzymology data that argue for SAM competition as the mechanism of action it is actually unknown whether EZH2 inhibitors bind precisely to the SAM-binding site of the enzyme. In the absence of an EZH2:compound co-crystal structure several alternatives have to be considered. The inhibitor might bind to a different region within the catalytic pocket, resulting in a conformational change that might render the enzyme inert to SAM binding. Alternatively, the compound could bind to one of the other PRC2 subunits, thereby inducing a conformational change that would render EZH2 less conducive to SAM binding. The recently published EZH2 SET domain crystal structure[150] represents good progress but the I-SET and

Table 6.2 EZH2 inhibitors.

	Structure	Name	Mechanism of action	*In vitro* IC_{50}[a] (nM)	Fold selectivity[a]	Cellular potency[a,b] (nM)	References
1	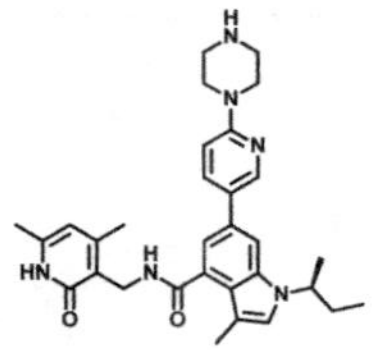	GSK-126	SAM competitive	<1	>150 (EZH1) >10 000	7–252 (DLBCL)	McCabe *et al.*[30]
2	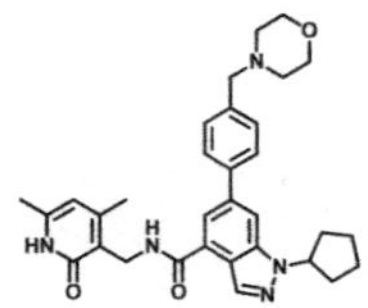	EPZ005687	SAM competitive	24 ± 7	~50 (EZH1) >500	80 ± 30 (OCI-LY19)	Knutson *et al.*[144]
3	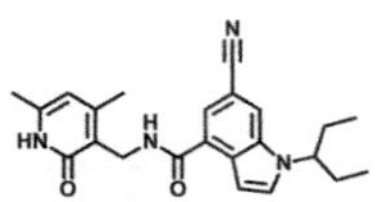	Novartis-EI 1	SAM competitive	15 ± 2	~90 (EZH1) >10 000	~500 (SUDHL6)	Qi *et al.*[147]
4	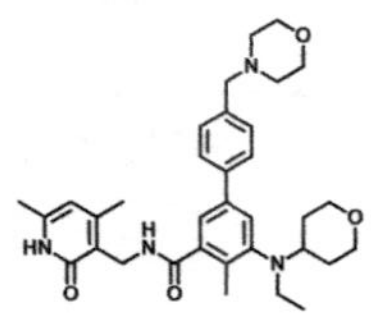	EPZ-6438	SAM competitive	2.5 ± 0.5	~35 (EZH1) >4500	~300 (G401) ~90 (KARPAS-422)	Knutson *et al.*[143,154]

(*continued*)

Table 6.2 (*continued*)

	Structure	Name	Mechanism of action	*In vitro* IC_{50}[a] (nM)	Fold selectivity[a]	Cellular potency[a,b] (nM)	References
5	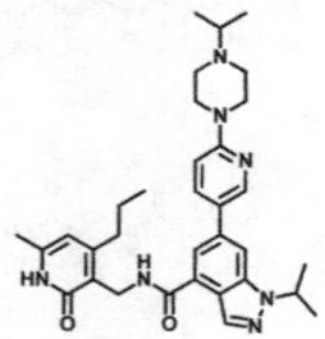	UNC-1999	SAM competitive	4.6 ± 0.8	~10 (EZH1) >10 000	124 ± 11 (MCF10A)	Konze *et al.*[145]
6	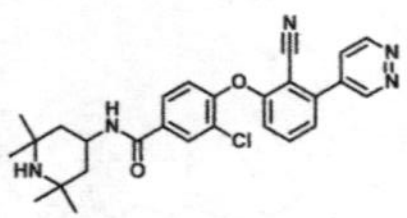	Compound 3	SAM competitive	32 ± 19	~10 (EZH1) >3800	~6700 (Pfeiffer)	Garapaty-Rao *et al.*[142]
7	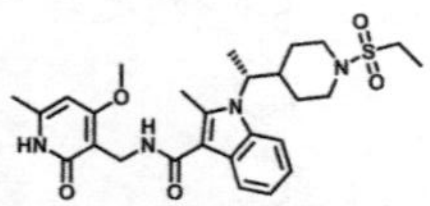	CPI-169	SAM competitive	0.24 ± 0.06	~100 (EZH1) >10 000	~41 (KARPAS-422)	Bradley *et al.*[148]

[a]As indicated in the primary reference.
[b]Global H3K27me3 levels were monitored as a surrogate of cellular target engagement; the cell line used is indicated in brackets.

post-SET domain conformation and the inability to bind SAM or substrate limits its utility for structure-guided drug design.

The various EZH2 inhibitors consistently achieve dose-dependent induction of gene expression, confirming the role of the PRC2 complex in gene silencing. There is no evidence that the protein levels of the PRC2 subunits are affected by the compounds, suggesting that the functional consequences observed upon compound treatment result from suppression of the complex's catalytic activity. Thus, EZH2 small-molecule inhibitors serve as exquisite tools to dissect PRC2 catalytic *versus* non-catalytic function(s). In addition, EZH2 inhibitors provided a wealth of data to support the idea that lymphomas harbouring activating EZH2 mutations are dependent on them.[142,144,146–148,151]

Wild-type EZH2 is required for the formation of germinal centre B-cells and transiently suppresses their differentiation. As B-cells exit the germinal centre reaction, EZH2 levels usually decrease, promoting the expression of genes that ensure terminal differentiation.[152] Conditional expression of an EZH2 mutant allele promotes hyperplasia and lymphomagenesis by aberrantly repressing B-cell differentiation genes through promoter H3K27 trimethylation,[151] providing a mechanistic explanation for the oncogenic role of mutant EZH2 in GCB-DLBCL.

EZH2 inhibitors achieved complete regression in various GCB-DLBCL subcutaneous xenograft models[146,148,153,154] and SMARCB1-deletion driven malignant rhabdoid tumour models[143] without causing any overt adverse effects. A good correlation between the administered dose of compound, reduction in H3K27me3 levels in tumour tissue and an effect on tumour growth was observed. It is noteworthy, however, that high doses and high systemic and tumour concentrations of inhibitor were required despite the exquisite potency of the compounds, perhaps suggesting that complete target engagement is required and needs to be maintained to achieve anti-tumour effects. It is encouraging that the suppression of tumour growth seems durable as no regrowth was observed after treatment was discontinued.[146,154] First-generation EZH2 inhibitors have recently entered phase I clinical trials (http://www.clinicaltrials.gov/) with a plan to focus on patients with germinal centre lymphomas and advanced solid tumours.

6.4 Pharmacological Inhibition of EHMT1 and EHMT2

EHMT1 (GLP, KMT1D) and EHMT2 (G9a, KMT1C) are homologous SET domain-containing, H3K9-specific KMTs (Figure 6.1). In contrast to the H3K9 trimethyl-specific KMT SUV39H1, EHMTs catalyse H3K9 and H1.4K26 mono- and dimethylation.[155,156] Both EHMT1 and EHMT2 function as part of a heterodimeric (EHMT) complex in the regulation of both transcriptional repression and activation, thereby impacting important cellular and developmental processes such as germ cell formation, embryogenesis and cardiac morphogenesis.[157,158] While EHMT co-repressor activity largely depends on its catalytic activity, the co-activator role of EHMTs does not require it.[159,160] EHMTs also contribute to transcriptional repression by physical interaction with DNA methyltransferases (DNMTs) and PRC2 and their recruitment to chromatin.[161,162] Interestingly, the EHMT complex also controls the repression of non-neural genes in neuronal tissues and thus seems to control adaptive behaviour and memory formation.[163–165] This role is corroborated by genomic data that linked inactivating EHMT1 mutations to Kleefstra syndrome, a neurodevelopmental disorder characterised by intellectual disability, among other features.[166] EHMT2 has a role in neurons of the nucleus accumbens, a region of the brain with important roles in reward, learning reinforcement and addiction. Cocaine and morphine addiction correlate with reduced H3K9me2 levels in the nucleus accumbens. Reduction of EHMT2 increases the preference for cocaine and morphine while its increased expression counteracts these effects.[167,168]

EHMTs are not significantly mutated in cancer,[22] but EHMT2 is overexpressed in lung and bladder cancers,[169,170] and in some instances is correlated with a poor prognosis. Reduction of EHMT2 levels by RNAi suppressed cell proliferation,[171] inhibited lung and breast cancer cell migration and invasion *in vitro* and impaired primary tumour growth and metastasis *in vivo*.[169,172] EHMTs are also implicated in cancer metabolism, promoting glycolysis in basal-like breast cancer cells and maintaining the activity of the serine–glycine biosynthetic pathway which is required for cancer cell proliferation.[173,174]

The first selective small-molecule inhibitor of EHMT1 and EHMT2 was identified through biochemical screening.[175] At micromolar concentrations the described compound, BIX-01294 (Table 6.3), reduced global and gene-specific H3K9me2 levels in mammalian cells[156,167,175] but also showed a fair amount of off-target toxicity in cell proliferation experiments. Interestingly, BIX-01294 inhibits EHMT1 and EHMT2 in a substrate-competitive manner,[176] while most other validated KMT inhibitors to date are SAM-competitive. Medicinal chemistry efforts improved the potency of this chemotype,[81,177–181] consequently achieving better separation of on-target mediated functional consequences of EHMT inhibition and off-target toxicities. UNC0638 (Table 6.3) is a highly potent and specific EHMT inhibitor that suppresses global H3K9me2 but not H3K9me3 levels in a dose- and time-dependent manner. Even at fairly high compound concentrations, H3K9me2 levels decreased gradually over the course of 4 days but were never completely depleted, suggesting that other KMTs contribute to global H3K9me2 levels. The effects persisted for several days after compound removal, indicating either slow H3K9me2 turnover or a slow compound off-rate. No impact on global DNA methylation levels was observed, even after long-term treatment. However, treatment with UNC0638 resulted in increased expression of a number of genes, which correlated with a reduction of gene-specific H3K9me2 and DNA methylation levels.[181] Continued chemistry efforts generated UNC0642 (Table 6.3), a compound with improved pharmacokinetic properties that was exquisitely non-toxic in 2D cell growth assays but which reduced colony formation of pancreatic adenocarcinoma cells.[178]

Table 6.3 EHMT inhibitors.

	Structure	Name	Mechanism of action	*In vitro* $IC_{50}{}^{a}$ (nM)	Fold selectivity[a]	Cellular potency[a,b] (nM)	References
1	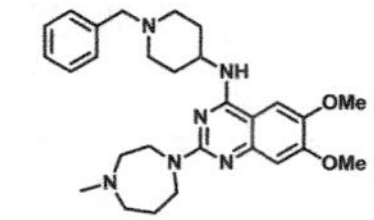	BIX-01294	Substrate-competitive	~180[c]	>1000	500 ± 43[c] (MDA-MB231)	Kubicek *et al.*[175]
2	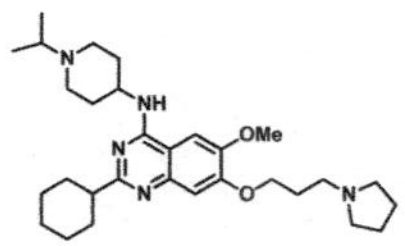	UNC0638	Substrate-competitive	<15	>10 000	81 ± 9 (MDA-MB231)	Vedadi *et al.*[181]
3	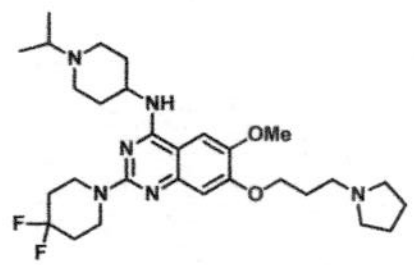	UNC0642	Substrate-competitive	<2.5	>2000	~110 (MDA-MB231)	Liu *et al.*[178]
4	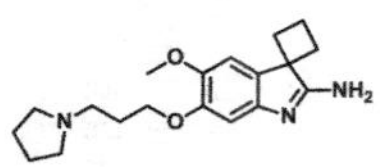	A-366	Substrate-competitive	~3.3	>10 000	~3000[d] (PC3)	Sweis *et al.*[182]
5	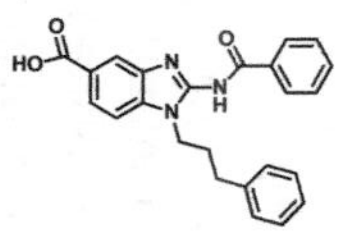	BRD9539	SAM-competitive	~6300	>6	~5000[e] (PANC-1)	Yuan *et al.*[183]

[a]As indicated in the primary reference.
[b]Global H3K9me2 levels were used as a surrogate of cellular target engagement; the cell line used is indicated in brackets.
[c]The values are taken from Vedadi *et al.*[181] to allow for direct comparison between BIX-01294 and UNC0638.
[d]Likely to be more potent because UNC0638 in this study also had an average H3K9me2 IC_{50} of 3000 nM.
[e]Determined with the cell-permeable pro-drug BRD4770.

A structurally unrelated chemotype was recently identified by screening of a diversity-oriented compound collection. A-366 (Table 6.3) is a potent, selective and cell-active inhibitor[182] that also functions on the basis of substrate competition. SAM-competitive inhibitors for EHMT2 have been identified as well[175,183] but these are less selective and potent, have not been validated by biophysical methods or crystallography and lack cellular activity. However, a methylester pro-drug of one of these compounds, BRD9539 (Table 6.3), exhibited cellular activity, modestly decreasing H3K9me2 and H3K9me3 levels at micromolar concentrations. This compound did not induce apoptosis but rather senescence in short-term 2D growth assays and caused modest reduction in colony formation in long-term soft agar assays.[183] However, a tight correlation between cellular target engagement and the observed phenotypic effects remains to be established.

The EHMT inhibitor BIX-01294 suppressed cancer cell growth in various contexts; however, potential off-target toxicity at high compound concentrations complicates the interpretation of cancer cell phenotypic data and it is not yet established to what extent EHMT catalytic activities are required in normal cells. Given these open questions, EHMT small-molecule inhibitors have not yet revealed clear applications for human cancer therapy. In one instance, BIX-01294 was shown to impact serine biosynthesis within 24 hours and down-regulated the expression of serine biosynthetic pathway genes within 6 hours of treatment.[173] Serine deprivation consequently resulted in defects in cell cycle progression, ribosome biogenesis and induced autophagy in cancer cells. Interestingly, addition of serine bypassed the growth-detrimental effects of BIX-01294. However, it is unclear whether the BIX-01294-induced gene expression changes were directly mediated by inhibition of EHMTs directly bound to the genes that undergo transcriptional changes. Neither was a correlation between the induction of gene expression and local changes in H3K9 methylation levels established. Despite the fact that H3K9me1 and H3K9me2 marks are turned over rapidly on a global scale,[16] the kinetics of local H3K9 methylation decreases on serine biosynthetic pathway genes needed to be even faster to directly account for the induction of gene expression.

It is also possible that the full potential of EHMT inhibitors to interfere with aberrant gene expression programmes in cancer requires their combination with DNMT inhibitors because gene silencing is often maintained by both H3K9me2 and DNA methylation and interfering with either one does not relieve repression maintained by the other.[184] Interestingly, EHMT inhibitors as single agents facilitated the generation of induced pluripotent cells,[185] prevented aminoglycoside-induced ototoxicity in auditory hair cells[186] and exhibited clear antimalarial effects;[187] however, in each case a mechanistic explanation for the observed effects is still lacking. Further improvement of EHMT inhibitors is necessary to carry out *in vivo* efficacy experiments that will also address open questions regarding the therapeutic window, especially given potential cardiac-specific toxicities. Moreover, the currently most advanced compounds partition into the brain, which may be undesirable given the well documented neurological function of EHMTs. A potential progression of EHMT inhibitors to clinical development will rest on the ability to further improve the pharmacological properties as well as to find a specific cancer context that is exquisitely dependent on EHMT catalytic activities.

6.5 Progress Towards Developing Small-Molecule Inhibitors against other Methyltransferases

Early drug discovery studies were confounded by a lack of methods to adequately validate the chemical matter and correctly assess the specificity and mechanism of action of potential KMT inhibitors. In 2005 a high-throughput screening campaign led to the identification of the fungal metabolite chaetocin as what was presumed to be the first selective inhibitor of the H3K9-specific methyltransferase SUV39H1 and its *Drosophila* homologue Su(var)3–9.[188] Years later, in-depth molecular studies proved chaetocin to be a non-specific, non-competitive and time-dependent KMT inhibitor whose mechanism of action is essentially dependent on its disulfide functionality and irreversible adduct formation with the enzyme.[189,190] In general, screening of large small-molecule compound libraries – despite the above-mentioned successful campaigns – has not uncovered a wealth of KMT inhibitor chemotypes. For the majority of KMTs there are still no small-molecule probe compounds available, with the exception of the few examples summarised below.

SMYD proteins belong to the KMT subfamily with a divergent SET domain (Figure 6.1). The SET domain is split by an intervening MYND zinc finger motif that mediates interactions with proteins and DNA but does not contribute residues to either substrate or cofactor binding. Another distinctive structural feature of SMYD proteins is the presence of an unusual C-terminal domain, a tetratrico-peptide repeat (TPR) motif that participates in the formation of the substrate binding channel and is thus required for the enzyme to be enzymatically competent.[191–193] Interestingly, the TPR domain shows striking conformational flexibility, either packing tightly against the SET domain and closing over the substrate binding pocket adopting an auto-inhibited conformation or promoting an open, catalytically active state through a hinge-bending motion.[194,195] The substrate specificity of SMYDs is not as clear as for other KMTs because H3K4, H4K20 and H4K5 have been claimed for SMYD3 and H3K4 and H3K36 as target sites for SMYD2.[196–200] In several instances it was suggested that the chaperone HSP90 alters the substrate specificity or promotes catalytic activity by stabilising the open conformation of the TPR domain. SMYD2 also methylates non-histone substrates such as HSP90, TP53, RB1 and ER.[196,201,202]

Given that both SMYD2 and SMYD3 are amplified and over-expressed in various cancer types[203–205] and functional experiments suggest a requirement for these enzymes for cancer cell proliferation in certain contexts,[199,204] these enzymes might be considered as candidate oncology targets. SMYD2 small-molecule inhibitors have been identified by HTS of a large compound library.[206] AZ505 (Table 6.4) was the most potent among several identified compounds whose target binding was confirmed by biophysical methods. This compound is highly selective, substrate-competitive and uncompetitive for SAM in enzymatic assays and its co-crystallisation with SMYD2 confirmed that it occupied the substrate peptide groove. AZ505 at a concentration of 10 μM significantly suppressed SMYD2-induced TP53 K370 monomethylation in the human osteosarcoma cell line U2OS within 24 hours. These data are consistent with the idea that TP53K370me1 is turned over relatively fast but the impact of SMYD2 inhibition on changes in the histone lysine methylation landscape or insight into the kinetics of such events were not examined. Despite the encouraging cellular data these compounds still need to be significantly improved before phenotypic consequences due to SMYD2 inhibition can be assessed *in vitro* and *in vivo*. A structurally similar compound, A-893, that is significantly more potent in biochemical assays did not markedly improve cellular potency with respect to TP53K370 methylation.[207] Collaborative efforts between the Structural Genomics Consortium (SGC) and Eli Lilly and Company have resulted in the discovery of LLY-507 (Table 6.4), another SMYD2 selective inhibitor.[208] LLY-507 is a potent inhibitor of SMYD2 with 100-fold selectivity over other methyltransferases and other non-epigenetic targets. LLY-507 treatment significantly reduces TP53K370me1 levels at sub-micromolar concentrations within 18 hours of treatment. No impact on histone methylation levels were observed, raising some questions regarding SMYD2's ability to methylate histone lysine residues. The compound affected cancer cell proliferation in a time-dependent manner. The underlying molecular effects of LLY-507 treatment in these cancer cells remain to be determined.

SETD7 (SET7, SET9) is a SET domain-containing enzyme (Figure 6.1) that was originally identified as a histone H3K4 monomethylase.[209,210] Since then an ever increasing number of transcription factors, chromatin modifiers and signalling proteins have been described as SETD7 substrates. These include TP53, TAF10, ER, DNMT1, PCAF, P65, E2F1, STAT3, AR (reviewed in Del Rizzo and Trievel[211]), MYPT1,[212] SUV39H1,[213] FXR,[214] FOXO3,[215] YAP[216] and IFITM3,[217] always resulting in SETD7-mediated monomethylation of the respective proteins. Initial reports claiming that TP53 methylation by SETD7 was important for its function in response to DNA damage were subsequently disproven by genetic studies.[218,219] Overall, there is evidence for SETD7 being involved in transcriptional regulation, differentiation, protein stability, genome integrity and Hippo signalling.[213,216,220–222] In many instances, the true functional impact of SETD7-mediated protein methylation remains elusive. Cancer genomic data do not reveal SETD7 as being significantly mutated[22] or over-expressed in cancer.

Notwithstanding the absence of strong disease rationale, collaborative efforts between the SGC and Pfizer, as part of a large chemical biology initiative in the chromatin space, has led to the discovery of (*R*)-PFI-2 (Table 6.4), a potent, selective and cell-active inhibitor of SETD7.[223] A structural analysis demonstrated that (*R*)-PFI-2 occupies the peptide binding groove and thus

Table 6.4 Other validated KMT inhibitors.

	Structure	Name	Mechanism of action	*In vitro* IC_{50}[a] (nM)	Fold selectivity[a]	Cellular potency[a] (nM)	References
1	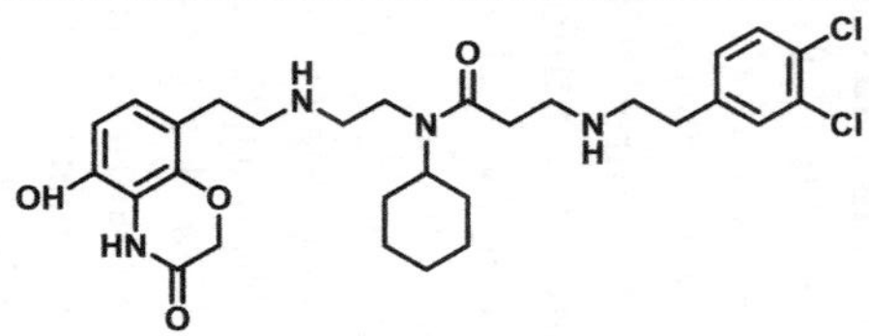	AZ505	Substrate-competitive	120	>690-fold	No dose response	Ferguson *et al.*[206]
	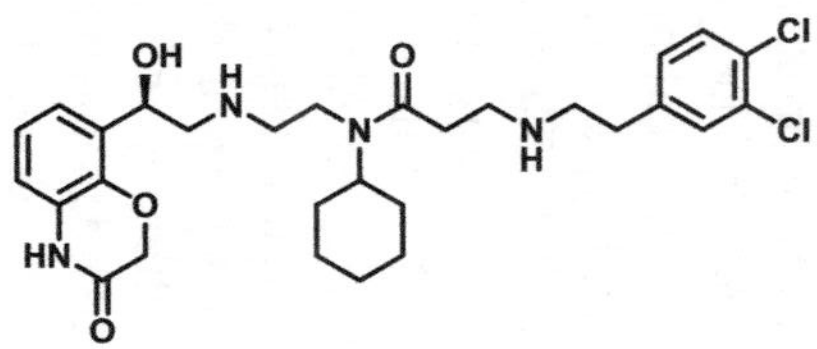	A-893	Substrate-competitive	2.8	>300	No dose response	Sweis *et al.*[207]
2	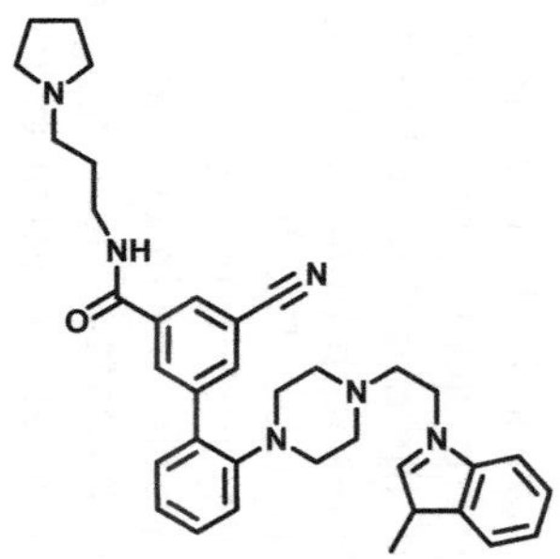	LLY-507	Substrate-competitive	<15	>100	~600	Nguyen *et al.*[208]
3	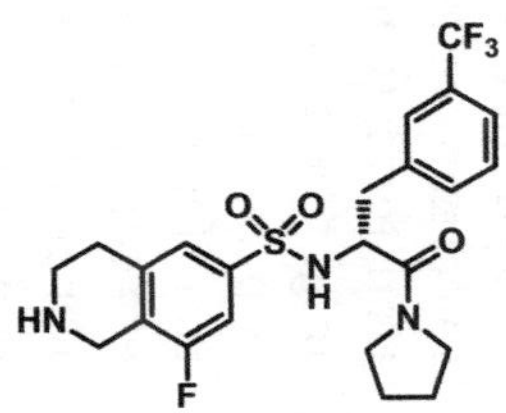	(*R*)-PFI-2	Substrate-competitive	<10	>100	Unknown	Barsyte-Lovejoy *et al.*[223]

[a]As indicated in the primary reference.

suggested a peptide-competitive mode of inhibition. Interestingly, the compound's pyrrolidine amide makes direct contact with SAM. Biophysical assessment of (*R*)-PFI-2-binding to SETD7 revealed that SAM is required for efficient target engagement, which is consistent with the structural data and suggested a binding mode that is uncompetitive for SAM. Favourable physicochemical properties make (*R*)-PFI-2 a suitable tool compound for cellular assays. SETD7 was previously proposed to have a role in the Hippo pathway, an evolutionarily conserved signalling cascade regulating cell proliferation and controlling organ size. (*R*)-PFI-2 promoted nuclear localisation of the transcriptional co-activator YAP, consistent with earlier findings that SETD7 KMT activity was required for YAP subcellular localisation.[216] Because Hippo activation and nuclear YAP-driven transcription are a means by which cancer cells bypass contact inhibition,[224] (*R*)-PFI-2 may actually promote cancer cell growth. The biological impact of these SETD7 inhibitors in other disease contexts has yet to be evaluated.

6.6 Conclusions and Outlook

Despite the excellent progress that has been made in drugging individual KMTs there are still many of the 51 SET domain-containing proteins that have not been pursued in drug discovery, either due to a lack of biological interest or concerns about these proteins perhaps being enzymatically inactive. For several KMTs with an important role in human disease drug discovery campaigns have so far been unsuccessful. For instance, the field continues to wait for high-quality inhibitors of the NSD subfamily of H3K36-specific KMTs given that NSD1 is translocated in a small subset of AML,[225] WHSC1 is translocated in 15% of multiple myeloma and WHSC1L1 (NSD3) is found as part of the prevalent 8p11 copy number gain in breast, lung and pancreatic cancers.[226,227] Similarly, inhibitors targeting the intrinsic catalytic activity of the MLL subfamily of H3K4-specifc methyltransferases have not yet been discovered. MLL is frequently translocated in myeloid and lymphoid leukaemias, where such rearrangements account for approximately 80% of infant leukaemias and 5–10% of adult leukaemias.[228–230] Despite the fact that the catalytic domain is lost in AML, with MLL translocations it was shown that in certain contexts the activity of the MLL wild-type allele is required to drive transformation of the translocated MLL allele.[231] Therefore, MLL inhibitors might be efficacious in human malignancies where a wild-type MLL allele is present and required. Finally, the entire PR domain-containing methyltransferase (PRDM) subfamily has been largely ignored by the drug discovery community despite the interesting biological aspects of these enzymes. For instance, the PRDM14-encoding gene is part of the focal 1q22 amplicon found in breast and ovarian cancer whereby the PRDM14 expression correlates with gene copy number.[232–234] PRDM14 over-expression promotes malignant transformation of fibroblasts and induces lymphoblastic leukaemia in bone marrow cells, and RNAi-mediated loss of PRDM14 causes proliferation defects in breast cancer cells.[234,235] MECOM (PRDM3, EVI1, MDS1) is frequently translocated in cancer; however, the N-terminally located catalytic domain is often lost in these translocation events,[236] suggesting rather a tumour-suppressive function. In fact, multiple other PRDMs have ascribed tumour suppressor roles, which might explain the reluctance the field has had for targeting this KMT subfamily. Surprisingly, recent data argue that the PR domain of MECOM is required for malignant transformation of myeloid progenitors by select MLL fusion proteins such as MLL-AF9 and MLL-ENL.[237] These findings indicate that MECOM, similar to other KMTs, might be implicated in either promoting or suppressing tumour growth depending on the given molecular context.

Overall, we still lack a comprehensive mechanistic understanding about KMT function in cancer, and we have yet to discover the defining parameters of those contexts in which cancer cells have become dependent upon the catalytic activity of individual KMTs. Frequent inactivating mutations observed in UTX, EZH2, MLL2 and MLL3, along with activating mutations in EZH2, translocations of NSD1, WHSC1, MLL, MECOM and the many KMT gene amplification events, provide the basis for the continued interest in KMTs as a potentially rich space for next-generation oncology targets.

The excitement in this target space and the initial enthusiasm have been tempered by certain KMT-targeted drug discovery challenges. For example, biochemical assay development involved reconstitution of multi-subunit protein complexes and the generation of complex substrates such as oligonucleosomes to recapitulate physiologically relevant activity and substrate specificity. Hit identification and lead finding phases were tedious because existing small-molecule compound libraries are not enriched for KMT-targeting chemotypes. Thus, the HTS campaigns frequently yielded weak inhibitors that required sophisticated hit triage to separate valid KMT inhibitors from false positives. Implementation of biophysical methods proved invaluable on several occasions to identify relevant chemistry starting points. Surprisingly, all the different approaches to identify relevant chemotypes have led to completely selective molecules that effectively suppress the activity of a particular KMT but do not cross over to other KMTs. Thus, initial hopes to generate and evolve compound libraries of unselective or partially selective KMT inhibitor chemotypes to accelerate future KMT drug discovery campaigns have so far not materialised. Perhaps the exceptions are rational design approaches around SAM analogue libraries that have already yielded very potent and selective lead chemotypes. However, the utility of SAM analogue libraries to broadly prosecute KMTs in a platform-type manner remains to be shown.

Several inhibitors discussed in this review result in efficient target engagement but show delayed histone methylation turnover and slow phenotypic responses. Adequate assessment of on-target activities requires more sophisticated cellular assays with respect to incubation period and model complexity to better predict *in vivo* responses. Because DOT1L and EZH2 inhibitors only show phenotypic activity in longer-term growth assays one might expect that this is a common phenomenon for KMT inhibitors. However, it is probably premature to make this generalisation, and for additional KMT inhibitors the kinetics of the downstream functional consequences of target engagement will have to be understood before one can predict the timing of the phenotypic responses. Chromatin-based therapeutics vary widely in their kinetics and phenotypic responses. While sub-toxic DNMT inhibitors seem to cause similarly slow response kinetics, HDAC inhibitors show a rapid induction of anti-growth phenotypes. Moreover, inhibitors that target the bromodomains of the chromatin readers BRD2, BRD3, BRD4 and BRDT reduce *MYC* transcript levels within hours of treatment and elicit phenotypic responses in 1–2 days.

Given the substantial efforts that continue to be made in the KMT field it is anticipated that progress in drugging these unique enzymes will continue. And now that KMT inhibitors have entered the clinic one can expect that the early experience with these first compounds in patients will inform the direction of future developments in the field.

Acknowledgements

I would like to thank Drs Michael R. Cooper and James E. Audia for critical reading of the manuscript and Dr Christopher Nasveschuk for help with the organisation of the tables.

References

1. S. Rea, F. Eisenhaber, D. O'Carroll, B. D. Strahl, Z. W. Sun, M. Schmid, S. Opravil, K. Mechtler, C. P. Ponting, C. D. Allis and T. Jenuwein, *Nature*, 2000, **406**, 593–599.
2. E. L. Greer and Y. Shi, *Nat. Rev. Genet.*, 2012, **13**, 343–357.
3. H. M. Herz, A. Garruss and A. Shilatifard, *Trends Biochem. Sci.*, 2013, **38**, 621–639.
4. C. Martin and Y. Zhang, *Nat. Rev. Mol. Cell Biol.*, 2005, **6**, 838–849.
5. T. C. Petrossian and S. G. Clarke, *Mol. Cell. Proteomics*, 2011, **10**, M110 000976.
6. A. T. Nguyen and Y. Zhang, *Genes Dev.*, 2011, **25**, 1345–1358.
7. B. D. Strahl and C. D. Allis, *Nature*, 2000, **403**, 41–45.
8. S. D. Taverna, H. Li, A. J. Ruthenburg, C. D. Allis and D. J. Patel, *Nat. Struct. Mol. Biol.*, 2007, **14**, 1025–1040.
9. S. G. Clarke, *Trends Biochem. Sci.*, 2013, **38**, 243–252.
10. K. E. Moore and O. Gozani, *Biochim. Biophys. Acta*, 2014, **1839**, 1395–1403.
11. C. Das and J. K. Tyler, *Biochim. Biophys. Acta*, 2013, **1819**, 332–342.
12. E. M. Duncan, T. L. Muratore-Schroeder, R. G. Cook, B. A. Garcia, J. Shabanowitz, D. F. Hunt and C. D. Allis, *Cell*, 2008, **135**, 284–294.
13. H. Santos-Rosa, A. Kirmizis, C. Nelson, T. Bartke, N. Saksouk, J. Cote and T. Kouzarides, *Nat. Struct. Mol. Biol.*, 2009, **16**, 17–22.
14. G. E. Zentner and S. Henikoff, *Nat. Struct. Mol. Biol.*, 2013, **20**, 259–266.
15. B. M. Zee, L. M. Britton, D. Wolle, D. M. Haberman and B. A. Garcia, *Mol. Cell. Biol.*, 2012, **32**, 2503–2514.
16. B. M. Zee, R. S. Levin, B. Xu, G. LeRoy, N. S. Wingreen and B. A. Garcia, *J. Biol. Chem.*, 2010, **285**, 3341–3350.

17. E. Apostolou and K. Hochedlinger, *Nature*, 2013, **502**, 462–471.
18. A. I. Badeaux and Y. Shi, *Nat. Rev. Mol. Cell Biol.*, 2013, **14**, 211–224.
19. P. Gut and E. Verdin, *Nature*, 2013, **502**, 489–498.
20. A. Gonzalez-Perez, A. Jene-Sanz and N. Lopez-Bigas, *Genome Biol.*, 2013, **14**, r106.
21. C. Kandoth, M. D. McLellan, F. Vandin, K. Ye, B. Niu, C. Lu, M. Xie, Q. Zhang, J. F. McMichael, M. A. Wyczalkowski, M. D. Leiserson, C. A. Miller, J. S. Welch, M. J. Walter, M. C. Wendl, T. J. Ley, R. K. Wilson, B. J. Raphael and L. Ding, *Nature*, 2013, **502**, 333–339.
22. M. S. Lawrence, P. Stojanov, C. H. Mermel, J. T. Robinson, L. A. Garraway, T. R. Golub, M. Meyerson, S. B. Gabriel, E. S. Lander and G. Getz, *Nature*, 2014, **505**, 495–501.
23. T. I. Zack, S. E. Schumacher, S. L. Carter, A. D. Cherniack, G. Saksena, B. Tabak, M. S. Lawrence, C. Z. Zhang, J. Wala, C. H. Mermel, C. Sougnez, S. B. Gabriel, B. Hernandez, H. Shen, P. W. Laird, G. Getz, M. Meyerson and R. Beroukhim, *Nat. Genet.*, 2013, **45**, 1134–1140.
24. P. Chi, C. D. Allis and G. G. Wang, *Nat. Rev. Cancer*, 2010, **10**, 457–469.
25. J. C. Black, C. Van Rechem and J. R. Whetstine, *Mol. Cell*, 2012, **48**, 491–507.
26. J. G. Lohr, P. Stojanov, M. S. Lawrence, D. Auclair, B. Chapuy, C. Sougnez, P. Cruz-Gordillo, B. Knoechel, Y. W. Asmann, S. L. Slager, A. J. Novak, A. Dogan, S. M. Ansell, B. K. Link, L. Zou, J. Gould, G. Saksena, N. Stransky, C. Rangel-Escareno, J. C. Fernandez-Lopez, A. Hidalgo-Miranda, J. Melendez-Zajgla, E. Hernandez-Lemus, A. Schwarz-Cruz y Celis, I. Imaz-Rosshandler, A. I. Ojesina, J. Jung, C. S. Pedamallu, E. S. Lander, T. M. Habermann, J. R. Cerhan, M. A. Shipp, G. Getz and T. R. Golub, *Proc. Natl. Acad. Sci. U. S. A.*, 2012, **109**, 3879–3884.
27. R. D. Morin, N. A. Johnson, T. M. Severson, A. J. Mungall, J. An, R. Goya, J. E. Paul, M. Boyle, B. W. Woolcock, F. Kuchenbauer, D. Yap, R. K. Humphries, O. L. Griffith, S. Shah, H. Zhu, M. Kimbara, P. Shashkin, J. F. Charlot, M. Tcherpakov, R. Corbett, A. Tam, R. Varhol, D. Smailus, M. Moksa, Y. Zhao, A. Delaney, H. Qian, I. Birol, J. Schein, R. Moore, R. Holt, D. E. Horsman, J. M. Connors, S. Jones, S. Aparicio, M. Hirst, R. D. Gascoyne and M. A. Marra, *Nat. Genet.*, 2010, **42**, 181–185.
28. R. D. Morin, M. Mendez-Lago, A. J. Mungall, R. Goya, K. L. Mungall, R. D. Corbett, N. A. Johnson, T. M. Severson, R. Chiu, M. Field, S. Jackman, M. Krzywinski, D. W. Scott, D. L. Trinh, J. Tamura-Wells, S. Li, M. R. Firme, S. Rogic, M. Griffith, S. Chan, O. Yakovenko, I. M. Meyer, E. Y. Zhao, D. Smailus, M. Moksa, S. Chittaranjan, L. Rimsza, A. Brooks-Wilson, J. J. Spinelli, S. Ben-Neriah, B. Meissner, B. Woolcock, M. Boyle, H. McDonald, A. Tam, Y. Zhao, A. Delaney, T. Zeng, K. Tse, Y. Butterfield, I. Birol, R. Holt, J. Schein, D. E. Horsman, R. Moore, S. J. Jones, J. M. Connors, M. Hirst, R. D. Gascoyne and M. A. Marra, *Nature*, 2011, **476**, 298–303.
29. C. R. Majer, L. Jin, M. P. Scott, S. K. Knutson, K. W. Kuntz, H. Keilhack, J. J. Smith, M. P. Moyer, V. M. Richon, R. A. Copeland and T. J. Wigle, *FEBS Lett.*, 2012, **586**, 3448–3451.

30. M. T. McCabe, A. P. Graves, G. Ganji, E. Diaz, W. S. Halsey, Y. Jiang, K. N. Smitheman, H. M. Ott, M. B. Pappalardi, K. E. Allen, S. B. Chen, A. Della Pietra, 3rd, E. Dul, A. M. Hughes, S. A. Gilbert, S. H. Thrall, P. J. Tummino, R. G. Kruger, M. Brandt, B. Schwartz and C. L. Creasy, *Proc. Natl. Acad. Sci. U. S. A.*, 2012, **109**, 2989–2994.
31. C. J. Sneeringer, M. P. Scott, K. W. Kuntz, S. K. Knutson, R. M. Pollock, V. M. Richon and R. A. Copeland, *Proc. Natl. Acad. Sci. U. S. A.*, 2010, **107**, 20980–20985.
32. T. J. Wigle, S. K. Knutson, L. Jin, K. W. Kuntz, R. M. Pollock, V. M. Richon, R. A. Copeland and M. P. Scott, *FEBS Lett.*, 2011, **585**, 3011–3014.
33. D. B. Yap, J. Chu, T. Berg, M. Schapira, S. W. Cheng, A. Moradian, R. D. Morin, A. J. Mungall, B. Meissner, M. Boyle, V. E. Marquez, M. A. Marra, R. D. Gascoyne, R. K. Humphries, C. H. Arrowsmith, G. B. Morin and S. A. Aparicio, *Blood*, 2011, **117**, 2451–2459.
34. P. L. Bergsagel and W. M. Kuehl, *J. Clin. Oncol.*, 2005, **23**, 6333–6338.
35. M. Chesi, E. Nardini, R. S. Lim, K. D. Smith, W. M. Kuehl and P. L. Bergsagel, *Blood*, 1998, **92**, 3025–3034.
36. A. J. Kuo, P. Cheung, K. Chen, B. M. Zee, M. Kioi, J. Lauring, Y. Xi, B. H. Park, X. Shi, B. A. Garcia, W. Li and O. Gozani, *Mol. Cell*, 2011, **44**, 609–620.
37. E. Martinez-Garcia, R. Popovic, D. J. Min, S. M. Sweet, P. M. Thomas, L. Zamdborg, A. Heffner, C. Will, L. Lamy, L. M. Staudt, D. L. Levens, N. L. Kelleher and J. D. Licht, *Blood*, 2011, **117**, 211–220.
38. A. M. Dubuc, M. Remke, A. Korshunov, P. A. Northcott, S. H. Zhan, M. Mendez-Lago, M. Kool, D. T. Jones, A. Unterberger, A. S. Morrissy, D. Shih, J. Peacock, V. Ramaswamy, A. Rolider, X. Wang, H. Witt, T. Hielscher, C. Hawkins, R. Vibhakar, S. Croul, J. T. Rutka, W. A. Weiss, S. J. Jones, C. G. Eberhart, M. A. Marra, S. M. Pfister and M. D. Taylor, *Acta Neuropathol.*, 2013, **125**, 373–384.
39. G. van Haaften, G. L. Dalgliesh, H. Davies, L. Chen, G. Bignell, C. Greenman, S. Edkins, C. Hardy, S. O'Meara, J. Teague, A. Butler, J. Hinton, C. Latimer, J. Andrews, S. Barthorpe, D. Beare, G. Buck, P. J. Campbell, J. Cole, S. Forbes, M. Jia, D. Jones, C. Y. Kok, C. Leroy, M. L. Lin, D. J. McBride, M. Maddison, S. Maquire, K. McLay, A. Menzies, T. Mironenko, L. Mulderrig, L. Mudie, E. Pleasance, R. Shepherd, R. Smith, L. Stebbings, P. Stephens, G. Tang, P. S. Tarpey, R. Turner, K. Turrell, J. Varian, S. West, S. Widaa, P. Wray, V. P. Collins, K. Ichimura, S. Law, J. Wong, S. T. Yuen, S. Y. Leung, G. Tonon, R. A. DePinho, Y. T. Tai, K. C. Anderson, R. J. Kahnoski, A. Massie, S. K. Khoo, B. T. Teh, M. R. Stratton and P. A. Futreal, *Nat. Genet.*, 2009, **41**, 521–523.
40. E. M. Jackson, A. J. Sievert, X. Gai, H. Hakonarson, A. R. Judkins, L. Tooke, J. C. Perin, H. Xie, T. H. Shaikh and J. A. Biegel, *Clin. Cancer Res.*, 2009, **15**, 1923–1930.
41. B. G. Wilson, X. Wang, X. Shen, E. S. McKenna, M. E. Lemieux, Y. J. Cho, E. C. Koellhoffer, S. L. Pomeroy, S. H. Orkin and C. W. Roberts, *Cancer Cell*, 2010, **18**, 316–328.

42. C. W. Brennan, R. G. Verhaak, A. McKenna, B. Campos, H. Noushmehr, S. R. Salama, S. Zheng, D. Chakravarty, J. Z. Sanborn, S. H. Berman, R. Beroukhim, B. Bernard, C. J. Wu, G. Genovese, I. Shmulevich, J. Barnholtz-Sloan, L. Zou, R. Vegesna, S. A. Shukla, G. Ciriello, W. K. Yung, W. Zhang, C. Sougnez, T. Mikkelsen, K. Aldape, D. D. Bigner, E. G. Van Meir, M. Prados, A. Sloan, K. L. Black, J. Eschbacher, G. Finocchiaro, W. Friedman, D. W. Andrews, A. Guha, M. Iacocca, B. P. O'Neill, G. Foltz, J. Myers, D. J. Weisenberger, R. Penny, R. Kucherlapati, C. M. Perou, D. N. Hayes, R. Gibbs, M. Marra, G. B. Mills, E. Lander, P. Spellman, R. Wilson, C. Sander, J. Weinstein, M. Meyerson, S. Gabriel, P. W. Laird, D. Haussler, G. Getz and L. Chin, *Cell*, 2013, **155**, 462–477.
43. H. Shen and P. W. Laird, *Cell*, 2013, **153**, 38–55.
44. S. C. Dillon, X. Zhang, R. C. Trievel and X. Cheng, *Genome Biol.*, 2005, **6**, 227.
45. Q. Feng, H. Wang, H. H. Ng, H. Erdjument-Bromage, P. Tempst, K. Struhl and Y. Zhang, *Curr. Biol.*, 2002, **12**, 1052–1058.
46. N. Lacoste, R. T. Utley, J. M. Hunter, G. G. Poirier and J. Cote, *J. Biol. Chem.*, 2002, **277**, 30421–30424.
47. F. van Leeuwen, P. R. Gafken and D. E. Gottschling, *Cell*, 2002, **109**, 745–756.
48. J. Min, Q. Feng, Z. Li, Y. Zhang and R. M. Xu, *Cell*, 2003, **112**, 711–723.
49. J. Vidgren, L. A. Svensson and A. Liljas, *Nature*, 1994, **368**, 354–358.
50. K. Luger, A. W. Mader, R. K. Richmond, D. F. Sargent and T. J. Richmond, *Nature*, 1997, **389**, 251–260.
51. S. D. Briggs, T. Xiao, Z. W. Sun, J. A. Caldwell, J. Shabanowitz, D. F. Hunt, C. D. Allis and B. D. Strahl, *Nature*, 2002, **418**, 498.
52. H. H. Ng, R. M. Xu, Y. Zhang and K. Struhl, *J. Biol. Chem.*, 2002, **277**, 34655–34657.
53. R. K. McGinty, J. Kim, C. Chatterjee, R. G. Roeder and T. W. Muir, *Nature*, 2008, **453**, 812–816.
54. B. Jones, H. Su, A. Bhat, H. Lei, J. Bajko, S. Hevi, G. A. Baltus, S. Kadam, H. Zhai, R. Valdez, S. Gonzalo, Y. Zhang, E. Li and T. Chen, *PLoS Genet.*, 2008, **4**, e1000190.
55. Y. Feng, Y. Yang, M. M. Ortega, J. N. Copeland, M. Zhang, J. B. Jacob, T. A. Fields, J. L. Vivian and P. E. Fields, *Blood*, 2010, **116**, 4483–4491.
56. H. H. Ng, Q. Feng, H. Wang, H. Erdjument-Bromage, P. Tempst, Y. Zhang and K. Struhl, *Genes Dev.*, 2002, **16**, 1518–1527.
57. M. S. Singer, A. Kahana, A. J. Wolf, L. L. Meisinger, S. E. Peterson, C. Goggin, M. Mahowald and D. E. Gottschling, *Genetics*, 1998, **150**, 613–632.
58. D. Ontoso, I. Acosta, F. van Leeuwen, R. Freire and P. A. San-Segundo, *PLoS Genet.*, 2013, **9**, e1003262.
59. P. A. San-Segundo and G. S. Roeder, *Mol. Biol. Cell*, 2000, **11**, 3601–3615.
60. F. Conde, E. Refolio, V. Cordon-Preciado, F. Cortes-Ledesma, L. Aragon, A. Aguilera and P. A. San-Segundo, *Genetics*, 2009, **182**, 437–446.
61. H. Fu, A. K. Maunakea, M. M. Martin, L. Huang, Y. Zhang, M. Ryan, R. Kim, C. M. Lin, K. Zhao and M. I. Aladjem, *PLoS Genet.*, 2013, **9**, e1003542.

62. W. Kim, R. Kim, G. Park, J. W. Park and J. E. Kim, *J. Biol. Chem.*, 2012, **287**, 5588–5599.
63. D. J. Steger, M. I. Lefterova, L. Ying, A. J. Stonestrom, M. Schupp, D. Zhuo, A. L. Vakoc, J. E. Kim, J. Chen, M. A. Lazar, G. A. Blobel and C. R. Vakoc, *Mol. Cell. Biol.*, 2008, **28**, 2825–2839.
64. A. Kouskouti and I. Talianidis, *EMBO J.*, 2005, **24**, 347–357.
65. D. Schubeler, D. M. MacAlpine, D. Scalzo, C. Wirbelauer, C. Kooperberg, F. van Leeuwen, D. E. Gottschling, L. P. O'Neill, B. M. Turner, J. Delrow, S. P. Bell and M. Groudine, *Genes Dev.*, 2004, **18**, 1263–1271.
66. S. K. Kim, I. Jung, H. Lee, K. Kang, M. Kim, K. Jeong, C. S. Kwon, Y. M. Han, Y. S. Kim, D. Kim and D. Lee, *J. Biol. Chem.*, 2012, **287**, 39698–39709.
67. V. Oksenych, A. Zhovmer, S. Ziani, P. O. Mari, J. Eberova, T. Nardo, M. Stefanini, G. Giglia-Mari, J. M. Egly and F. Coin, *PLoS Genet.*, 2013, **9**, e1003611.
68. T. T. Onder, N. Kara, A. Cherry, A. U. Sinha, N. Zhu, K. M. Bernt, P. Cahan, B. O. Marcarci, J. Unternaehrer, P. B. Gupta, E. S. Lander, S. A. Armstrong and G. Q. Daley, *Nature*, 2012, **483**, 598–602.
69. A. V. Krivtsov, Z. Feng, M. E. Lemieux, J. Faber, S. Vempati, A. U. Sinha, X. Xia, J. Jesneck, A. P. Bracken, L. B. Silverman, J. L. Kutok, A. L. Kung and S. A. Armstrong, *Cancer Cell*, 2008, **14**, 355–368.
70. A. T. Nguyen, O. Taranova, J. He and Y. Zhang, *Blood*, 2011, **117**, 6912–6922.
71. Y. Okada, Q. Feng, Y. Lin, Q. Jiang, Y. Li, V. M. Coffield, L. Su, G. Xu and Y. Zhang, *Cell*, 2005, **121**, 167–178.
72. Y. Okada, Q. Jiang, M. Lemieux, L. Jeannotte, L. Su and Y. Zhang, *Nat. Cell Biol.*, 2006, **8**, 1017–1024.
73. E. R. Barry, G. N. Corry and T. P. Rasmussen, *Expert Opin. Ther. Targets*, 2010, **14**, 405–418.
74. K. M. Bernt and S. A. Armstrong, *Epigenomics*, 2011, **3**, 667–670.
75. M. Mohan, H. M. Herz, Y. H. Takahashi, C. Lin, K. C. Lai, Y. Zhang, M. P. Washburn, L. Florens and A. Shilatifard, *Genes Dev.*, 2010, **24**, 574–589.
76. D. Biswas, T. A. Milne, V. Basrur, J. Kim, K. S. Elenitoba-Johnson, C. D. Allis and R. G. Roeder, *Proc. Natl. Acad. Sci. U. S. A.*, 2011, **108**, 15751–15756.
77. E. Bitoun, P. L. Oliver and K. E. Davies, *Hum. Mol. Genet.*, 2007, **16**, 92–106.
78. S. C. Monroe, S. Y. Jo, D. S. Sanders, V. Basrur, K. S. Elenitoba-Johnson, R. K. Slany and J. L. Hess, *Exp. Hematol.*, 2011, **39**, 77–86.
79. D. Mueller, C. Bach, D. Zeisig, M. P. Garcia-Cuellar, S. Monroe, A. Sreekumar, R. Zhou, A. Nesvizhskii, A. Chinnaiyan, J. L. Hess and R. K. Slany, *Blood*, 2007, **110**, 4445–4454.
80. D. Mueller, M. P. Garcia-Cuellar, C. Bach, S. Buhl, E. Maethner and R. K. Slany, *PLoS Biol.*, 2009, **7**, e1000249.
81. M. J. Chang, H. Wu, N. J. Achille, M. R. Reisenauer, C. W. Chou, N. J. Zeleznik-Le, C. S. Hemenway and W. Zhang, *Cancer Res.*, 2010, **70**, 10234–10242.

82. A. J. Deshpande, L. Chen, M. Fazio, A. U. Sinha, K. M. Bernt, D. Banka, S. Dias, J. Chang, E. J. Olhava, S. R. Daigle, V. M. Richon, R. M. Pollock and S. A. Armstrong, *Blood*, 2013, **121**, 2533–2541.
83. Y. Zhang, A. Chen, X. M. Yan and G. Huang, *Int. J. Hematol.*, 2012, **96**, 428–437.
84. M. C. Castano Betancourt, F. Cailotto, H. J. Kerkhof, F. M. Cornelis, S. A. Doherty, D. J. Hart, A. Hofman, F. P. Luyten, R. A. Maciewicz, M. Mangino, S. Metrustry, K. Muir, M. J. Peters, F. Rivadeneira, M. Wheeler, W. Zhang, N. Arden, T. D. Spector, A. G. Uitterlinden, M. Doherty, R. J. Lories, A. M. Valdes and J. B. van Meurs, *Proc. Natl. Acad. Sci. U. S. A.*, 2012, **109**, 8218–8223.
85. T. Mahmoudi, S. F. Boj, P. Hatzis, V. S. Li, N. Taouatas, R. G. Vries, H. Teunissen, H. Begthel, J. Korving, S. Mohammed, A. J. Heck and H. Clevers, *PLoS Biol.*, 2010, **8**, e1000539.
86. L. L. Ho, A. Sinha, M. Verzi, K. M. Bernt, S. A. Armstrong and R. A. Shivdasani, *Mol. Cell. Biol.*, 2013, **33**, 1735–1745.
87. S. R. Daigle, E. J. Olhava, C. A. Therkelsen, C. R. Majer, C. J. Sneeringer, J. Song, L. D. Johnston, M. P. Scott, J. J. Smith, Y. Xiao, L. Jin, K. W. Kuntz, R. Chesworth, M. P. Moyer, K. M. Bernt, J. C. Tseng, A. L. Kung, S. A. Armstrong, R. A. Copeland, V. M. Richon and R. M. Pollock, *Cancer Cell*, 2011, **20**, 53–65.
88. A. Basavapathruni, L. Jin, S. R. Daigle, C. R. Majer, C. A. Therkelsen, T. J. Wigle, K. W. Kuntz, R. Chesworth, R. M. Pollock, M. P. Scott, M. P. Moyer, V. M. Richon, R. A. Copeland and E. J. Olhava, *Chem. Biol. Drug Des.*, 2012, **80**, 971–980.
89. R. A. Copeland, M. E. Solomon and V. M. Richon, *Nat. Rev. Drug Discovery*, 2009, **8**, 724–732.
90. Y. Yao, P. Chen, J. Diao, G. Cheng, L. Deng, J. L. Anglin, B. V. Prasad and Y. Song, *J. Am. Chem. Soc.*, 2011, **133**, 16746–16749.
91. J. L. Anglin, L. Deng, Y. Yao, G. Cai, Z. Liu, H. Jiang, G. Cheng, P. Chen, S. Dong and Y. Song, *J. Med. Chem.*, 2012, **55**, 8066–8074.
92. W. Yu, E. J. Chory, A. K. Wernimont, W. Tempel, A. Scopton, A. Federation, J. J. Marineau, J. Qi, D. Barsyte-Lovejoy, J. Yi, R. Marcellus, R. E. Iacob, J. R. Engen, C. Griffin, A. Aman, E. Wienholds, F. Li, J. Pineda, G. Estiu, T. Shatseva, T. Hajian, R. Al-Awar, J. E. Dick, M. Vedadi, P. J. Brown, C. H. Arrowsmith, J. E. Bradner and M. Schapira, *Nat. Commun.*, 2012, **3**, 1288.
93. W. Yu, D. Smil, F. Li, W. Tempel, O. Fedorov, K. T. Nguyen, Y. Bolshan, R. Al-Awar, S. Knapp, C. H. Arrowsmith, M. Vedadi, P. J. Brown and M. Schapira, *Bioorg. Med. Chem.*, 2013, **21**, 1787–1794.
94. L. Deng, L. Zhang, Y. Yao, C. Wang, M. S. Redell, S. Dong and Y. Song, *MedChemComm*, 2013, **4**, 822–826.
95. S. R. Daigle, E. J. Olhava, C. A. Therkelsen, A. Basavapathruni, L. Jin, P. A. Boriack-Sjodin, C. J. Allain, C. R. Klaus, A. Raimondi, M. P. Scott, N. J. Waters, R. Chesworth, M. P. Moyer, R. A. Copeland, V. M. Richon and R. M. Pollock, *Blood*, 2013, **122**, 1017–1025.

96. L. Chen, A. J. Deshpande, D. Banka, K. M. Bernt, S. Dias, C. Buske, E. J. Olhava, S. R. Daigle, V. M. Richon, R. M. Pollock and S. A. Armstrong, *Leukemia*, 2013, **27**, 813–822.
97. K. M. Bernt, N. Zhu, A. U. Sinha, S. Vempati, J. Faber, A. V. Krivtsov, Z. Feng, N. Punt, A. Daigle, L. Bullinger, R. M. Pollock, V. M. Richon, A. L. Kung and S. A. Armstrong, *Cancer Cell*, 2011, **20**, 66–78.
98. M. W. M. Kühn, M. Hadler, S. R. Daigle, C.-W. Chen, A. U. Sinha, A. V. Krivtsov, E. J. Olhava, M. A. Caligiuri, J. E. Bradner, R. M. Pollock and S. A. Armstrong, in *The American Society of Hematology, 2013 Annual Meeting, Poster Sessions - 602. Disordered Gene Expression in Hematologic Malignancy, including Disordered Epigenetic Regulation: Poster I*, The American Society of Hematology, 2013.
99. A. Basavapathruni, E. J. Olhava, S. R. Daigle, C. A. Therkelsen, L. Jin, P. A. Boriack-Sjodin, C. J. Allain, C. R. Klaus, A. Raimondi, M. P. Scott, A. Dovletoglou, V. M. Richon, R. M. Pollock, R. A. Copeland, M. P. Moyer, R. Chesworth, P. G. Pearson and N. J. Waters, *Biopharm. Drug Dispos.*, 2014, **35**, 237–352.
100. C. R. Klaus, S. R. Daigle, D. Iwanowics, L. D. Johnston, C. A. Therkelsen, J. J. Smith, M. P. Moyer, R. A. Copeland, E. J. Olhava, M. Porter Scott, R. M. Pollock and A. Raimondi, in *The American Society of Hematology, 2013 Annual Meeting, Poster Sessions - 615. Acute Myeloid Leukemia: Therapy, excluding Transplantation: Poster III*, 2013.
101. R. Cao, L. Wang, H. Wang, L. Xia, H. Erdjument-Bromage, P. Tempst, R. S. Jones and Y. Zhang, *Science*, 2002, **298**, 1039–1043.
102. A. Kuzmichev, K. Nishioka, H. Erdjument-Bromage, P. Tempst and D. Reinberg, *Genes Dev.*, 2002, **16**, 2893–2905.
103. J. Muller, C. M. Hart, N. J. Francis, M. L. Vargas, A. Sengupta, B. Wild, E. L. Miller, M. B. O'Connor, R. E. Kingston and J. A. Simon, *Cell*, 2002, **111**, 197–208.
104. D. O'Carroll, S. Erhardt, M. Pagani, S. C. Barton, M. A. Surani and T. Jenuwein, *Mol. Cell. Biol.*, 2001, **21**, 4330–4336.
105. R. Cao and Y. Zhang, *Mol. Cell*, 2004, **15**, 57–67.
106. R. Margueron and D. Reinberg, *Nature*, 2011, **469**, 343–349.
107. S. L. Squazzo, H. O'Geen, V. M. Komashko, S. R. Krig, V. X. Jin, S. W. Jang, R. Margueron, D. Reinberg, R. Green and P. J. Farnham, *Genome Res.*, 2006, **16**, 890–900.
108. K. H. Hansen, A. P. Bracken, D. Pasini, N. Dietrich, S. S. Gehani, A. Monrad, J. Rappsilber, M. Lerdrup and K. Helin, *Nat. Cell Biol.*, 2008, **10**, 1291–1300.
109. R. Margueron, G. Li, K. Sarma, A. Blais, J. Zavadil, C. L. Woodcock, B. D. Dynlacht and D. Reinberg, *Mol. Cell*, 2008, **32**, 503–518.
110. J. van der Vlag and A. P. Otte, *Nat. Genet.*, 1999, **23**, 474–478.
111. N. D. Montgomery, D. Yee, A. Chen, S. Kalantry, S. J. Chamberlain, A. P. Otte and T. Magnuson, *Curr. Biol.*, 2005, **15**, 942–947.
112. D. Pasini, A. P. Bracken, M. R. Jensen, E. Lazzerini Denchi and K. Helin, *EMBO J.*, 2004, **23**, 4061–4071.

113. G. Li, R. Margueron, M. Ku, P. Chambon, B. E. Bernstein and D. Reinberg, *Genes Dev.*, 2010, **24**, 368–380.
114. D. Pasini, P. A. Cloos, J. Walfridsson, L. Olsson, J. P. Bukowski, J. V. Johansen, M. Bak, N. Tommerup, J. Rappsilber and K. Helin, *Nature*, 2010, **464**, 306–310.
115. J. C. Peng, A. Valouev, T. Swigut, J. Zhang, Y. Zhao, A. Sidow and J. Wysocka, *Cell*, 2009, **139**, 1290–1302.
116. J. Tan, X. Yang, L. Zhuang, X. Jiang, W. Chen, P. L. Lee, R. K. Karuturi, P. B. Tan, E. T. Liu and Q. Yu, *Genes Dev.*, 2007, **21**, 1050–1063.
117. E. Ezhkova, W. H. Lien, N. Stokes, H. A. Pasolli, J. M. Silva and E. Fuchs, *Genes Dev.*, 2011, **25**, 485–498.
118. X. Shen, Y. Liu, Y. J. Hsu, Y. Fujiwara, J. Kim, X. Mao, G. C. Yuan and S. H. Orkin, *Mol. Cell*, 2008, **32**, 491–502.
119. H. Xie, J. Xu, J. H. Hsu, M. Nguyen, Y. Fujiwara, C. Peng and S. H. Orkin, *Cell Stem Cell*, 2014, **14**, 68–80.
120. H. Chen, X. Gu, I. H. Su, R. Bottino, J. L. Contreras, A. Tarakhovsky and S. K. Kim, *Genes Dev.*, 2009, **23**, 975–985.
121. P. Delgado-Olguin, Y. Huang, X. Li, D. Christodoulou, C. E. Seidman, J. G. Seidman, A. Tarakhovsky and B. G. Bruneau, *Nat. Genet.*, 2012, **44**, 343–347.
122. E. Ezhkova, H. A. Pasolli, J. S. Parker, N. Stokes, I. H. Su, G. Hannon, A. Tarakhovsky and E. Fuchs, *Cell*, 2009, **136**, 1122–1135.
123. J. D. Pereira, S. N. Sansom, J. Smith, M. W. Dobenecker, A. Tarakhovsky and F. J. Livesey, *Proc. Natl. Acad. Sci. U. S. A.*, 2010, **107**, 15957–15962.
124. I. H. Su, A. Basavaraj, A. N. Krutchinsky, O. Hobert, A. Ullrich, B. T. Chait and A. Tarakhovsky, *Nat. Immunol.*, 2003, **4**, 124–131.
125. I. H. Su, M. W. Dobenecker, E. Dickinson, M. Oser, A. Basavaraj, R. Margueron, A. Viale, D. Reinberg, C. Wulfing and A. Tarakhovsky, *Cell*, 2005, **121**, 425–436.
126. S. Erhardt, I. H. Su, R. Schneider, S. Barton, A. J. Bannister, L. Perez-Burgos, T. Jenuwein, T. Kouzarides, A. Tarakhovsky and M. A. Surani, *Development*, 2003, **130**, 4235–4248.
127. K. Mousavi, H. Zare, A. H. Wang and V. Sartorelli, *Mol. Cell*, 2012, **45**, 255–262.
128. C. G. Kleer, Q. Cao, S. Varambally, R. Shen, I. Ota, S. A. Tomlins, D. Ghosh, R. G. Sewalt, A. P. Otte, D. F. Hayes, M. S. Sabel, D. Livant, S. J. Weiss, M. A. Rubin and A. M. Chinnaiyan, *Proc. Natl. Acad. Sci. U. S. A.*, 2003, **100**, 11606–11611.
129. S. Varambally, S. M. Dhanasekaran, M. Zhou, T. R. Barrette, C. Kumar-Sinha, M. G. Sanda, D. Ghosh, K. J. Pienta, R. G. Sewalt, A. P. Otte, M. A. Rubin and A. M. Chinnaiyan, *Nature*, 2002, **419**, 624–629.
130. S. Weikert, F. Christoph, J. Kollermann, M. Muller, M. Schrader, K. Miller and H. Krause, *Int. J. Mol. Med.*, 2005, **16**, 349–353.
131. C. Lu, H. D. Han, L. S. Mangala, R. Ali-Fehmi, C. S. Newton, L. Ozbun, G. N. Armaiz-Pena, W. Hu, R. L. Stone, A. Munkarah, M. K. Ravoori, M. M. Shahzad, J. W. Lee, E. Mora, R. R. Langley, A. R. Carroll, K. Matsuo, W.

A. Spannuth, R. Schmandt, N. B. Jennings, B. W. Goodman, R. B. Jaffe, A. M. Nick, H. S. Kim, E. O. Guven, Y. H. Chen, L. Y. Li, M. C. Hsu, R. L. Coleman, G. A. Calin, E. B. Denkbas, J. Y. Lim, J. S. Lee, V. Kundra, M. J. Birrer, M. C. Hung, G. Lopez-Berestein and A. K. Sood, *Cancer Cell*, 2010, **18**, 185–197.

132. J. Min, A. Zaslavsky, G. Fedele, S. K. McLaughlin, E. E. Reczek, T. De Raedt, I. Guney, D. E. Strochlic, L. E. Macconaill, R. Beroukhim, R. T. Bronson, S. Ryeom, W. C. Hahn, M. Loda and K. Cichowski, *Nat. Med.*, 2010, **16**, 286–294.

133. J. Shi, E. Wang, J. Zuber, A. Rappaport, M. Taylor, C. Johns, S. W. Lowe and C. R. Vakoc, *Oncogene*, 2013, **32**, 930–938.

134. M. L. Suva, N. Riggi, M. Janiszewska, I. Radovanovic, P. Provero, J. C. Stehle, K. Baumer, M. A. Le Bitoux, D. Marino, L. Cironi, V. E. Marquez, V. Clement and I. Stamenkovic, *Cancer Res.*, 2009, **69**, 9211–9218.

135. E. Hodis, I. R. Watson, G. V. Kryukov, S. T. Arold, M. Imielinski, J. P. Theurillat, E. Nickerson, D. Auclair, L. Li, C. Place, D. Dicara, A. H. Ramos, M. S. Lawrence, K. Cibulskis, A. Sivachenko, D. Voet, G. Saksena, N. Stransky, R. C. Onofrio, W. Winckler, K. Ardlie, N. Wagle, J. Wargo, K. Chong, D. L. Morton, K. Stemke-Hale, G. Chen, M. Noble, M. Meyerson, J. E. Ladbury, M. A. Davies, J. E. Gershenwald, S. N. Wagner, D. S. Hoon, D. Schadendorf, E. S. Lander, S. B. Gabriel, G. Getz, L. A. Garraway and L. Chin, *Cell*, 2012, **150**, 251–263.

136. T. Ernst, A. J. Chase, J. Score, C. E. Hidalgo-Curtis, C. Bryant, A. V. Jones, K. Waghorn, K. Zoi, F. M. Ross, A. Reiter, A. Hochhaus, H. G. Drexler, A. Duncombe, F. Cervantes, D. Oscier, J. Boultwood, F. H. Grand and N. C. Cross, *Nat. Genet.*, 2010, **42**, 722–726.

137. A. M. Jankowska, H. Makishima, R. V. Tiu, H. Szpurka, Y. Huang, F. Traina, V. Visconte, Y. Sugimoto, C. Prince, C. O'Keefe, E. D. Hsi, A. List, M. A. Sekeres, A. Rao, M. A. McDevitt and J. P. Maciejewski, *Blood*, 2011, **118**, 3932–3941.

138. H. Makishima, A. M. Jankowska, R. V. Tiu, H. Szpurka, Y. Sugimoto, Z. Hu, Y. Saunthararajah, K. Guinta, M. A. Keddache, P. Putnam, M. A. Sekeres, A. R. Moliterno, A. F. List, M. A. McDevitt and J. P. Maciejewski, *Leukemia*, 2010, **24**, 1799–1804.

139. J. Score, C. Hidalgo-Curtis, A. V. Jones, N. Winkelmann, A. Skinner, D. Ward, K. Zoi, T. Ernst, F. Stegelmann, K. Dohner, A. Chase and N. C. Cross, *Blood*, 2012, **119**, 1208–1213.

140. J. Zhang, L. Ding, L. Holmfeldt, G. Wu, S. L. Heatley, D. Payne-Turner, J. Easton, X. Chen, J. Wang, M. Rusch, C. Lu, S. C. Chen, L. Wei, J. R. Collins-Underwood, J. Ma, K. G. Roberts, S. B. Pounds, A. Ulyanov, J. Becksfort, P. Gupta, R. Huether, R. W. Kriwacki, M. Parker, D. J. McGoldrick, D. Zhao, D. Alford, S. Espy, K. C. Bobba, G. Song, D. Pei, C. Cheng, S. Roberts, M. I. Barbato, D. Campana, E. Coustan-Smith, S. A. Shurtleff, S. C. Raimondi, M. Kleppe, J. Cools, K. A. Shimano, M. L. Hermiston, S. Doulatov, K. Eppert, E. Laurenti, F. Notta, J. E. Dick, G. Basso, S. P.

Hunger, M. L. Loh, M. Devidas, B. Wood, S. Winter, K. P. Dunsmore, R. S. Fulton, L. L. Fulton, X. Hong, C. C. Harris, D. J. Dooling, K. Ochoa, K. J. Johnson, J. C. Obenauer, W. E. Evans, C. H. Pui, C. W. Naeve, T. J. Ley, E. R. Mardis, R. K. Wilson, J. R. Downing and C. G. Mullighan, *Nature*, 2012, **481**, 157–163.

141. E. Diaz, C. A. Machutta, S. Chen, Y. Jiang, C. Nixon, G. Hofmann, D. Key, S. Sweitzer, M. Patel, Z. Wu, C. L. Creasy, R. G. Kruger, L. Lafrance, S. K. Verma, M. B. Pappalardi, B. Le, G. S. Van Aller, M. T. McCabe, P. J. Tummino, A. J. Pope, S. H. Thrall, B. Schwartz and M. Brandt, *J. Biomol. Screening*, 2012, **17**, 1279–1292.
142. S. Garapaty-Rao, C. Nasveschuk, A. Gagnon, E. Y. Chan, P. Sandy, J. Busby, S. Balasubramanian, R. Campbell, F. Zhao, L. Bergeron, J. E. Audia, B. K. Albrecht, J. C. Harmange, R. Cummings and P. Trojer, *Chem. Biol.*, 2013, **20**, 1329–1339.
143. S. K. Knutson, N. M. Warholic, T. J. Wigle, C. R. Klaus, C. J. Allain, A. Raimondi, M. Porter Scott, R. Chesworth, M. P. Moyer, R. A. Copeland, V. M. Richon, R. M. Pollock, K. W. Kuntz and H. Keilhack, *Proc. Natl. Acad. Sci. U. S. A.*, 2013, **110**, 7922–7927.
144. S. K. Knutson, T. J. Wigle, N. M. Warholic, C. J. Sneeringer, C. J. Allain, C. R. Klaus, J. D. Sacks, A. Raimondi, C. R. Majer, J. Song, M. P. Scott, L. Jin, J. J. Smith, E. J. Olhava, R. Chesworth, M. P. Moyer, V. M. Richon, R. A. Copeland, H. Keilhack, R. M. Pollock and K. W. Kuntz, *Nat. Chem. Biol.*, 2012, **8**, 890–896.
145. K. D. Konze, A. Ma, F. Li, D. Barsyte-Lovejoy, T. Parton, C. J. Macnevin, F. Liu, C. Gao, X. P. Huang, E. Kuznetsova, M. Rougie, A. Jiang, S. G. Pattenden, J. L. Norris, L. I. James, B. L. Roth, P. J. Brown, S. V. Frye, C. H. Arrowsmith, K. M. Hahn, G. G. Wang, M. Vedadi and J. Jin, *ACS Chem. Biol.*, 2013, **8**, 1324–1334.
146. M. T. McCabe, H. M. Ott, G. Ganji, S. Korenchuk, C. Thompson, G. S. Van Aller, Y. Liu, A. P. Graves, A. D. Iii, E. Diaz, L. V. Lafrance, M. Mellinger, C. Duquenne, X. Tian, R. G. Kruger, C. F. McHugh, M. Brandt, W. H. Miller, D. Dhanak, S. K. Verma, P. J. Tummino and C. L. Creasy, *Nature*, 2012, **492**, 108–112.
147. W. Qi, H. Chan, L. Teng, L. Li, S. Chuai, R. Zhang, J. Zeng, M. Li, H. Fan, Y. Lin, J. Gu, O. Ardayfio, J. H. Zhang, X. Yan, J. Fang, Y. Mi, M. Zhang, T. Zhou, G. Feng, Z. Chen, G. Li, T. Yang, K. Zhao, X. Liu, Z. Yu, C. X. Lu, P. Atadja and E. Li, *Proc. Natl. Acad. Sci. U. S. A.*, 2012, **109**, 21360–21365.
148. W. D. Bradley, S. Arora, J. Busby, S. Balasubramanian, V. S. Gehling, C. G. Nasveschuk, R. G. Vaswani, C. C. Yuan, C. Hatton, F. Zhao, K. E. Williamson, P. Iyer, J. Mendez, R. Campbell, N. Cantone, S. Garapaty-Rao, J. E. Audia, A. S. Cook, L. A. Dakin, B. K. Albrecht, J. C. Harmange, D. L. Daniels, R. T. Cummings, B. M. Bryant, E. Normant and P. Trojer, *Chem. Biol.*, 2014, **21**, 1463–1475.
149. C. G. Nasveschuk, A. Gagnon, S. Garapaty-Rao, S. Balasubramanian, R. Campbell, C. Lee, F. Zhao, L. Bergeron, R. Cummings, P. Trojer, J. E.

Audia, B. K. Albrecht and J. C. Harmange, *ACS Med. Chem. Lett.*, 2014, **5**, 378–383.

150. H. Wu, H. Zeng, A. Dong, F. Li, H. He, G. Senisterra, A. Seitova, S. Duan, P. J. Brown, M. Vedadi, C. H. Arrowsmith and M. Schapira, *PLoS One*, 2013, **8**, e83737.
151. W. Beguelin, R. Popovic, M. Teater, Y. Jiang, K. L. Bunting, M. Rosen, H. Shen, S. N. Yang, L. Wang, T. Ezponda, E. Martinez-Garcia, H. Zhang, Y. Zheng, S. K. Verma, M. T. McCabe, H. M. Ott, G. S. Van Aller, R. G. Kruger, Y. Liu, C. F. McHugh, D. W. Scott, Y. R. Chung, N. Kelleher, R. Shaknovich, C. L. Creasy, R. D. Gascoyne, K. K. Wong, L. Cerchietti, R. L. Levine, O. Abdel-Wahab, J. D. Licht, O. Elemento and A. M. Melnick, *Cancer Cell*, 2013, **23**, 677–692.
152. I. Velichutina, R. Shaknovich, H. Geng, N. A. Johnson, R. D. Gascoyne, A. M. Melnick and O. Elemento, *Blood*, 2010, **116**, 5247–5255.
153. W. D. Bradley, S. Arora, J. Busby, S. Balasubramanian, V. S. Gehling, C. G. Nasveschuk, R. G. Vaswani, C. C. Yuan, C. Hatton, F. Zhao, *et al.*, *Chem. Biol.,* 2014, **21**, 1463–1475.
154. S. K. Knutson, S. Kawano, Y. Minoshima, N. M. Warholic, K. C. Huang, Y. Xiao, T. Kadowaki, M. Uesugi, G. Kuznetsov, N. Kumar, T. J. Wigle, C. R. Klaus, C. J. Allain, A. Raimondi, N. J. Waters, J. J. Smith, M. Porter-Scott, R. Chesworth, M. P. Moyer, R. A. Copeland, V. M. Richon, T. Uenaka, R. Pollock, K. W. Kuntz, A. Yokoi and H. Keilhack, *Mol. Cancer Ther.*, 2014, **13**, 842–854.
155. M. Tachibana, K. Sugimoto, M. Nozaki, J. Ueda, T. Ohta, M. Ohki, M. Fukuda, N. Takeda, H. Niida, H. Kato and Y. Shinkai, *Genes Dev.*, 2002, **16**, 1779–1791.
156. P. Trojer, J. Zhang, M. Yonezawa, A. Schmidt, H. Zheng, T. Jenuwein and D. Reinberg, *J. Biol. Chem.*, 2009, **284**, 8395–8405.
157. M. Inagawa, K. Nakajima, T. Makino, S. Ogawa, M. Kojima, S. Ito, A. Ikenishi, T. Hayashi, R. J. Schwartz, K. Nakamura, T. Obayashi, M. Tachibana, Y. Shinkai, K. Maeda, S. Miyagawa-Tomita and T. Takeuchi, *Mech. Dev.*, 2013, **130**, 519–531.
158. Y. Shinkai and M. Tachibana, *Genes Dev.*, 2011, **25**, 781–788.
159. D. Y. Lee, J. P. Northrop, M. H. Kuo and M. R. Stallcup, *J. Biol. Chem.*, 2006, **281**, 8476–8485.
160. D. J. Purcell, K. W. Jeong, D. Bittencourt, D. S. Gerke and M. R. Stallcup, *J. Biol. Chem.*, 2011, **286**, 41963–41971.
161. P. O. Esteve, H. G. Chin, A. Smallwood, G. R. Feehery, O. Gangisetty, A. R. Karpf, M. F. Carey and S. Pradhan, *Genes Dev.*, 2006, **20**, 3089–3103.
162. C. Mozzetta, J. Pontis, L. Fritsch, P. Robin, M. Portoso, C. Proux, R. Margueron and S. Ait-Si-Ali, *Mol. Cell*, 2014, **53**, 277–289.
163. S. Gupta-Agarwal, A. V. Franklin, T. Deramus, M. Wheelock, R. L. Davis, L. L. McMahon and F. D. Lubin, *J. Neurosci.*, 2012, **32**, 5440–5453.
164. J. M. Kramer, K. Kochinke, M. A. Oortveld, H. Marks, D. Kramer, E. K. de Jong, Z. Asztalos, J. T. Westwood, H. G. Stunnenberg, M. B. Sokolowski,

K. Keleman, H. Zhou, H. van Bokhoven and A. Schenck, *PLoS Biol.*, 2011, **9**, e1000569.
165. A. Schaefer, S. C. Sampath, A. Intrator, A. Min, T. S. Gertler, D. J. Surmeier, A. Tarakhovsky and P. Greengard, *Neuron*, 2009, **64**, 678–691.
166. M. H. Willemsen, A. T. Vulto-van Silfhout, W. M. Nillesen, W. M. Wissink-Lindhout, H. van Bokhoven, N. Philip, E. M. Berry-Kravis, U. Kini, C. M. van Ravenswaaij-Arts, B. Delle Chiaie, A. M. Innes, G. Houge, T. Kosonen, K. Cremer, M. Fannemel, A. Stray-Pedersen, W. Reardon, J. Ignatius, K. Lachlan, C. Mircher, P. T. Helderman van den Enden, M. Mastebroek, P. E. Cohn-Hokke, H. G. Yntema, S. Drunat and T. Kleefstra, *Mol. Syndromol.*, 2012, **2**, 202–212.
167. I. Maze, H. E. Covington, 3rd, D. M. Dietz, Q. LaPlant, W. Renthal, S. J. Russo, M. Mechanic, E. Mouzon, R. L. Neve, S. J. Haggarty, Y. Ren, S. C. Sampath, Y. L. Hurd, P. Greengard, A. Tarakhovsky, A. Schaefer and E. J. Nestler, *Science*, 2010, **327**, 213–216.
168. H. Sun, I. Maze, D. M. Dietz, K. N. Scobie, P. J. Kennedy, D. Damez-Werno, R. L. Neve, V. Zachariou, L. Shen and E. J. Nestler, *J. Neurosci.*, 2012, **32**, 17454–17464.
169. M. W. Chen, K. T. Hua, H. J. Kao, C. C. Chi, L. H. Wei, G. Johansson, S. G. Shiah, P. S. Chen, Y. M. Jeng, T. Y. Cheng, T. C. Lai, J. S. Chang, Y. H. Jan, M. H. Chien, C. J. Yang, M. S. Huang, M. Hsiao and M. L. Kuo, *Cancer Res.*, 2010, **70**, 7830–7840.
170. H. S. Cho, J. D. Kelly, S. Hayami, G. Toyokawa, M. Takawa, M. Yoshimatsu, T. Tsunoda, H. I. Field, D. E. Neal, B. A. Ponder, Y. Nakamura and R. Hamamoto, *Neoplasia*, 2011, **13**, 676–684.
171. J. Huang, J. Dorsey, S. Chuikov, L. Perez-Burgos, X. Zhang, T. Jenuwein, D. Reinberg and S. L. Berger, *J. Biol. Chem.*, 2010, **285**, 9636–9641.
172. C. Dong, Y. Wu, J. Yao, Y. Wang, Y. Yu, P. G. Rychahou, B. M. Evers and B. P. Zhou, *J. Clin. Invest.*, 2012, **122**, 1469–1486.
173. J. Ding, T. Li, X. Wang, E. Zhao, J. H. Choi, L. Yang, Y. Zha, Z. Dong, S. Huang, J. M. Asara, H. Cui and H. F. Ding, *Cell Metab.*, 2013, **18**, 896–907.
174. C. Dong, T. Yuan, Y. Wu, Y. Wang, T. W. Fan, S. Miriyala, Y. Lin, J. Yao, J. Shi, T. Kang, P. Lorkiewicz, D. St Clair, M. C. Hung, B. M. Evers and B. P. Zhou, *Cancer Cell*, 2013, **23**, 316–331.
175. S. Kubicek, R. J. O'Sullivan, E. M. August, E. R. Hickey, Q. Zhang, M. L. Teodoro, S. Rea, K. Mechtler, J. A. Kowalski, C. A. Homon, T. A. Kelly and T. Jenuwein, *Mol. Cell*, 2007, **25**, 473–481.
176. Y. Chang, X. Zhang, J. R. Horton, A. K. Upadhyay, A. Spannhoff, J. Liu, J. P. Snyder, M. T. Bedford and X. Cheng, *Nat. Struct. Mol. Biol.*, 2009, **16**, 312–317.
177. F. Liu, D. Barsyte-Lovejoy, A. Allali-Hassani, Y. He, J. M. Herold, X. Chen, C. M. Yates, S. V. Frye, P. J. Brown, J. Huang, M. Vedadi, C. H. Arrowsmith and J. Jin, *J. Med. Chem.*, 2011, **54**, 6139–6150.
178. F. Liu, D. Barsyte-Lovejoy, F. Li, Y. Xiong, V. Korboukh, X. P. Huang, A. Allali-Hassani, W. P. Janzen, B. L. Roth, S. V. Frye, C. H. Arrowsmith, P. J. Brown, M. Vedadi and J. Jin, *J. Med. Chem.*, 2013, **56**, 8931–8942.

179. F. Liu, X. Chen, A. Allali-Hassani, A. M. Quinn, G. A. Wasney, A. Dong, D. Barsyte, I. Kozieradzki, G. Senisterra, I. Chau, A. Siarheyeva, D. B. Kireev, A. Jadhav, J. M. Herold, S. V. Frye, C. H. Arrowsmith, P. J. Brown, A. Simeonov, M. Vedadi and J. Jin, *J. Med. Chem.*, 2009, **52**, 7950–7953.
180. F. Liu, X. Chen, A. Allali-Hassani, A. M. Quinn, T. J. Wigle, G. A. Wasney, A. Dong, G. Senisterra, I. Chau, A. Siarheyeva, J. L. Norris, D. B. Kireev, A. Jadhav, J. M. Herold, W. P. Janzen, C. H. Arrowsmith, S. V. Frye, P. J. Brown, A. Simeonov, M. Vedadi and J. Jin, *J. Med. Chem.*, 2010, **53**, 5844–5857.
181. M. Vedadi, D. Barsyte-Lovejoy, F. Liu, S. Rival-Gervier, A. Allali-Hassani, V. Labrie, T. J. Wigle, P. A. Dimaggio, G. A. Wasney, A. Siarheyeva, A. Dong, W. Tempel, S. C. Wang, X. Chen, I. Chau, T. J. Mangano, X. P. Huang, C. D. Simpson, S. G. Pattenden, J. L. Norris, D. B. Kireev, A. Tripathy, A. Edwards, B. L. Roth, W. P. Janzen, B. A. Garcia, A. Petronis, J. Ellis, P. J. Brown, S. V. Frye, C. H. Arrowsmith and J. Jin, *Nat. Chem. Biol.*, 2011, **7**, 566–574.
182. R. F. Sweis, M. Pliushchev, P. J. Brown, J. Guo, F. Li, D. Maag, A. M. Petros, N. B. Soni, C. Tse, M. Vedadi, M. R. Michaelides, G. G. Chiang and W. N. Pappano, *ACS Med. Chem. Lett.*, 2014, **5**, 205–209.
183. Y. Yuan, Q. Wang, J. Paulk, S. Kubicek, M. M. Kemp, D. J. Adams, A. F. Shamji, B. K. Wagner and S. L. Schreiber, *ACS Chem. Biol.*, 2012, **7**, 1152–1157.
184. M. Tachibana, Y. Matsumura, M. Fukuda, H. Kimura and Y. Shinkai, *EMBO J.*, 2008, **27**, 2681–2690.
185. Y. Shi, J. T. Do, C. Desponts, H. S. Hahm, H. R. Scholer and S. Ding, *Cell Stem Cell*, 2008, **2**, 525–528.
186. H. Yu, Q. Lin, Y. Wang, Y. He, S. Fu, H. Jiang, Y. Yu, S. Sun, Y. Chen, J. Shou and H. Li, *Cell Death Dis.*, 2013, **4**, e506.
187. N. A. Malmquist, T. A. Moss, S. Mecheri, A. Scherf and M. J. Fuchter, *Proc. Natl. Acad. Sci. U. S. A.*, 2012, **109**, 16708–16713.
188. D. Greiner, T. Bonaldi, R. Eskeland, E. Roemer and A. Imhof, *Nat. Chem. Biol.*, 2005, **1**, 143–145.
189. F. L. Cherblanc, K. L. Chapman, R. Brown and M. J. Fuchter, *Nat. Chem. Biol.*, 2013, **9**, 136–137.
190. F. L. Cherblanc, K. L. Chapman, J. Reid, A. J. Borg, S. Sundriyal, L. Alcazar-Fuoli, E. Bignell, M. Demetriades, C. J. Schofield, P. A. Dimaggio Jr., R. Brown and M. J. Fuchter, *J. Med. Chem.*, 2013, **56**, 8616–8625.
191. N. Sirinupong, J. Brunzelle, J. Ye, A. Pirzada, L. Nico and Z. Yang, *J. Biol. Chem.*, 2010, **285**, 40635–40644.
192. S. Xu, J. Wu, B. Sun, C. Zhong and J. Ding, *Nucleic Acids Res.*, 2011, **39**, 4438–4449.
193. S. Xu, C. Zhong, T. Zhang and J. Ding, *J. Mol. Cell Biol.*, 2011, **3**, 293–300.
194. Y. Jiang, N. Sirinupong, J. Brunzelle and Z. Yang, *PLoS One*, 2011, **6**, e21640.
195. N. Sirinupong, J. Brunzelle, E. Doko and Z. Yang, *J. Mol. Biol.*, 2011, **406**, 149–159.

196. M. Abu-Farha, J. P. Lambert, A. S. Al-Madhoun, F. Elisma, I. S. Skerjanc and D. Figeys, *Mol. Cell. Proteomics*, 2008, **7**, 560–572.
197. M. A. Brown, R. J. Sims 3rd, P. D. Gottlieb and P. W. Tucker, *Mol. Cancer*, 2006, **5**, 26.
198. K. W. Foreman, M. Brown, F. Park, S. Emtage, J. Harriss, C. Das, L. Zhu, A. Crew, L. Arnold, S. Shaaban and P. Tucker, *PLoS One*, 2011, **6**, e22290.
199. R. Hamamoto, Y. Furukawa, M. Morita, Y. Iimura, F. P. Silva, M. Li, R. Yagyu and Y. Nakamura, *Nat. Cell Biol.*, 2004, **6**, 731–740.
200. G. S. Van Aller, N. Reynoird, O. Barbash, M. Huddleston, S. Liu, A. F. Zmoos, P. McDevitt, R. Sinnamon, B. Le, G. Mas, R. Annan, J. Sage, B. A. Garcia, P. J. Tummino, O. Gozani and R. G. Kruger, *Epigenetics*, 2012, **7**, 340–343.
201. H. S. Cho, S. Hayami, G. Toyokawa, K. Maejima, Y. Yamane, T. Suzuki, N. Dohmae, M. Kogure, D. Kang, D. E. Neal, B. A. Ponder, H. Yamaue, Y. Nakamura and R. Hamamoto, *Neoplasia*, 2012, **14**, 476–486.
202. J. Huang, L. Perez-Burgos, B. J. Placek, R. Sengupta, M. Richter, J. A. Dorsey, S. Kubicek, S. Opravil, T. Jenuwein and S. L. Berger, *Nature*, 2006, **444**, 629–632.
203. K. R. Kalari, S. J. Hebbring, H. S. Chai, L. Li, J. P. Kocher, L. Wang and R. M. Weinshilboum, *BMC Genomics*, 2010, **11**, 357.
204. S. Komatsu, I. Imoto, H. Tsuda, K. I. Kozaki, T. Muramatsu, Y. Shimada, S. Aiko, Y. Yoshizumi, D. Ichikawa, E. Otsuji and J. Inazawa, *Carcinogenesis*, 2009, **30**, 1139–1146.
205. P. H. Rooney, G. I. Murray, D. A. Stevenson, N. E. Haites, J. Cassidy and H. L. McLeod, *Br. J. Cancer*, 1999, **80**, 862–873.
206. A. D. Ferguson, N. A. Larsen, T. Howard, H. Pollard, I. Green, C. Grande, T. Cheung, R. Garcia-Arenas, S. Cowen, J. Wu, R. Godin, H. Chen and N. Keen, *Structure*, 2011, **19**, 1262–1273.
207. R. F. Sweis, *et al.*, *ACS Med. Chem. Lett.*, 2015, **6**, 695–700.
208. H. Nguyen, *et al.*, *J. Biol. Chem.*, 2015, **290**, 13641–13653.
209. K. Nishioka, S. Chuikov, K. Sarma, H. Erdjument-Bromage, C. D. Allis, P. Tempst and D. Reinberg, *Genes Dev.*, 2002, **16**, 479–489.
210. H. Wang, R. Cao, L. Xia, H. Erdjument-Bromage, C. Borchers, P. Tempst and Y. Zhang, *Mol. Cell*, 2001, **8**, 1207–1217.
211. P. A. Del Rizzo and R. C. Trievel, *Epigenetics*, 2011, **6**, 1059–1067.
212. H. S. Cho, T. Suzuki, N. Dohmae, S. Hayami, M. Unoki, M. Yoshimatsu, G. Toyokawa, M. Takawa, T. Chen, J. K. Kurash, H. I. Field, B. A. Ponder, Y. Nakamura and R. Hamamoto, *Cancer Res.*, 2011, **71**, 655–660.
213. D. Wang, J. Zhou, X. Liu, D. Lu, C. Shen, Y. Du, F. Z. Wei, B. Song, X. Lu, Y. Yu, L. Wang, Y. Zhao, H. Wang, Y. Yang, Y. Akiyama, H. Zhang and W. G. Zhu, *Proc. Natl. Acad. Sci. U. S. A.*, 2013, **110**, 5516–5521.
214. N. Balasubramaniyan, M. Ananthanarayanan and F. J. Suchy, *Am. J. Physiol.: Gastrointest. Liver Physiol.*, 2012, **302**, G937–G947.
215. D. R. Calnan, A. E. Webb, J. L. White, T. R. Stowe, T. Goswami, X. Shi, A. Espejo, M. T. Bedford, O. Gozani, S. P. Gygi and A. Brunet, *Aging*, 2012, **4**, 462–479.

216. M. J. Oudhoff, S. A. Freeman, A. L. Couzens, F. Antignano, E. Kuznetsova, P. H. Min, J. P. Northrop, B. Lehnertz, D. Barsyte-Lovejoy, M. Vedadi, C. H. Arrowsmith, H. Nishina, M. R. Gold, F. M. Rossi, A. C. Gingras and C. Zaph, *Dev. Cell*, 2013, **26**, 188–194.
217. Z. Shan, Q. Han, J. Nie, X. Cao, Z. Chen, S. Yin, Y. Gao, F. Lin, X. Zhou, K. Xu, H. Fan, Z. Qian, B. Sun, J. Zhong, B. Li and A. Tsun, *J. Biol. Chem.*, 2013, **288**, 35093–35103.
218. S. Campaner, F. Spreafico, T. Burgold, M. Doni, U. Rosato, B. Amati and G. Testa, *Mol. Cell*, 2011, **43**, 681–688.
219. B. Lehnertz, J. C. Rogalski, F. M. Schulze, L. Yi, S. Lin, J. Kast and F. M. Rossi, *Mol. Cell*, 2011, **43**, 673–680.
220. H. Kontaki and I. Talianidis, *Mol. Cell*, 2010, **39**, 152–160.
221. A. Kouskouti, E. Scheer, A. Staub, L. Tora and I. Talianidis, *Mol. Cell*, 2004, **14**, 175–182.
222. Y. Tao, R. L. Neppl, Z. P. Huang, J. Chen, R. H. Tang, R. Cao, Y. Zhang, S. W. Jin and D. Z. Wang, *J. Cell Biol.*, 2011, **194**, 551–565.
223. D. Barsyte-Lovejoy, F. Li, M. J. Oudhoff, J. H. Tatlock, A. Dong, H. Zeng, H. Wu, S. A. Freeman, M. Schapira, G. A. Senisterra, E. Kuznetsova, R. Marcellus, A. Allali-Hassani, S. Kennedy, J. P. Lambert, A. L. Couzens, A. Aman, A. C. Gingras, R. Al-Awar, P. V. Fish, B. S. Gerstenberger, L. Roberts, C. L. Benn, R. L. Grimley, M. J. Braam, F. M. Rossi, M. Sudol, P. J. Brown, M. E. Bunnage, D. R. Owen, C. Zaph, M. Vedadi and C. H. Arrowsmith, *Proc. Natl. Acad. Sci. U. S. A.*, 2014, **111**, 12853–12858.
224. B. Zhao, X. Wei, W. Li, R. S. Udan, Q. Yang, J. Kim, J. Xie, T. Ikenoue, J. Yu, L. Li, P. Zheng, K. Ye, A. Chinnaiyan, G. Halder, Z. C. Lai and K. L. Guan, *Genes Dev.*, 2007, **21**, 2747–2761.
225. F. Thol, B. Kolking, I. H. Hollink, F. Damm, M. M. van den Heuvel-Eibrink, C. Michel Zwaan, G. Bug, O. Ottmann, K. Wagner, M. Morgan, W. K. Hofmann, G. Gohring, B. Schlegelberger, J. Krauter, A. Ganser and M. Heuser, *Leukemia*, 2013, **27**, 750–754.
226. P. O. Angrand, F. Apiou, A. F. Stewart, B. Dutrillaux, R. Losson and P. Chambon, *Genomics*, 2001, **74**, 79–88.
227. S. F. Mahmood, N. Gruel, R. Nicolle, E. Chapeaublanc, O. Delattre, F. Radvanyi and I. Bernard-Pierrot, *Am. J. Pathol.*, 2013, **183**, 1634–1644.
228. E. Smith, C. Lin and A. Shilatifard, *Genes Dev.*, 2011, **25**, 661–672.
229. Y. Zhang, A. Chen, X. M. Yan and G. Huang, *Int. J. Hematol.*, 2012, **96**, 428–437.
230. W. D. Li, Q. R. Li, S. N. Xu, F. J. Wei, Z. J. Ye, J. K. Cheng and J. P. Chen, *Blood*, 2013, **121**, 1478–1479.
231. A. M. Dorrance, S. Liu, W. Yuan, B. Becknell, K. J. Arnoczky, M. Guimond, M. P. Strout, L. Feng, T. Nakamura, L. Yu, L. J. Rush, M. Weinstein, G. Leone, L. Wu, A. Ferketich, S. P. Whitman, G. Marcucci and M. A. Caligiuri, *J. Clin. Invest.*, 2006, **116**, 2707–2716.
232. M. Hu, J. Yao, L. Cai, K. E. Bachman, F. van den Brule, V. Velculescu and K. Polyak, *Nat. Genet.*, 2005, **37**, 899–905.
233. C. B. Moelans, R. A. de Weger, H. N. Monsuur, R. Vijzelaar and P. J. van Diest, *Mod. Pathol.*, 2010, **23**, 1029–1039.

234. N. Nishikawa, M. Toyota, H. Suzuki, T. Honma, T. Fujikane, T. Ohmura, T. Nishidate, M. Ohe-Toyota, R. Maruyama, T. Sonoda, Y. Sasaki, T. Urano, K. Imai, K. Hirata and T. Tokino, *Cancer Res.*, 2007, **67**, 9649–9657.
235. E. J. Dettman, S. J. Simko, B. Ayanga, B. L. Carofino, J. F. Margolin, H. C. Morse, 3rd and M. J. Justice, *Oncogene*, 2011, **30**, 2859–2873.
236. K. Maki, T. Yamagata and K. Mitani, *Cancer Sci.*, 2008, **99**, 1878–1883.
237. Y. Zhang, K. Owens, L. Hatem, C. H. Glass, K. Karuppaiah, F. Camargo and A. S. Perkins, *Blood*, 2013, **122**, 2888–2892.
238. L. Liu, X. T. Zhen, E. Denton, B. D. Marsden and M. Schapira, *Bioinformatics*, 2012, **28**, 2205–2206.
239. E. Cerami, J. Gao, U. Dogrusoz, B. E. Gross, S. O. Sumer, B. A. Aksoy, A. Jacobsen, C. J. Byrne, M. L. Heuer, E. Larsson, Y. Antipin, B. Reva, A. P. Goldberg, C. Sander and N. Schultz, *Cancer Discovery*, 2012, **2**, 401–404.
240. J. Gao, B. A. Aksoy, U. Dogrusoz, G. Dresdner, B. Gross, S. O. Sumer, Y. Sun, A. Jacobsen, R. Sinha, E. Larsson, E. Cerami, C. Sander and N. Schultz, *Sci. Signaling*, 2013, **6**, l1.

CHAPTER 7

Targeting Non-Acetylation Histone Erasers

FEI LU[a] AND HUI ZHANG*[b]

[a]School of Chemical Biology and Biotechnology, Peking University Shenzhen Graduate School, Shenzhen, Guangdong 518055, China;
[b]Department of Chemistry, University of Nevada, Las Vegas, NV 89154, USA
*E-mail: hui.zhang@unlv.edu

7.1 Introduction

Alterations in the genome and the epigenome are frequent events in human cancers and various other diseases.[1,2] A major epigenetic mechanism in the dynamic regulation of gene expression is the reversible post-translational modification of histones.[3] Histone modifications store important chromatin regulatory information to control gene expression and development.[4,5] The major histone modifications include acetylation, methylation, phosphorylation and ubiquitination.[3] While much progress has been made in understanding and targeting the processes of phosphorylation and acetylation of histones, emerging evidence indicates that reversible histone methylation is a key regulatory mechanism that defines chromatin structure, the patterns of gene expression, and consequently the identity of a specific cell.[4,6]

Multiple methylations have been identified on the lysine residues at the amino-terminal tails of core histones *in vivo*, although arginine residues on histones are also often methylated.[7] The major methylations occur on histone H3 at the ε-amine group of lysines 4 (H3K4), 9 (H3K9), 27 (H3K27), 36 (H3K36)

RSC Drug Discovery Series No. 48
Epigenetics for Drug Discovery
Edited by Nessa Carey

Published by the Royal Society of Chemistry, www.rsc.org

and 79 (H3K79) and histone H4 at lysine 20 (H4K20), even though other histones are also methylated.[4] These lysines on histone H3 and H4 can be mono- (me1), di- (me2) and trimethylated (me3). While methylations of H3K4, H3K36 and H3K79, especially that of trimethylation of H3K4 (H3K4me3), H3K36 (H3K36me3) and H3K79 (H3K79me3) are associated with active transcriptional activity, di- and trimethylations of H3K9 (H3K9me2/me3) and trimethylations on H4K20 and H3K27 (H3K20me3 and H3K27me3) are generally considered as repressive for transcriptional activities.[4,6] Recent studies show that the location and degree of methylation (mono-, di- and trimethylations) on these histone tails are associated with differential gene expression status. While H3K4 trimethylation is generally located in the promoter regions of active genes or genes poised to be activated, the monomethylated H3K4 is often linked with enhancer function. Methylated H3K4 and H3K27 also form bivalent histone modification markers on silenced but poised to be activated genes during development. The dimethylation of H3K79 is functionally linked to cell cycle regulation, whereas the trimethylated H3K79 is associated with the Wnt pathway. These location-specific methylations of histones are catalysed by histone methyl transferases (HMTs) that transfer the methyl group from the *S*-adenosyl-methionine to the lysines of histones.[4] Many of these SET domain-containing methyltransferases form complexes with other proteins that help define the substrate specificity.

Unlike histone deacetylation or dephosphorylation, which involves simple hydrolysis of an amide or an ester bond, methylated histones were thought to be irreversibly altered because of the stable nature of the N–CH_3 bond.[7–9] This view was supported by some studies showing that the half-life of histone methyl groups was approximately equal to that of histones themselves.[8,9] However, many lines of evidence also appeared to suggest that histone methylation patterns change with gene expression in a dynamic and reversible way, indicating that the removal of histone methylation is associated with gene activity.[9,10] However, until the discovery of the first histone demethylase, lysine-specific demethylase 1 (LSD1/KDM1A/AOF2), in 2004,[11] histone methylation was considered as a stable epigenetic mark that can pass from one generation to the next during cell divisions.

7.2 Function of Histone Demethylases

LSD1 was originally found in several protein complexes that contain transcriptional regulatory activity such as the CtBP, NuRD, CoREST and/or HDAC1/2 complexes.[12–15] Shi *et al.*[11] were the first to show that LSD1 can specifically oxidise and remove the methyl group from methylated H3K4 through a conserved amine oxidase domain, which classifies it in the flavin adenine dinucleotide (FAD)-dependent amine oxidase family. However, LSD1 can only catalyse the demethylation of the mono- and dimethylated H3K4 through amine oxidation (Figure 7.1).[11,16,17] This is because demethylation by LSD1 requires a protonated nitrogen in the methylated histone, precluding it from removing the

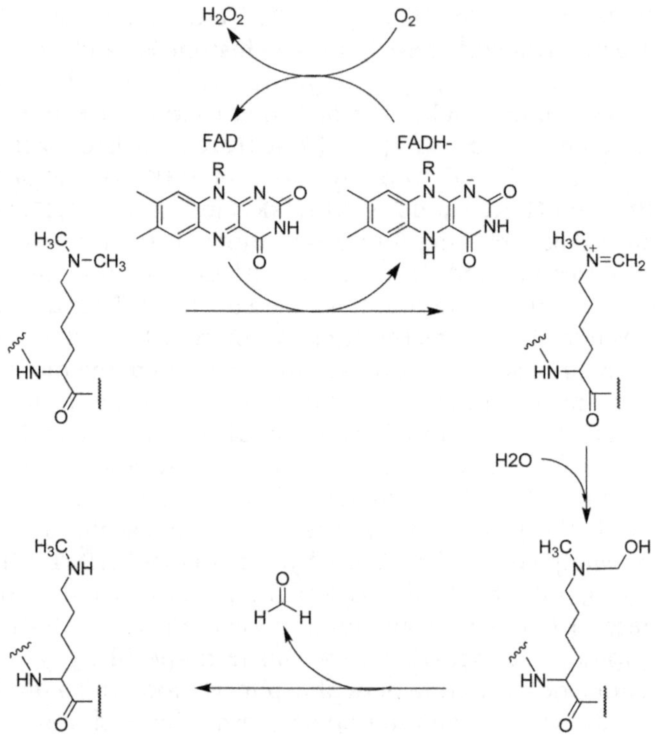

Figure 7.1 The flavin adenosine dinucleotide (FAD)-dependent amine oxidation reaction catalysed by LSD1 or LSD2 to remove mono- and dimethylated substrates.

methyl group from trimethylated H3K4.[11,16,18] In certain prostate cancer cells, LSD1 was found to associate with the androgen receptor to demethylate the mono- and dimethylated H3K9 in a ligand-dependent manner.[19] Thus, the specificity of the LSD1 demethylase towards the methylated H3K4 or H3K9 can be modulated by its interaction with other proteins. LSD1 is highly conserved from *S. pombe*, *C. elegans* and *D. melanogaster* to human, but is absent from *S. cerevisiae*.[6] In *S. pombe*, LSD1 has a weak intrinsic demethylase activity on H3K9me1/me2 but not H3K4me1/me2, suggesting that the H3K9 demethylase activity of LSD1 might be evolutionarily important.[20] Our recent studies indicate that the H3K9me1/me2-specific demethylation activity of LSD1 is also present in many cancer cells that express pluripotent stem cell protein Sox2. *In vivo*, LSD1 is often found as a component of multi-protein complexes such as the CoREST or NuRD complexes.[13–15] It has been demonstrated that the presence of CoREST in the LSD1 complexes allows LSD1 to demethylate the nucleosomal substrates.[14,15] These studies suggest that the ability of LSD1 to create both permissive and repressive epigenetic histone modifications depends on its interaction with other proteins in a cell-specific manner.

A homologue of LSD1, LSD2 (KDN1B/AOF1), which shares 38% of identical amino acid residues and 56% of amino acid similarity with LSD1 in the conserved amine oxidase domain, also exists. It also contains the H3K4me1/me2 demethylase activity.[21,22] However, the function of LSD1 appears to be unique in development, as the mouse null mutant of LSD1 exhibited early embryonic lethality.[23] In contrast, LSD2 deficiency had no effects on mouse development and oogenesis. However, oocytes from the LSD2-deficient female mice showed increased levels of H3K4me1/me2 and these oocytes failed to set up DNA methylation at specific imprinted genes (see Table 7.1).[22]

Because there were only two LSD1 homologues in the mammalian genome and both could only catalyse the removal of mono- and dimethylated histones but not trimethylated substrates, it suggested that additional demethylases might exist to remove the methyl group from other known histone methylation sites and from the trimethylated histones. It was found that the bacterial AlkB protein contained an iron-dependent and α-ketoglutarate-dependent oxidation mechanism that is capable of demethylating DNA, producing formaldehyde as the reaction product. Protein sequence homologies between the AlkB catalytic domain and the Jumonji C (JmjC) domain in eukaryotic cells and many lines of evidence for the involvement of JmjC domain proteins in transcriptional processes suggested that JmjC domain-containing proteins might act as demethylases.[7,24] Indeed, a large family of JmjC domain-containing demethylases was demonstrated to contain the demethylase activities for histones in a manner similar to that of dioxygenases.[6,16,24,25] Thus, this family of histone demethylases removes the methyl group from mono-, di- and trimethylated core histone proteins H3 and H4 using iron (Fe(II)) and α-ketoglutarate as cofactors in the dioxygenase reaction that produces hydroxymethyl lysine, succinate and CO_2.[7] The hydroxymethyl group is then spontaneously lost as formaldehyde to liberate a methyl group (Figure 7.2).

Like LSD1 and LSD2, the JmjC domain-containing demethylases are site-specific enzymes (Figure 7.3 and Table 7.1). Although LSD1 demethylates H3K4me1/me2,[11] the removal of H3K4me2/me3 is catalysed by the JmjC domain-containing KDM5A–D (JARID1A–1D) demethylases,[26–28] although reports suggest that NO66 also acts on H3K4me1/me3 and has a weaker activity on H3K4me2.[6,29] KDM2B (JHDM1B/FBXL10) was also reported to demethylate H3K4me3 but this specificity was not confirmed.[30] In *Drosophila*, the KDM5 homologue LID was identified as the Trithorax group protein that interacts with *Drosophila* Myc (dMyc) and is required for dMyc-induced cell growth.[31] This function of LID appears to be conserved, as human homologues of LID, KDM5A and 5B were also found to interact with Myc.[31]

KDM2A (JHDM1A/FBXL11) and KDM2B were the first JmjC domain-containing demethylases that were shown to specifically remove the methyl group from H3K36me1/me2.[24] The mammalian KDM2B interacts with the proteins in the Polycomb repressive complex-1 (PRC-1) and the BCL-6 co-repressor (BCOR) complexes.[32] In addition to the JmjC domain, mammalian KDM2 proteins contain an F-box, a hallmark of SCF ubiquitin E3-ligase activity, and a CXXC zinc finger domain that can interact with unmethylated

Table 7.1 Demethylases and their function.

Name	Synonyms	Histone substrates	Non-histone substrates and/or reported functions
KDM1A	AOF2, LSD1, KIAA0601	H3K4me1/me2, H3K9me1/me2	Essential for mouse ES cells. Demethylates p53/TP53 and DNMT1
KDM1B	AOF1, LSD2, C6orf193	H3K4me1/me2	Increased levels of H3K4me1/me2 in oocytes in deficient female mice with altered DNA methylation at specific imprinted genes
KDM2A	CXXC8, FBL7, FBXL11, JHDM1A, KIAA1004	H3K36me1/me2	Contains F-box and CXXC zinc finger domains; the latter interacts with unmethylated CpG islands
KDM2B	CXXC2, FBL10, FBXL10, JHDM1B, PCCX2	H3K4me3, H3K36me1/me2	Contains F-box and CXXC zinc finger domains. Interacts with PRC-1 and BCOR complexes
KDM3A	JHDM2A, JMJD1, JMJD1A, KIAA0742, TSGA	H3K9me1/me2	Essential for mouse ES cells. Highly expressed in testis and facilitates the androgen receptor-mediated transcription activation. Mouse deficiency causes adult onset obesity, metabolic syndromes, and defects in spermatogenesis
KDM3B	C5orf7, JHDM2B, JMJD1B, KIAA1082	H3K9me	
KDM4A	JHDM3A, JMJD2, JMJD2A, KIAA0677	H3K9me3, H3K36me2/me3, H1.4K26me2/me3	*Drosophila* homologue is stimulated by the heterochromatin protein 1a (HP1a) to remove H3K36me2/me3
KDM4B	JHDM3B, JMJD2B, KIAA0876	H3K9me2/me3, H1.4K26me2/me3	Demethylates the H3K9me3 in the pericentric heterochromatin
KDM4C	GASC1, JHDM3C, JMJD2C, KIAA0780	H3K9me3, H3K36me2/me3, H1.4K26me2/me3	Essential in mouse ES cells. Cooperates with LSD1 to remove H3K9me3 in prostate cancer cells
KDM4D	JHDM3D, JMJD2D	H3K9me2/me3, H1.4K26me2/me3	

KDM4E	JMJD2E, KDM4DL	H3K9me2/me3	
KDM5A	JARID1A, RBBP2, RBP2	H3K4me2/me3	*Drosophila* (LID) and human KDM5A interact with Myc. LID is required for Myc-induced growth
KDM5B	JARID1B, PLU1, RBBP2H1	H3K4me1/me2/me3	Interacts with Myc
KDM5C	DXS1272E, JARID1C, SMCX, XE169	H3K4me2/me3	
KDM5D	HY, HYA, JARID1D, KIAA0234, SMCY	H3K4me2/me3	
KDM6A	UTX	H3K27me2/me3	Required in development. Regulates Hox gene expression. Component of MLL-H3K4 methyltransferase complexes
KDM6B	JMJD3, KIAA0346	H3K27me2/me3	Required for neural commitment and differentiation of self-renewal tissues and Polycomb regulation. Required for cell cycle progression, rDNA transcription, and brain development
KDM7A	JHDM1DA	H3K9me1/me2, H3K27me1/me2	
KDM7B	JHDM1DB	H3K9me1/me2, H3K27me1/me2	
KDM8	JMJD5	H3K36me2	
PHF2	JHDM1E	H3K9me2	Demethylates ARID5B; recruited to H3K4me3 at the rDNA promoters to regulate rDNA expression
PHF8	JHDM1F	H3K9me1/me2, H3K27me2, H4K20me1	Co-activator of RARα fusions for the retinoic acid response in acute promyelocytic leukaemia
NO66	C14orf169, MAPJD	H3K4me1/me2/me3, H3K36me2/me3	Catalyses the hydroxylation of L8 on histidine 216 residue
JMJD6	PSR/PTDSR	H3R2me2, H4R3me2	Acts as a lysyl hydroxylase that catalyses 5-hydroxylation of U2AF2/U2AF65 and LUC7L2

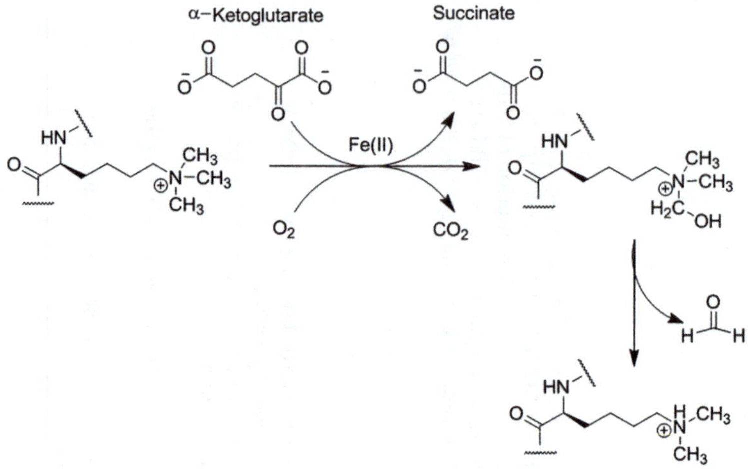

Figure 7.2 The Fe(II)- and α-ketoglutarate-dependent oxidation reaction catalysed by the JmjC domain histone demethylases to demethylate mono-, di- and trimethylated substrates.

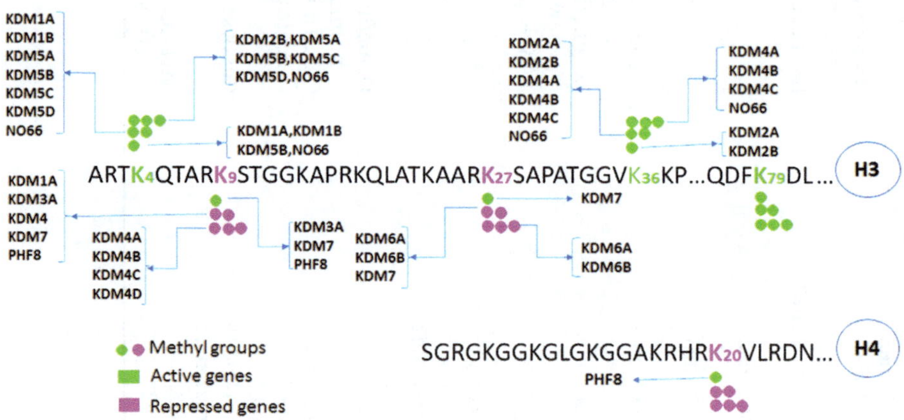

Figure 7.3 Histone demethylases and their histone substrates.

CpG islands.[32,33] Indeed, recent studies showed that KDM2A binds to a large majority of CpG islands in mouse embryonic stem (ES) cells.[33] In addition to KDM2, NO66 also acts on H3K36me2/me3, whereas KDM8 (JMJD5) was recently shown to be a demethylase for H3K36me2.[6,29,34]

KDM3A (JHDM2A/JMJD1A/TSGA), KDM7 (JHDM1D/KIAA1718) and PHF8 (JHDM1F) are demethylases for H3K9me1/me2.[34–36] KDM3A was found to be highly expressed in testis and it functions to facilitate the androgen receptor-mediated transcription activation.[37] Loss of KDM3A in the mouse caused adult-onset obesity and metabolic syndromes, as well as defects in spermatogenesis.[37–39]

KDM4A–4C (JHDM3A–D/JMJD2A–C) were found to remove the methyl groups from di- and trimethylated H3K9 and H3K36 while KDM4D and 4E only acted on H3K9me2/me3.[40,41] KDM4A–D were the first JmjC domain-containing demethylases that are capable of removing the trimethylated histones.[25] Recent studies also demonstrated that KDM7 is capable of demethylating H3K9me1/2 and H3K27me1/me2 while PHF2 acts on H3K9me2.[34,42] PHF2 is recruited to H3K4me3 at rDNA promoters to regulate the expression of rDNA.[42] In addition, PHF8 was also shown to catalyse the removal of H3K9me1/me2.[43] KDM4B was found to demethylate the H3K9me3 in the peri-centric heterochromatin,[44] and in *Drosophila*, KDM4A was stimulated by the heterochromatin protein 1a (HP1a) to remove H3K36me2/me3.[45,46] Moreover, it was shown that KDM4C cooperates with LSD1 to promote androgen receptor-dependent transcription in prostate cancer cells by removing the methyl group from H3K9me3.[47] KDM4C, as well as KDM3A, is essential for the self-renewal of mouse ES cells.[48]

KDM6A (UTX) and KDM6B (JMJD3) are specific demethylases for H3K27me2/me3.[49–52] KDM7 also functions as a dual specific demethylase to demethylate H3K27me1/me2 and PHF8 also catalyses the removal of H3K27me2.[34,36] Many lines of evidence indicate that KDM6s are conserved demethylases that play important roles in development.[6,49,52] KDM6A was shown to regulate Hox gene expression and KDM6B was required for neural commitment and differentiation of self-renewal tissues and Polycomb regulation.[49,51–54] In addition, KDM6A is a component of MLL-H3K4 methyltransferase complexes and coordinates the H3K4 methylation with the methylated H3K27.[55,56]

PHF8 was demonstrated to contain the demethylase activity for H4K20me1 and acted as a co-activator of RARα fusions to govern retinoic acid response in acute promyelocytic leukaemia.[57,58] It also plays a key role in cell cycle progression, rDNA transcription and brain development.[36,43,57,59] While additional JmjC domain-containing demethylases exist that have unknown functions, it is likely that these and more demethylases will be assigned to the specific methylation sites on histones.[4,6] For example, JMJD6 (PSR or PTDSR) was initially identified as a demethylase that specifically removes the methyl groups from the methylated histone H3 at arginine 2 (H3R2) and histone arginine 4 (H4R3),[60] although later evidence cast doubt on this finding. Instead, recent studies have shown that JMJD6 catalyses the lysyl hydroxylation of U28F65, a protein associated with RNA splicing.[61] The important biological activities of various KDMs described in this chapter are summarised in Table 7.1.

In addition to histones, it was found that some demethylases can also remove methyl groups from non-histone substrates. For example, LSD1 removes the mono- and dimethyl groups from lysine 370 in p53, lysine 185 in E2F1, and lysine 1096 in DNMT1, whereas KDM2A demethylates p65 and NF-κB.[23,62–65] The demethylation of p53 by LSD1 is reported to keep p53 in an inactive state whereas demethylation of DNMT1 and E2F1 by LSD1 leads to destabilisation of these proteins. It is likely that additional non-histone substrates will be identified for various demethylases that are involved in a diverse array of biological processes independent of their role in histone demethylation. For example, PHF2 has been shown to act on methylated ARID5B.[66]

7.3 Histone Demethylases and Human Diseases

Because of the importance of histone demethylases in regulating gene expression and cell specificity, many of them have been implicated in human cancers and other diseases, as summarised in Table 7.2. Several reports showed that the levels of LSD1 are elevated in various cancers, including germ cell tumours, prostate and bladder cancers, oestrogen receptor-negative breast cancers, undifferentiated malignant neuroblastomas, oesophageal squamous cell carcinomas, hepatocellular carcinomas, lung and colorectal carcinomas.[67–75] LSD1 also triggers Myc-induced transcription.[76] The gene encoding LSD2 is amplified and over-expressed in urothelial carcinomas.[77] Loss of KDM2B induces p15-dependent senescence and it is over-expressed in various lymphomas and adenocarcinomas.[78,79] However, the expression of KDM2A or KDM2B was reported to be lower in prostate cancers or glioblastomas, respectively, suggesting they might act as potential tumour suppressors in these conditions.[30,80] Several lines of evidence suggest that KDM2 might have cell-specific pro- or anti-oncogenic activities.[81,82] The gene encoding KDM4C, located on 9p23–24, is often amplified in oesophageal squamous cell carcinomas, medulloblastomas and breast carcinomas.[40,83–88] KDM4A, 4B and 4C are also over-expressed in prostate cancers.[40] High expression of KDM5B was identified in breast and testis cancers.[89–93] KDM6A is frequently mutated in myelomas, oesophageal squamous cell carcinomas, and renal cell carcinomas and acts as a tumour suppressor.[94–97] KDM6A was also found to regulate RB-dependent cell fate control.[53] The H3K27me3 demethylase KDM6B is required for the activation of the INK4A-ARF locus in response to oncogene- and stress-induced senescence such as that induced by oncogenic Ras.[98,99] In addition to KDM5A and 5B, which are involved in Myc function,[31] two other JmjC domain-containing proteins, Mina53 and NO66, are related to Myc activities. While Mina53 is highly expressed in oesophageal squamous cell carcinomas, early stage lung cancer, hepatocellular carcinomas and colon cancers, NO66 is over-expressed in non-small cell lung carcinomas.[100–104] PHF8 is over-expressed in prostate cancer with an impact on cell proliferation, migration and invasion.[57,105] Furthermore, PHF8 governs the retinoic acid response in acute promyelocytic leukaemia.[58]

Mutations of histone demethylases have also been linked to other diseases. For instance, mutations of KDM5C and PHF8 were found in patients who had certain X-linked mental retardation syndromes.[106–109] KDM3A, on the other hand, appears to be involved in regulating metabolism and obesity resistance.[38,39] Thus, histone demethylases are considered as excellent potential targets for drug development for the treatment of various human diseases. The diseases that are associated with altered functions of various KDMs are summarised in Table 7.2.

Table 7.2 Demethylases and diseases.

Name	Tissue specificity and associated diseases
KDM1A	Over-expression in germ cell tumours, prostate and bladder cancers, oestrogen receptor-negative breast cancers, undifferentiated malignant neuroblastomas, hepatocellular carcinomas, oesophageal squamous cell carcinomas, and lung and colorectal carcinomas
KDM1B	Amplification and over-expression in urothelial carcinoma
KDM2B	Loss of KDM2B induces p15-dependent senescence. Over-expression in leukaemias and bladder carcinomas
KDM3A	Important for spermatogenesis and involved in regulating metabolism and obesity
KDM4A	Over-expression in breast and prostate cancers
KDM4B	Over-expression in malignant peripheral nerve sheath tumours and prostate cancers
	Amplification in medulloblastomas
KDM4C	Over-expressed in oesophageal squamous cell carcinomas. Amplification in oesophageal squamous carcinomas, medulloblastomas and breast cancers. Translocation in lymphoma
KDM5A	Links to regulation of differentiation, cell cycle progression and mitochondrial function
	Silencing/down-regulation or deletion in melanomas, translocation in acute leukaemias and mutation in ankylosing spondylitis
KDM5B	Over-expression in bladder, prostate and breast cancers
KDM5C	High expression in brain and skeletal muscle. Mutations in X-linked mental retardation in autism and in renal carcinomas
KDM5D	Deletion in prostate cancers
KDM6A	Mutation in multiple tumour types including multiple myelomas, oesophageal squamous cell carcinomas, renal clear cell carcinomas, transitional cell carcinomas, chronic myelomonocytic leukaemias, and deletion in the Kabuki syndrome
KDM6B	Required for activation of INK4A-ARF locus by oncogenic stresses
PHF2	Mutation or silencing/down-regulation in breast carcinomas and head and neck squamous cell carcinomas.
PHF8	Over-expression in prostate cancers. Mutation and deletion associated with X-linked mental retardation and cleft lip/palate
Mina53	Highly expressed in oesophageal squamous cell carcinomas, early stage lung cancers, hepatocellular carcinomas and colon cancers
NO66	Over-expression in non-small cell lung cancers
JMJD6	Over-expression in chronic pancreatitis

7.4 Targeting Histone Demethylases for Drug Development

7.4.1 Development of Specific LSD1 Inhibitors that Non-Covalently Bind to LSD1

As LSD1 regulation is altered in various cancers, LSD1 may serve as a good target for therapeutic intervention in these cancers. Because LSD1 belongs to the FAD-dependent amine oxidase family, which catalyses the demethylation of di- and monomethylated H3K4 through amine oxidation,[11] most pharmacological investigations of LSD1 function in cancers and other cells involved the use of non-selective monoamine oxidase inhibitors (MAOIs).[11,18,110] These chemical inhibitors were previously developed against two major isoforms of monoamine oxidases, MAO-A and MAO-B, as drugs for the treatment of depression.[111] These MAOs remove the amino groups from neuron transmitters in a FAD-dependent oxidation process similar to the demethylation activity of LSD1 on methylated histones. Because FAD is essential for the oxidation of amines by the amine oxidases, these MAOIs act through the formation of irreversible covalent adducts with the enzyme-bound FAD to block the activity of MAOs.[18,110,112] Several of these MAOIs, such as *trans*-2-phenylcyclopropylamine (tranylcypromine, 2-PCPA, parnate and TCP) (1), pargyline (2) and phenelzine (3), were found to inhibit the demethylation activity of LSD1 (Figure 7.4).[18,111,113,114] These MAOIs, especially tranylcypromine, have been derivatised and a significant number of derivatives or analogues were reported with improved potency against LSD1

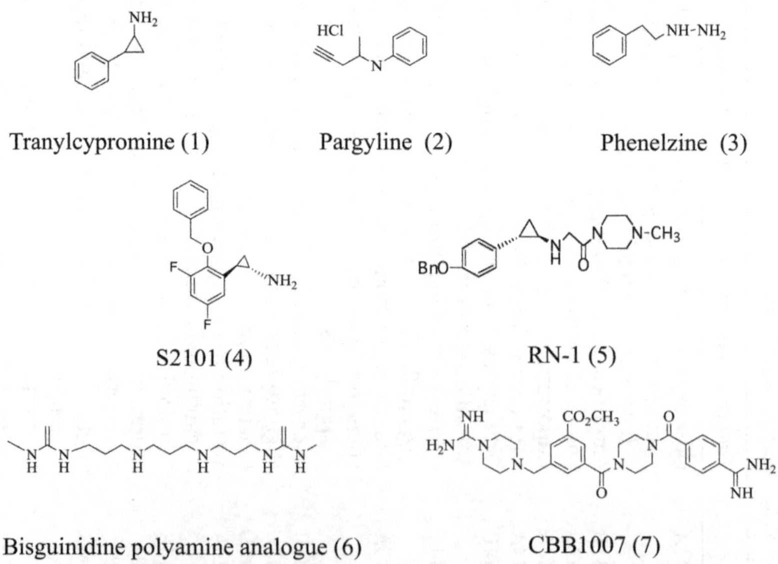

Figure 7.4 Examples of LSD1 and LSD2 inhibitors.

(Figure 7.4), including S2101 (4) and RN-1 (5).[115–117] The active site of LSD1 shares homologies with the FAD-dependent polyamine oxidases and LSD1 can also use polyamines as substrates for oxidation. It was reported that bisguanide and bisguanidine polyamine analogues (6), as well as (bis)urea and (bis)thiourea analogues, can also inhibit LSD1 demethylase activities.[118] Although these inhibitors and analogues can penetrate cells and inhibit the activity of LSD1, one concern is that these compounds nevertheless may have substantial off-target effects on other FAD-dependent amine oxidases.[119,120] Thus, it is often difficult to cross-examine the cytotoxicity or other physiological effects of these reported LSD1 or LSD2 inhibitors.

Several groups have resolved the crystal structures of LSD1 protein or the co-crystal structures of LSD1 and its substrates or tranylcypromine and its analogues.[112,121–124] These studies suggest that with respect to histone substrates, the binding of tranylcypromine or its analogues to the active site of LSD1 is not optimised. This is in part because tranylcypromine is a suicide inhibitor that forms a covalent linkage with FAD so there are more stringent structural requirements on tranylcypromine and its analogues to interact with LSD1 to facilitate such a covalent modification.

Our group has attempted to employ the crystal structure of LSD1 protein to develop novel chemical compounds that will selectively interact with LSD1.[71] It was reported that a peptide pseudo-substrate, derived from the amino-terminal 21 amino acid residues of histone H3 peptide, in which lysine 4 (K4) is replaced by methionine (H3K4M), can bind to LSD1 with high binding affinity (K_i = 0.05 μM).[112,125,126] This pseudo-substrate established favourable electrostatic interactions with a cluster of negatively-charged residues on the LSD1 surface and blocked the access to FAD. We used the co-crystal structure of LSD1 and H3K4M as a template to design specific LSD1 inhibitors.[112,121,122] Utilising the structural features of the LSD1 active site, especially the highly acidic properties of the surface around the active site, we designed a non-peptide chemical scaffold that binds to LSD1 with similar non-covalent binding mode to H3K4M. Unlike the previous MAOIs, these new LSD1 inhibitors specifically interact with the LSD1 protein and inhibit its demethylase activity without forming a covalent bond with FAD.

By screening a small synthetic chemical library, our initial attempt has identified two promising LSD1 inhibitors, CBB1003 and CBB1007 (7). These specifically inhibited the demethylation activity of LSD1, but not that of other histone demethylases such as KDM5A and LSD2, both *in vitro* and *in vivo*.[71] The development of the CBB series of LSD1 inhibitory compounds thus represents a new strategy to selectively inhibit the activity of LSD1, and to modulate the epigenetic regulation of chromatin structure and gene expression without interfering with the activity of other amine oxidases.

The development of new LSD1 inhibitors helped clarify the physiological and pathological roles of LSD1. Although the mouse null mutant of LSD1 caused early embryonic lethality,[23] the precise physiological function of LSD1 remained unclear. In sharp contrast to previous reports using MAOIs as LSD1 inhibitors, we found that the effects of LSD1 inhibition on various human or mouse cells are highly selective and specific. The growth of many cancer cell lines or normal cells, such as human cervical carcinoma HeLa, colorectal carcinoma HCT116, adenovirus-transformed kidney 293, or immortalised mouse NIH3T3 cells, was not sensitive to our LSD1 inhibitors. However, one established cancer cell line, the mouse teratocarcinoma F9 cell, was sensitive to CBB inhibitors.[71] The F9 cells were derived from a teratocarcinoma that belongs to a class of pluripotent germ cell tumours that include teratoma/teratocarcinomas, embryonic carcinomas and seminomas, which are caused by abnormal development of ES cells.[127] Like ES cells, these tumour cells usually display pluripotent stem cell properties, express pluripotent stem cell proteins Oct4, Sox2, Lin28 and Nanog, and are capable of differentiating into various tissues.[128] Consistently, we found that our LSD1 inhibitors blocked the growth of mouse embryonic stem (mES) and human teratocarcinoma/embryonic carcinoma cells.[71] Similarly, another group also independently showed that ablation of LSD1 by RNA interference in mES cells caused these pluripotent stem cells to arrest their cell cycle and to undergo cellular differentiation.[129] Recent studies demonstrated that LSD1 is essential for the maintenance of the oncogenic potential of MLL-AF9 leukaemia stem cells and acute myeloid leukaemia (AML).[117,130] Thus, these studies indicate that LSD1 is a key regulator of cells that exhibit pluripotent or multipotent stem cell properties.

7.4.2 Cellular Mechanism for Sensitivity Towards LSD1 Inhibitors in Cancer Cells

LSD1 inhibitors only selectively target cancer cells with pluripotent or multipotent stem cell properties, but the underlying mechanism for such selectivity is unclear. We found that the levels of LSD1 protein are elevated in germ cell tumours, and our most recent investigation showed that LSD1 is often co-over-expressed with pluripotent stem cell protein Sox2 in lung squamous cell carcinomas.[131] Lung squamous cell carcinoma is a major form of frequent and aggressive lung cancer.[132] Several studies showed that the gene amplification of *Sox2* at 3q26.33 is a common event in squamous cell carcinomas of the lung, oesophagus and oral cavity.[133–135] The Sox2 gene is also reported to be amplified in small cell lung carcinomas[136] and glioblastoma multiforme (GBM).[137] While Sox2 is a key regulator of pluripotent embryonic stem cells (ESCs) and adult neural stem cells,[138,139] it was also reported to act as a lineage-survival oncogene for the expression of pluripotent stem cell signatures and the lineage-specific genes in lung squamous cell carcinomas.[133] Ectopic expression of Sox2 is sufficient to cause the oncogenic transformation of normal tracheobronchial epithelial cells.[133] Sox2 is frequently expressed in other types of poorly differentiated and aggressive human cancers,[140] including breast and ovarian carcinomas, and the cells that express Sox2 often display stem cell-like properties. We found that LSD1 inhibitors specifically and selectively inhibited the growth of lung, breast and ovarian cancer cells that express Sox2 but not that of Sox2-negative cells.[131] Treatment of these cancer cells with our LSD1 inhibitors caused the down-regulation of Sox2 expression, promoted the G1 cell cycle arrest, suppressed Sox2-dependent lineage-survival oncogenic potential, and induced cellular differentiation. Inactivation of LSD1 by specific small RNA interference (siRNA) phenocopied all the selective effects of our LSD1 inhibitors only in Sox2-expressing cancer cells. These studies suggest that LSD1 may serve as a selective and specific demethylase target for the therapy of Sox2-expressing cancer cells.

7.4.3 Chemical Inhibitors for JmjC Domain-Containing Histone Demethylases

Because JmjC domain histone demethylases utilise the Fe(II)- and α-ketoglutarate-dependent oxidation reaction to demethylate methylated histones,[6] several studies employed known inhibitors of other Fe(II)- and α-ketoglutarate-dependent oxidases to determine whether they inhibit the activities of JmjC domain-containing histone demethylases. It was reported that *N*-oxalylglycine (NOG, 8, Figure 7.5), an amide that resembles α-ketoglutarate, inhibited the activity of JMJD2E (KDM4E).[141,142] As NOG and JMJD2A (KDM2A) can form a complex which can be co-crystallised if Fe(II) is replaced by Ni(II), additional NOG analogues were designed and analysed.[143] An improved NOG-like compound (9) that inhibited the activity of KDM4E (IC_{50} = 37.1 μM) was identified based on the presence of an adjacent and unique hydrophobic subpocket to the

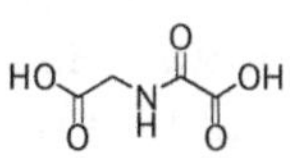
NOG (8)

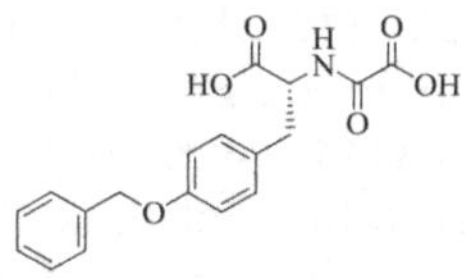
NOG-like (9)

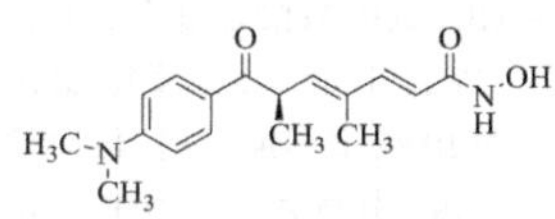
TSA (10)

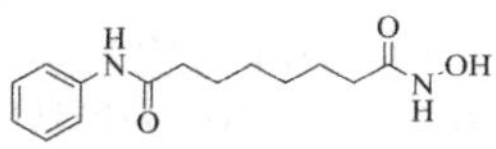
SAHA (11)

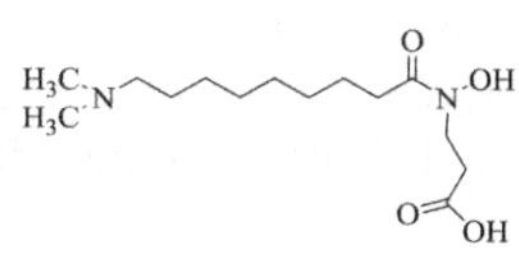
Internal hydroxamic acid (12)

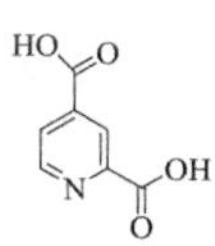
2,4-PDCA (13)

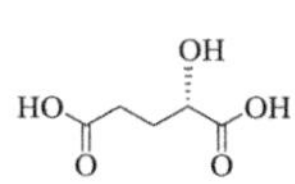
2-hydroxyglutarate (14)

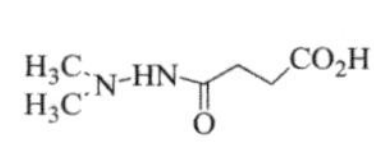
Daminozide (15)

Figure 7.5 Examples of chemical compounds that inhibit the JmjC domain histone demethylases.

active site of JmjC domain histone demethylases.[144] These studies also revealed that hydroxamic acid compounds that were used for inhibiting histone deacetylases (HDACs) such as trichostatin A (TSA, 10) and SAHA (11) can also inhibit KDM4E *via* their metal binding capacities (Figure 7.5).[142] Such an inhibitory activity may be due to the ability of these compounds to chelate Fe(II) in the active site of JmjC domain histone demethylases. The concept of chelating Fe(II) was further explored to generate a structure-based optimisation of inhibitory compounds (12) that exhibits an *in vitro* IC_{50} about 1 μM against JMJD2C (KDM4C) and 3 μM on JMJD2A (KDM4A).[145] On the other hand, pyridine dicarboxylic acids were also examined and one of them, pyridine-2,4-dicarboxylic acid (2,4-PDCA) (13), displayed a potent inhibitory activity towards KDM4E with an IC_{50} of 1.4 μM.[142] It was also found that 2-hydroxyglutarate (2-HG, 14), an oncometabolite that shares high structural similarity to α-ketoglutarate, often accumulates in gliomas, glioblastoma multiforme, and AMLs that contain familial mutations on the gene encoding isocitrate dehydrogenase 1 or 2 (IDH1/2). 2-HG acts as a competitive inhibitor for the binding of α-ketoglutarate to the JmjC domain-containing histone demethylases and other α-ketoglutarate-dependent dioxygenases at physiologically relevant high concentrations.[146,147] More recently, daminozide (15, Figure 7.5), a plant growth regulator, was shown to act as a selective inhibitor of human KDM2 and KDM7, probably due to the metal chelating capability of this compound *via* its hydrazide carbonyl and dimethylamino groups.[146]

As JmjC domain histone demethylases share structural similarities, it is important to obtain subtype-specific inhibitors to prevent cross-reactivity. It was demonstrated that BIX-01294 (16) (Figure 7.6), an inhibitor of H3K9 methyltransferase G9a, and its analogue E67-2 (17), can inhibit H3K9me2 demethylase KDM7 (JHDM1D/KIAA1718) with an IC_{50} of 3.7 μM for E67-2.[148] E67-2 was also an inhibitor of PHF8 but not KDM5C. This inhibitory effect is probably due to the fact that both G9a and KDM7/PHF8 can recognise the methylated H3K9 as products or substrates. In addition, a specific inhibitor, GSK-J1 (18), was recently reported that was designed based on the crystal structure of the mouse KDM6B protein in complex with cofactor analogue N-oxalylglycine (NOG) and metal (Ni2+).[149] It was demonstrated that this compound selectively inhibits the activity of JmjC domain KDM6B (JMJD3) and

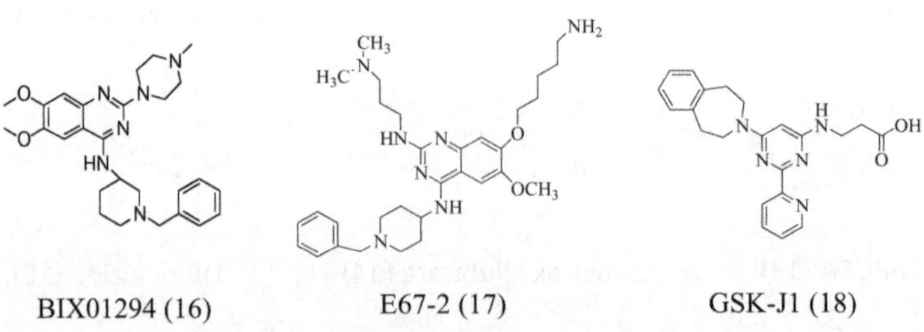

BIX01294 (16) E67-2 (17) GSK-J1 (18)

Figure 7.6 Inhibitors of the JmjC domain histone demethylases.

KDM6A (UTX) demethylases, and thereby modulates the levels of trimethylated H3K27 in cells. It was also shown that this inhibitor reduces the lipopolysaccharide-induced proinflammatory cytokine production by human primary macrophages through inhibiting the demethylation of H3K27me3 by KDM6A/6B.

While these studies indicate that histone demethylases serve as a promising new target class of non-acetylation histone erasers, further work is warranted to understand the roles of these new histone modulators in various biological processes and to develop novel classes of chemical inhibitors and new therapeutic strategies for the treatment of various human diseases.

References

1. L. T. Smith, G. A. Otterson and C. Plass, *Trends Genet.*, 2007, **23**, 449–456.
2. Y. Shi and J. R. Whetstine, *Mol. Cell*, 2007, **25**, 1–14.
3. T. Suganuma and J. L. Workman, *Annu. Rev. Biochem.*, 2011, **80**, 473–499.
4. E. L. Greer and Y. Shi, *Nat. Rev. Genet.*, 2012, **13**, 343–357.
5. R. Margueron and D. Reinberg, *Nat. Rev. Genet.*, 2010, **11**, 285–296.
6. S. M. Kooistra and K. Helin, *Nat. Rev. Mol. Cell Biol.*, 2012, **13**, 297–311.
7. R. J. Klose and Y. Zhang, *Nat. Rev. Mol. Cell Biol.*, 2007, **8**, 307–318.
8. T. Jenuwein and C. D. Allis, *Science*, 2001, **293**, 1074–1080.
9. A. J. Bannister, R. Schneider and T. Kouzarides, *Cell*, 2002, **109**, 801–806.
10. T. Kouzarides, *Cell*, 2007, **128**, 693–705.
11. Y. Shi, F. Lan, C. Matson, P. Mulligan, J. R. Whetstine, P. A. Cole and R. A. Casero, *Cell*, 2004, **119**, 941–953.
12. Y. Shi, J. Sawada, G. Sui, B. Affar el, J. R. Whetstine, F. Lan, H. Ogawa, M. P. Luke and Y. Nakatani, *Nature*, 2003, **422**, 735–738.
13. Y. Wang, H. Zhang, Y. Chen, Y. Sun, F. Yang, W. Yu, J. Liang, L. Sun, X. Yang, L. Shi, R. Li, Y. Li, Y. Zhang, Q. Li, X. Yi and Y. Shang, *Cell*, 2009, **138**, 660–672.
14. Y. J. Shi, C. Matson, F. Lan, S. Iwase, T. Baba and Y. Shi, *Mol. Cell*, 2005, **19**, 857–864.
15. M. G. Lee, C. Wynder, N. Cooch and R. Shiekhattar, *Nature*, 2005, **437**, 432–435.
16. Y. Shi, *Nat. Rev. Genet.*, 2007, **8**, 829–833.
17. F. Forneris, C. Binda, E. Battaglioli and A. Mattevi, *Trends Biochem. Sci.*, 2008, **33**, 181–189.
18. J. C. Culhane and P. A. Cole, *Curr. Opin. Chem. Biol.*, 2007, **11**, 561–568.
19. E. Metzger, M. Wissmann, N. Yin, J. M. Muller, R. Schneider, A. H. Peters, T. Gunther, R. Buettner and R. Schule, *Nature*, 2005, **437**, 436–439.
20. F. Lan, M. Zaratiegui, J. Villen, M. W. Vaughn, A. Verdel, M. Huarte, Y. Shi, S. P. Gygi, D. Moazed, R. A. Martienssen and Y. Shi, *Mol. Cell*, 2007, **26**, 89–101.
21. R. Fang, A. J. Barbera, Y. Xu, M. Rutenberg, T. Leonor, Q. Bi, F. Lan, P. Mei, G. C. Yuan, C. Lian, J. Peng, D. Cheng, G. Sui, U. B. Kaiser, Y. Shi and Y. G. Shi, *Mol. Cell*, 2010, **39**, 222–233.

22. D. N. Ciccone, H. Su, S. Hevi, F. Gay, H. Lei, J. Bajko, G. Xu, E. Li and T. Chen, *Nature*, 2009, **461**, 415–418.
23. J. Wang, S. Hevi, J. K. Kurash, H. Lei, F. Gay, J. Bajko, H. Su, W. Sun, H. Chang, G. Xu, F. Gaudet, E. Li and T. Chen, *Nat. Genet.*, 2009, **41**, 125–129.
24. Y. Tsukada, J. Fang, H. Erdjument-Bromage, M. E. Warren, C. H. Borchers, P. Tempst and Y. Zhang, *Nature*, 2006, **439**, 811–816.
25. R. J. Klose, E. M. Kallin and Y. Zhang, *Nat. Rev. Genet.*, 2006, **7**, 715–727.
26. R. J. Klose, Q. Yan, Z. Tothova, K. Yamane, H. Erdjument-Bromage, P. Tempst, D. G. Gilliland, Y. Zhang and W. G. Kaelin Jr, *Cell*, 2007, **128**, 889–900.
27. J. Christensen, K. Agger, P. A. Cloos, D. Pasini, S. Rose, L. Sennels, J. Rappsilber, K. H. Hansen, A. E. Salcini and K. Helin, *Cell*, 2007, **128**, 1063–1076.
28. S. Iwase, F. Lan, P. Bayliss, L. de la Torre-Ubieta, M. Huarte, H. H. Qi, J. R. Whetstine, A. Bonni, T. M. Roberts and Y. Shi, *Cell*, 2007, **128**, 1077–1088.
29. K. M. Sinha, H. Yasuda, M. M. Coombes, S. Y. Dent and B. de Crombrugghe, *EMBO J.*, 2010, **29**, 68–79.
30. D. Frescas, D. Guardavaccaro, F. Bassermann, R. Koyama-Nasu and M. Pagano, *Nature*, 2007, **450**, 309–313.
31. J. Secombe, L. Li, L. Carlos and R. N. Eisenman, *Genes Dev.*, 2007, **21**, 537–551.
32. M. D. Gearhart, C. M. Corcoran, J. A. Wamstad and V. J. Bardwell, *Mol. Cell. Biol.*, 2006, **26**, 6880–6889.
33. N. P. Blackledge, J. C. Zhou, M. Y. Tolstorukov, A. M. Farcas, P. J. Park and R. J. Klose, *Mol. Cell*, 2010, **38**, 179–190.
34. Y. Tsukada, T. Ishitani and K. I. Nakayama, *Genes Dev.*, 2010, **24**, 432–437.
35. K. Yamane, C. Toumazou, Y. Tsukada, H. Erdjument-Bromage, P. Tempst, J. Wong and Y. Zhang, *Cell*, 2006, **125**, 483–495.
36. H. H. Qi, M. Sarkissian, G. Q. Hu, Z. Wang, A. Bhattacharjee, D. B. Gordon, M. Gonzales, F. Lan, P. P. Ongusaha, M. Huarte, N. K. Yaghi, H. Lim, B. A. Garcia, L. Brizuela, K. Zhao, T. M. Roberts and Y. Shi, *Nature*, 2010, **466**, 503–507.
37. Y. Okada, G. Scott, M. K. Ray, Y. Mishina and Y. Zhang, *Nature*, 2007, **450**, 119–123.
38. K. Tateishi, Y. Okada, E. M. Kallin and Y. Zhang, *Nature*, 2009, **458**, 757–761.
39. T. Inagaki, M. Tachibana, K. Magoori, H. Kudo, T. Tanaka, M. Okamura, M. Naito, T. Kodama, Y. Shinkai and J. Sakai, *Genes Cells*, 2009, **14**, 991–1001.
40. P. A. Cloos, J. Christensen, K. Agger, A. Maiolica, J. Rappsilber, T. Antal, K. H. Hansen and K. Helin, *Nature*, 2006, **442**, 307–311.
41. R. J. Klose, K. Yamane, Y. Bae, D. Zhang, H. Erdjument-Bromage, P. Tempst, J. Wong and Y. Zhang, *Nature*, 2006, **442**, 312–316.
42. H. Wen, J. Li, T. Song, M. Lu, P. Y. Kan, M. G. Lee, B. Sha and X. Shi, *J. Biol. Chem.*, 2010, **285**, 9322–9326.

43. W. Feng, M. Yonezawa, J. Ye, T. Jenuwein and I. Grummt, *Nat. Struct. Mol. Biol.*, 2010, **17**, 445–450.
44. B. D. Fodor, S. Kubicek, M. Yonezawa, R. J. O'Sullivan, R. Sengupta, L. Perez-Burgos, S. Opravil, K. Mechtler, G. Schotta and T. Jenuwein, *Genes Dev.*, 2006, **20**, 1557–1562.
45. C. H. Lin, B. Li, S. Swanson, Y. Zhang, L. Florens, M. P. Washburn, S. M. Abmayr and J. L. Workman, *Mol. Cell*, 2008, **32**, 696–706.
46. L. L. Wallrath and S. C. Elgin, *Mol. Cell*, 2008, **32**, 601–602.
47. M. Wissmann, N. Yin, J. M. Muller, H. Greschik, B. D. Fodor, T. Jenuwein, C. Vogler, R. Schneider, T. Gunther, R. Buettner, E. Metzger and R. Schule, *Nat. Cell Biol.*, 2007, **9**, 347–353.
48. Y. H. Loh, W. Zhang, X. Chen, J. George and H. H. Ng, *Genes Dev.*, 2007, **21**, 2545–2557.
49. K. Agger, P. A. Cloos, J. Christensen, D. Pasini, S. Rose, J. Rappsilber, I. Issaeva, E. Canaani, A. E. Salcini and K. Helin, *Nature*, 2007, **449**, 731–734.
50. S. Hong, Y. W. Cho, L. R. Yu, H. Yu, T. D. Veenstra and K. Ge, *Proc. Natl. Acad. Sci. U. S. A.*, 2007, **104**, 18439–18444.
51. F. De Santa, M. G. Totaro, E. Prosperini, S. Notarbartolo, G. Testa and G. Natoli, *Cell*, 2007, **130**, 1083–1094.
52. F. Lan, P. E. Bayliss, J. L. Rinn, J. R. Whetstine, J. K. Wang, S. Chen, S. Iwase, R. Alpatov, I. Issaeva, E. Canaani, T. M. Roberts, H. Y. Chang and Y. Shi, *Nature*, 2007, **449**, 689–694.
53. J. K. Wang, M. C. Tsai, G. Poulin, A. S. Adler, S. Chen, H. Liu, Y. Shi and H. Y. Chang, *Genes Dev.*, 2010, **24**, 327–332.
54. S. A. Miller, S. E. Mohn and A. S. Weinmann, *Mol. Cell*, 2010, **40**, 594–605.
55. Y. W. Cho, T. Hong, S. Hong, H. Guo, H. Yu, D. Kim, T. Guszczynski, G. R. Dressler, T. D. Copeland, M. Kalkum and K. Ge, *J. Biol. Chem.*, 2007, **282**, 20395–20406.
56. I. Issaeva, Y. Zonis, T. Rozovskaia, K. Orlovsky, C. M. Croce, T. Nakamura, A. Mazo, L. Eisenbach and E. Canaani, *Mol. Cell. Biol.*, 2007, **27**, 1889–1903.
57. W. Liu, B. Tanasa, O. V. Tyurina, T. Y. Zhou, R. Gassmann, W. T. Liu, K. A. Ohgi, C. Benner, I. Garcia-Bassets, A. K. Aggarwal, A. Desai, P. C. Dorrestein, C. K. Glass and M. G. Rosenfeld, *Nature*, 2010, **466**, 508–512.
58. M. F. Arteaga, J. H. Mikesch, J. Qiu, J. Christensen, K. Helin, S. C. Kogan, S. Dong and C. W. So, *Cancer Cell*, 2013, **23**, 376–389.
59. Z. Zhu, Y. Wang, X. Li, Y. Wang, L. Xu, X. Wang, T. Sun, X. Dong, L. Chen, H. Mao, Y. Yu, J. Li, P. A. Chen and C. D. Chen, *Cell Res.*, 2010, **20**, 794–801.
60. B. Chang, Y. Chen, Y. Zhao and R. K. Bruick, *Science*, 2007, **318**, 444–447.
61. C. J. Webby, A. Wolf, N. Gromak, M. Dreger, H. Kramer, B. Kessler, M. L. Nielsen, C. Schmitz, D. S. Butler, J. R. Yates 3rd, C. M. Delahunty, P. Hahn, A. Lengeling, M. Mann, N. J. Proudfoot, C. J. Schofield and A. Bottger, *Science*, 2009, **325**, 90–93.

62. J. Huang, R. Sengupta, A. B. Espejo, M. G. Lee, J. A. Dorsey, M. Richter, S. Opravil, R. Shiekhattar, M. T. Bedford, T. Jenuwein and S. L. Berger, *Nature*, 2007, **449**, 105–108.
63. H. Kontaki and I. Talianidis, *Mol. Cell*, 2010, **39**, 152–160.
64. T. Lu, M. W. Jackson, A. D. Singhi, E. S. Kandel, M. Yang, Y. Zhang, A. V. Gudkov and G. R. Stark, *Proc. Natl. Acad. Sci. U. S. A.*, 2009, **106**, 16339–16344.
65. T. Lu, M. W. Jackson, B. Wang, M. Yang, M. R. Chance, M. Miyagi, A. V. Gudkov and G. R. Stark, *Proc. Natl. Acad. Sci. U. S. A.*, 2010, **107**, 46–51.
66. A. Baba, F. Ohtake, Y. Okuno, K. Yokota, M. Okada, Y. Imai, M. Ni, C. A. Meyer, K. Igarashi, J. Kanno, M. Brown and S. Kato, *Nat. Cell Biol.*, 2011, **13**, 668–675.
67. P. Kahl, L. Gullotti, L. C. Heukamp, S. Wolf, N. Friedrichs, R. Vorreuther, G. Solleder, P. J. Bastian, J. Ellinger, E. Metzger, R. Schule and R. Buettner, *Cancer Res.*, 2006, **66**, 11341–11347.
68. J. H. Schulte, S. Lim, A. Schramm, N. Friedrichs, J. Koster, R. Versteeg, I. Ora, K. Pajtler, L. Klein-Hitpass, S. Kuhfittig-Kulle, E. Metzger, R. Schule, A. Eggert, R. Buettner and J. Kirfel, *Cancer Res.*, 2009, **69**, 2065–2071.
69. S. Lim, A. Janzer, A. Becker, A. Zimmer, R. Schule, R. Buettner and J. Kirfel, *Carcinogenesis*, 2010, **31**, 512–520.
70. Y. Yu, B. Wang, K. Zhang, Z. Lei, Y. Guo, H. Xiao, J. Wang, L. Fan, C. Lan, Y. Wei, Q. Ma, L. Lin, C. Mao, X. Yang, X. Chen, Y. Li, Y. Bai and D. Chen, *Biochem. Biophys. Res. Commun.*, 2013, **437**, 192–198.
71. J. Wang, F. Lu, Q. Ren, H. Sun, Z. Xu, R. Lan, Y. Liu, D. Ward, J. Quan, T. Ye and H. Zhang, *Cancer Res.*, 2011, **71**, 7238–7249.
72. S. Hayami, J. D. Kelly, H. S. Cho, M. Yoshimatsu, M. Unoki, T. Tsunoda, H. I. Field, D. E. Neal, H. Yamaue, B. A. Ponder, Y. Nakamura and R. Hamamoto, *Int. J. Cancer*, 2011, **128**, 574–586.
73. E. C. Kauffman, B. D. Robinson, M. J. Downes, L. G. Powell, M. M. Lee, D. S. Scherr, L. J. Gudas and N. P. Mongan, *Mol. Carcinog.*, 2011, **50**, 931–944.
74. Z. K. Zhao, P. Dong, J. Gu, L. Chen, M. Zhuang, W. J. Lu, D. R. Wang and Y. B. Liu, *Tumour Biol.*, 2013, **34**, 173–180.
75. I. M. Bennani-Baiti, I. Machado, A. Llombart-Bosch and H. Kovar, *Hum. Pathol.*, 2012, **43**, 1300–1307.
76. S. Amente, A. Bertoni, A. Morano, L. Lania, E. V. Avvedimento and B. Majello, *Oncogene*, 2010, **29**, 3691–3702.
77. M. Heidenblad, D. Lindgren, T. Jonson, F. Liedberg, S. Veerla, G. Chebil, S. Gudjonsson, A. Borg, W. Mansson and M. Hoglund, *BMC Med. Genomics*, 2008, **1**, 3.
78. J. He, A. T. Nguyen and Y. Zhang, *Blood*, 2011, **117**, 3869–3880.
79. F. Kottakis, C. Polytarchou, P. Foltopoulou, I. Sanidas, S. C. Kampranis and P. N. Tsichlis, *Mol. Cell*, 2011, **43**, 285–298.
80. D. Frescas, D. Guardavaccaro, S. M. Kuchay, H. Kato, A. Poleshko, V. Basrur, K. S. Elenitoba-Johnson, R. A. Katz and M. Pagano, *Cell Cycle*, 2008, **7**, 3539–3547.

81. R. Pfau, A. Tzatsos, S. C. Kampranis, O. B. Serebrennikova, S. E. Bear and P. N. Tsichlis, *Proc. Natl. Acad. Sci. U. S. A.*, 2008, **105**, 1907–1912.
82. T. Suzuki, K. Minehata, K. Akagi, N. A. Jenkins and N. G. Copeland, *EMBO J.*, 2006, **25**, 3422–3431.
83. Z. Q. Yang, I. Imoto, Y. Fukuda, A. Pimkhaokham, Y. Shimada, M. Imamura, S. Sugano, Y. Nakamura and J. Inazawa, *Cancer Res.*, 2000, **60**, 4735–4739.
84. Z. Q. Yang, I. Imoto, A. Pimkhaokham, Y. Shimada, K. Sasaki, M. Oka and J. Inazawa, *Jpn. J. Cancer Res.*, 2001, **92**, 423–428.
85. G. Liu, A. Bollig-Fischer, B. Kreike, M. J. van de Vijver, J. Abrams, S. P. Ethier and Z. Q. Yang, *Oncogene*, 2009, **28**, 4491–4500.
86. B. Berdel, K. Nieminen, Y. Soini, M. Tengstrom, M. Malinen, V. M. Kosma, J. J. Palvimo and A. Mannermaa, *BMC Cancer*, 2012, **12**, 516.
87. A. Ehrbrecht, U. Muller, M. Wolter, A. Hoischen, A. Koch, B. Radlwimmer, B. Actor, A. Mincheva, T. Pietsch, P. Lichter, G. Reifenberger and R. G. Weber, *J. Pathol.*, 2006, **208**, 554–563.
88. P. A. Northcott, Y. Nakahara, X. Wu, L. Feuk, D. W. Ellison, S. Croul, S. Mack, P. N. Kongkham, J. Peacock, A. Dubuc, Y. S. Ra, K. Zilberberg, J. McLeod, S. W. Scherer, J. Sunil Rao, C. G. Eberhart, W. Grajkowska, Y. Gillespie, B. Lach, R. Grundy, I. F. Pollack, R. L. Hamilton, T. Van Meter, C. G. Carlotti, F. Boop, D. Bigner, R. J. Gilbertson, J. T. Rutka and M. D. Taylor, *Nat. Genet.*, 2009, **41**, 465–472.
89. B. Madsen, B. Spencer-Dene, R. Poulsom, D. Hall, P. J. Lu, K. Scott, A. T. Shaw, J. M. Burchell, P. Freemont and J. Taylor-Papadimitriou, *Mech. Dev.*, 2002, **119**(1), S239–S246.
90. P. J. Lu, K. Sundquist, D. Baeckstrom, R. Poulsom, A. Hanby, S. Meier-Ewert, T. Jones, M. Mitchell, P. Pitha-Rowe, P. Freemont and J. Taylor-Papadimitriou, *J. Biol. Chem.*, 1999, **274**, 15633–15645.
91. A. Izawa, D. Kobayashi, S. Nasu, K. Saito, R. Moriai, K. Asanuma, M. Nakamura, A. Yagihashi and N. Watanabe, *Anticancer Res.*, 2002, **22**, 2965–2969.
92. S. Catchpole, B. Spencer-Dene, D. Hall, S. Santangelo, I. Rosewell, M. Guenatri, R. Beatson, A. G. Scibetta, J. M. Burchell and J. Taylor-Papadimitriou, *Int. J. Oncol.*, 2011, **38**, 1267–1277.
93. A. Barrett, B. Madsen, J. Copier, P. J. Lu, L. Cooper, A. G. Scibetta, J. Burchell and J. Taylor-Papadimitriou, *Int. J. Cancer*, 2002, **101**, 581–588.
94. G. L. Dalgliesh, K. Furge, C. Greenman, L. Chen, G. Bignell, A. Butler, H. Davies, S. Edkins, C. Hardy, C. Latimer, J. Teague, J. Andrews, S. Barthorpe, D. Beare, G. Buck, P. J. Campbell, S. Forbes, M. Jia, D. Jones, H. Knott, C. Y. Kok, K. W. Lau, C. Leroy, M. L. Lin, D. J. McBride, M. Maddison, S. Maguire, K. McLay, A. Menzies, T. Mironenko, L. Mulderrig, L. Mudie, S. O'Meara, E. Pleasance, A. Rajasingham, R. Shepherd, R. Smith, L. Stebbings, P. Stephens, G. Tang, P. S. Tarpey, K. Turrell, K. J. Dykema, S. K. Khoo, D. Petillo, B. Wondergem, J. Anema, R. J. Kahnoski, B. T. Teh, M. R. Stratton and P. A. Futreal, *Nature*, 2010, **463**, 360–363.

95. G. van Haaften, G. L. Dalgliesh, H. Davies, L. Chen, G. Bignell, C. Greenman, S. Edkins, C. Hardy, S. O'Meara, J. Teague, A. Butler, J. Hinton, C. Latimer, J. Andrews, S. Barthorpe, D. Beare, G. Buck, P. J. Campbell, J. Cole, S. Forbes, M. Jia, D. Jones, C. Y. Kok, C. Leroy, M. L. Lin, D. J. McBride, M. Maddison, S. Maquire, K. McLay, A. Menzies, T. Mironenko, L. Mulderrig, L. Mudie, E. Pleasance, R. Shepherd, R. Smith, L. Stebbings, P. Stephens, G. Tang, P. S. Tarpey, R. Turner, K. Turrell, J. Varian, S. West, S. Widaa, P. Wray, V. P. Collins, K. Ichimura, S. Law, J. Wong, S. T. Yuen, S. Y. Leung, G. Tonon, R. A. DePinho, Y. T. Tai, K. C. Anderson, R. J. Kahnoski, A. Massie, S. K. Khoo, B. T. Teh, M. R. Stratton and P. A. Futreal, *Nat. Genet.*, 2009, **41**, 521–523.
96. M. C. Tsai, J. K. Wang and H. Y. Chang, *Cell Cycle*, 2010, **9**, 2043–2044.
97. A. M. Jankowska, H. Makishima, R. V. Tiu, H. Szpurka, Y. Huang, F. Traina, V. Visconte, Y. Sugimoto, C. Prince, C. O'Keefe, E. D. Hsi, A. List, M. A. Sekeres, A. Rao, M. A. McDevitt and J. P. Maciejewski, *Blood*, 2011, **118**, 3932–3941.
98. K. Agger, P. A. Cloos, L. Rudkjaer, K. Williams, G. Andersen, J. Christensen and K. Helin, *Genes Dev.*, 2009, **23**, 1171–1176.
99. M. Barradas, E. Anderton, J. C. Acosta, S. Li, A. Banito, M. Rodriguez-Niedenfuhr, G. Maertens, M. Banck, M. M. Zhou, M. J. Walsh, G. Peters and J. Gil, *Genes Dev.*, 2009, **23**, 1177–1182.
100. M. Tsuneoka, Y. Nishimune, K. Ohta, K. Teye, H. Tanaka, M. Soejima, H. Iida, T. Inokuchi, H. Kimura and Y. Koda, *Int. J. Androl.*, 2006, **29**, 323–330.
101. M. Tsuneoka, H. Fujita, N. Arima, K. Teye, T. Okamura, H. Inutsuka, Y. Koda, K. Shirouzu and H. Kimura, *Clin. Cancer Res.*, 2004, **10**, 7347–7356.
102. K. Teye, M. Tsuneoka, N. Arima, Y. Koda, Y. Nakamura, Y. Ueta, K. Shirouzu and H. Kimura, *Am. J. Pathol.*, 2004, **164**, 205–216.
103. K. Komiya, N. Sueoka-Aragane, A. Sato, T. Hisatomi, T. Sakuragi, M. Mitsuoka, T. Sato, S. Hayashi, H. Izumi, M. Tsuneoka and E. Sueoka, *Lung Cancer*, 2010, **69**, 232–238.
104. C. Suzuki, K. Takahashi, S. Hayama, N. Ishikawa, T. Kato, T. Ito, E. Tsuchiya, Y. Nakamura and Y. Daigo, *Mol. Cancer Ther.*, 2007, **6**, 542–551.
105. M. Bjorkman, P. Ostling, V. Harma, J. Virtanen, J. P. Mpindi, J. Rantala, T. Mirtti, T. Vesterinen, M. Lundin, A. Sankila, A. Rannikko, E. Kaivanto, P. Kohonen, O. Kallioniemi and M. Nees, *Oncogene*, 2012, **31**, 3444–3456.
106. C. Loenarz, W. Ge, M. L. Coleman, N. R. Rose, C. D. Cooper, R. J. Klose, P. J. Ratcliffe and C. J. Schofield, *Hum. Mol. Genet.*, 2010, **19**, 217–222.
107. D. Kleine-Kohlbrecher, J. Christensen, J. Vandamme, I. Abarrategui, M. Bak, N. Tommerup, X. Shi, O. Gozani, J. Rappsilber, A. E. Salcini and K. Helin, *Mol. Cell*, 2010, **38**, 165–178.
108. L. R. Jensen, M. Amende, U. Gurok, B. Moser, V. Gimmel, A. Tzschach, A. R. Janecke, G. Tariverdian, J. Chelly, J. P. Fryns, H. Van Esch, T. Kleefstra, B. Hamel, C. Moraine, J. Gecz, G. Turner, R. Reinhardt, V. M. Kalscheuer, H. H. Ropers and S. Lenzner, *Am. J. Hum. Genet.*, 2005, **76**, 227–236.

109. A. Tzschach, S. Lenzner, B. Moser, R. Reinhardt, J. Chelly, J. P. Fryns, T. Kleefstra, M. Raynaud, G. Turner, H. H. Ropers, A. Kuss and L. R. Jensen, *Hum. Mutat.*, 2006, **27**, 389.
110. L. M. Szewczuk, J. C. Culhane, M. Yang, A. Majumdar, H. Yu and P. A. Cole, *Biochemistry*, 2007, **46**, 6892–6902.
111. P. Stavropoulos and A. Hoelz, *Expert. Opin. Ther. Targets*, 2007, **11**, 809–820.
112. M. Yang, J. C. Culhane, L. M. Szewczuk, C. B. Gocke, C. A. Brautigam, D. R. Tomchick, M. Machius, P. A. Cole and H. Yu, *Nat. Struct. Mol. Biol.*, 2007, **14**, 535–539.
113. M. G. Lee, C. Wynder, D. M. Schmidt, D. G. McCafferty and R. Shiekhattar, *Chem. Biol.*, 2006, **13**, 563–567.
114. D. M. Schmidt and D. G. McCafferty, *Biochemistry*, 2007, **46**, 4408–4416.
115. R. Neelamegam, E. L. Ricq, M. Malvaez, D. Patnaik, S. Norton, S. M. Carlin, I. T. Hill, M. A. Wood, S. J. Haggarty and J. M. Hooker, *ACS Chem. Neurosci.*, 2012, **3**, 120–128.
116. S. Mimasu, N. Umezawa, S. Sato, T. Higuchi, T. Umehara and S. Yokoyama, *Biochemistry*, 2010, **49**, 6494–6503.
117. W. J. Harris, X. Huang, J. T. Lynch, G. J. Spencer, J. R. Hitchin, Y. Li, F. Ciceri, J. G. Blaser, B. F. Greystoke, A. M. Jordan, C. J. Miller, D. J. Ogilvie and T. C. Somervaille, *Cancer Cell*, 2012, **21**, 473–487.
118. Y. Huang, E. Greene, T. Murray Stewart, A. C. Goodwin, S. B. Baylin, P. M. Woster and R. A. Casero Jr, *Proc. Natl. Acad. Sci. U. S. A.*, 2007, **104**, 8023–8028.
119. B. Lohse, J. L. Kristensen, L. H. Kristensen, K. Agger, K. Helin, M. Gajhede and R. P. Clausen, *Bioorg. Med. Chem.*, 2011, **19**, 3625–3636.
120. I. Hoffmann, M. Roatsch, M. L. Schmitt, L. Carlino, M. Pippel, W. Sippl and M. Jung, *Mol. Oncol.*, 2012, **6**, 683–703.
121. Y. Chen, Y. Yang, F. Wang, K. Wan, K. Yamane, Y. Zhang and M. Lei, *Proc. Natl. Acad. Sci. U. S. A.*, 2006, **103**, 13956–13961.
122. M. Yang, C. B. Gocke, X. Luo, D. Borek, D. R. Tomchick, M. Machius, Z. Otwinowski and H. Yu, *Mol. Cell*, 2006, **23**, 377–387.
123. S. Mimasu, T. Sengoku, S. Fukuzawa, T. Umehara and S. Yokoyama, *Biochem. Biophys. Res. Commun.*, 2008, **366**, 15–22.
124. M. Yang, J. C. Culhane, L. M. Szewczuk, P. Jalili, H. L. Ball, M. Machius, P. A. Cole and H. Yu, *Biochemistry*, 2007, **46**, 8058–8065.
125. F. Forneris, C. Binda, A. Adamo, E. Battaglioli and A. Mattevi, *J. Biol. Chem.*, 2007, **282**, 20070–20074.
126. R. Baron, C. Binda, M. Tortorici, J. A. McCammon and A. Mattevi, *Structure*, 2011, **19**, 212–220.
127. S. Strickland, K. K. Smith and K. R. Marotti, *Cell*, 1980, **21**, 347–355.
128. L. Cheng, M. T. Sung, P. Cossu-Rocca, T. D. Jones, G. T. MacLennan, J. De Jong, A. Lopez-Beltran, R. Montironi and L. H. Looijenga, *J. Pathol.*, 2007, **211**, 1–9.
129. A. Adamo, B. Sese, S. Boue, J. Castano, I. Paramonov, M. J. Barrero and J. C. Izpisua Belmonte, *Nat. Cell Biol.*, 2011, **13**, 652–659.

130. T. Schenk, W. C. Chen, S. Gollner, L. Howell, L. Jin, K. Hebestreit, H. U. Klein, A. C. Popescu, A. Burnett, K. Mills, R. A. Casero Jr, L. Marton, P. Woster, M. D. Minden, M. Dugas, J. C. Wang, J. E. Dick, C. Muller-Tidow, K. Petrie and A. Zelent, *Nat. Med.*, 2012, **18**, 605–611.
131. X. Zhang, F. Lu, J. Wang, F. Yin, Z. Xu, D. Qi, X. Wu, Y. Cao, W. Liang, Y. Liu, H. Sun, T. Ye and H. Zhang, *Cell Rep.*, 2013, **5**, 445–457.
132. D. R. Jones, T. M. Daniel, C. E. Denlinger, B. K. Rundall, M. E. Smolkin and M. R. Wick, *Ann. Thorac. Surg.*, 2006, **81**, 1958–1962.
133. A. J. Bass, H. Watanabe, C. H. Mermel, S. Yu, S. Perner, R. G. Verhaak, S. Y. Kim, L. Wardwell, P. Tamayo, I. Gat-Viks, A. H. Ramos, M. S. Woo, B. A. Weir, G. Getz, R. Beroukhim, M. O'Kelly, A. Dutt, O. Rozenblatt-Rosen, P. Dziunycz, J. Komisarof, L. R. Chirieac, C. J. Lafargue, V. Scheble, T. Wilbertz, C. Ma, S. Rao, H. Nakagawa, D. B. Stairs, L. Lin, T. J. Giordano, P. Wagner, J. D. Minna, A. F. Gazdar, C. Q. Zhu, M. S. Brose, I. Cecconello, U. Ribeiro Jr, S. K. Marie, O. Dahl, R. A. Shivdasani, M. S. Tsao, M. A. Rubin, K. K. Wong, A. Regev, W. C. Hahn, D. G. Beer, A. K. Rustgi and M. Meyerson, *Nat. Genet.*, 2009, **41**, 1238–1242.
134. T. Hussenet, S. Dali, J. Exinger, B. Monga, B. Jost, D. Dembele, N. Martinet, C. Thibault, J. Huelsken, E. Brambilla and S. du Manoir, *PLoS One*, 2010, **5**, e8960.
135. S. Maier, T. Wilbertz, M. Braun, V. Scheble, M. Reischl, R. Mikut, R. Menon, P. Nikolov, K. Petersen, C. Beschorner, H. Moch, C. Kakies, C. Protzel, J. Bauer, A. Soltermann, F. Fend, A. Staebler, C. Lengerke and S. Perner, *Hum. Pathol.*, 2011, **42**, 1078–1088.
136. C. M. Rudin, S. Durinck, E. W. Stawiski, J. T. Poirier, Z. Modrusan, D. S. Shames, E. A. Bergbower, Y. Guan, J. Shin, J. Guillory, C. S. Rivers, C. K. Foo, D. Bhatt, J. Stinson, F. Gnad, P. M. Haverty, R. Gentleman, S. Chaudhuri, V. Janakiraman, B. S. Jaiswal, C. Parikh, W. Yuan, Z. Zhang, H. Koeppen, T. D. Wu, H. M. Stern, R. L. Yauch, K. E. Huffman, D. D. Paskulin, P. B. Illei, M. Varella-Garcia, A. F. Gazdar, F. J. de Sauvage, R. Bourgon, J. D. Minna, M. V. Brock and S. Seshagiri, *Nat. Genet.*, 2012, **44**, 1111–1116.
137. M. M. Alonso, R. Diez-Valle, L. Manterola, A. Rubio, D. Liu, N. Cortes-Santiago, L. Urquiza, P. Jauregi, A. Lopez de Munain, N. Sampron, A. Aramburu, S. Tejada-Solis, C. Vicente, M. D. Odero, E. Bandres, J. Garcia-Foncillas, M. A. Idoate, F. F. Lang, J. Fueyo and C. Gomez-Manzano, *PLoS One*, 2011, **6**, e26740.
138. V. Episkopou, *Trends Neurosci.*, 2005, **28**, 219–221.
139. S. Yamanaka, *Cell Stem Cell*, 2007, **1**, 39–49.
140. I. Ben-Porath, M. W. Thomson, V. J. Carey, R. Ge, G. W. Bell, A. Regev and R. A. Weinberg, *Nat. Genet.*, 2008, **40**, 499–507.
141. J. F. Couture, E. Collazo, P. A. Ortiz-Tello, J. S. Brunzelle and R. C. Trievel, *Nat. Struct. Mol. Biol.*, 2007, **14**, 689–695.
142. N. R. Rose, S. S. Ng, J. Mecinovic, B. M. Lienard, S. H. Bello, Z. Sun, M. A. McDonough, U. Oppermann and C. J. Schofield, *J. Med. Chem.*, 2008, **51**, 7053–7056.

143. S. S. Ng, K. L. Kavanagh, M. A. McDonough, D. Butler, E. S. Pilka, B. M. Lienard, J. E. Bray, P. Savitsky, O. Gileadi, F. von Delft, N. R. Rose, J. Offer, J. C. Scheinost, T. Borowski, M. Sundstrom, C. J. Schofield and U. Oppermann, *Nature*, 2007, **448**, 87–91.
144. N. R. Rose, E. C. Woon, G. L. Kingham, O. N. King, J. Mecinovic, I. J. Clifton, S. S. Ng, J. Talib-Hardy, U. Oppermann, M. A. McDonough and C. J. Schofield, *J. Med. Chem.*, 2010, **53**, 1810–1818.
145. S. Hamada, T. Suzuki, K. Mino, K. Koseki, F. Oehme, I. Flamme, H. Ozasa, Y. Itoh, D. Ogasawara, H. Komaarashi, A. Kato, H. Tsumoto, H. Nakagawa, M. Hasegawa, R. Sasaki, T. Mizukami and N. Miyata, *J. Med. Chem.*, 2010, **53**, 5629–5638.
146. N. R. Rose, E. C. Woon, A. Tumber, L. J. Walport, R. Chowdhury, X. S. Li, O. N. King, C. Lejeune, S. S. Ng, T. Krojer, M. C. Chan, A. M. Rydzik, R. J. Hopkinson, K. H. Che, M. Daniel, C. Strain-Damerell, C. Gileadi, G. Kochan, I. K. Leung, J. Dunford, K. K. Yeoh, P. J. Ratcliffe, N. Burgess-Brown, F. von Delft, S. Muller, B. Marsden, P. E. Brennan, M. A. McDonough, U. Oppermann, R. J. Klose, C. J. Schofield and A. Kawamura, *J. Med. Chem.*, 2012, **55**, 6639–6643.
147. W. Xu, H. Yang, Y. Liu, Y. Yang, P. Wang, S. H. Kim, S. Ito, C. Yang, P. Wang, M. T. Xiao, L. X. Liu, W. Q. Jiang, J. Liu, J. Y. Zhang, B. Wang, S. Frye, Y. Zhang, Y. H. Xu, Q. Y. Lei, K. L. Guan, S. M. Zhao and Y. Xiong, *Cancer Cell*, 2011, **19**, 17–30.
148. A. K. Upadhyay, D. Rotili, J. W. Han, R. Hu, Y. Chang, D. Labella, X. Zhang, Y. S. Yoon, A. Mai and X. Cheng, *J. Mol. Biol.*, 2012, **416**, 319–327.
149. L. Kruidenier, C. W. Chung, Z. Cheng, J. Liddle, K. Che, G. Joberty, M. Bantscheff, C. Bountra, A. Bridges, H. Diallo, D. Eberhard, S. Hutchinson, E. Jones, R. Katso, M. Leveridge, P. K. Mander, J. Mosley, C. Ramirez-Molina, P. Rowland, C. J. Schofield, R. J. Sheppard, J. E. Smith, C. Swales, R. Tanner, P. Thomas, A. Tumber, G. Drewes, U. Oppermann, D. J. Patel, K. Lee and D. M. Wilson, *Nature*, 2012, **488**, 404–408.

CHAPTER 8

Progress in Targeting Epigenetic Readers

CHUN-WA CHUNG*[a]

[a]GlaxoSmithKline R&D, Stevenage, UK
*E-mail: chun-wa.h.chung@gsk.com

8.1 Introduction

There is growing appreciation that epigenetic changes play critical roles in the onset and maintenance of many diseases, and that compounds targeting epigenetic mechanisms offer exciting new therapeutic opportunities. Other chapters within this book focus on traditional enzyme targets that catalyse the covalent attachment and removal of post-translational modifications (PTMs) of histones and modifications to DNA. This chapter describes progress towards the discovery of small molecules that target alternative classes of proteins, namely those involved in the protein–protein interactions (PPIs) that lie at the heart of the molecular complexes governing transcriptional regulation.

Here we outline the advances that have been made in, and the future prospects for, inhibition of two classes of reader domains (RDs), acetyl lysine (Kac) and methyl lysine (Kme) readers, that sense the presence or absence of Kac and Kme modifications on histones and other proteins. Upon recruitment to their respective histone marks, these reader module-containing proteins (RCPs) act as scaffolds, attracting additional members into protein complexes and altering the epigenetic architecture at specific locations, resulting in enhanced or repressed gene expression. For some time it was

RSC Drug Discovery Series No. 48
Epigenetics for Drug Discovery
Edited by Nessa Carey

Published by the Royal Society of Chemistry, www.rsc.org

assumed that these PPIs would be impossible to target with small molecules. However, this view has been overturned by the discovery of potent small-molecule inhibitors of bromodomains (BD), the readers of acetylated histones, and more recently by methyl lysine reader (MD) inhibitors of the 'Royal' and WD40 repeat (WD40r) protein families. Success has come from a variety of hit discovery approaches: phenotypic screening, fragment and structure-based drug design, and peptide mimicry. This allows us to survey the merits of these approaches and highlights the value of combining diverse discovery strategies where this is possible. The wealth of emerging and potential biology resulting from targeting RDs is not discussed as it is covered in many recent reviews.[1–4]

8.2 Why Target Epigenetic RDs?

A successful small-molecule drug target must be both chemically tractable and biologically relevant. Published analysis of the 'druggable' genome has consistently suggested that of the ~30 000 expressed human proteins only ~3000 will be susceptible to small-molecule modulation.[5,6] This prediction employed several reasoned assumptions based on historical experience. For example, it assumed that proteins whose native function involves small-molecule binding, *e.g.* enzymes, GPCRs, ion channels, nuclear receptors, are most likely to have well-defined and often enclosed cavities capable of supporting sufficiently high-affinity interactions with a drug-like molecule to yield functional consequences.[7] Furthermore, the expectation is that only a fraction (estimated at ~20–50%) of these 'chemically tractable' targets will possess biological activities that make them attractive for therapeutic intervention. Sequencing the human genome has therefore not opened up a wealth of untapped targets for pharmaceutical exploitation, as was hoped, but rather highlights that the pool of 'traditionally' successful targets, such as enzymes, is more modest in size than previously supposed. Expanding the druggable genome either by finding novel targets for small-molecule modulation[2] or by deploying a greater diversity of drug modalities has therefore received much attention.[8–10]

One mechanism of pharmacological intervention neglected in this initial analysis was the prospect of modulating regulatory pathways by interfering with the PPIs ubiquitous within biology. MetaBAse (Thomson Reuters), a curated database, documents there to be ~40 000 substrate–product reactions, but well over 600 000 PPIs.[11,12] The general avoidance of pursing PPIs is based on the impression that the interface between two proteins tends to be too large and featureless for small molecules to compete effectively with the endogenous interaction. Yet a systematic structural analysis of PPI interfaces suggests a rather diverse situation.[13] Some PPIs are mediated by a multitude of widely distributed weak interactions that can prove challenging for small-molecule competition. For these, alternative approaches have been proposed.[14] These include use of cyclic or stapled peptides;[15–17] small molecules that violate Lipinski's 'rule of five',[18,19] often with more 'three-dimensional' character;[20] compounds that do not attempt to compete orthosterically with the PPI directly but exploit an allosteric site.[21] There is, however, also an increased realisation that affinity may not be uniformly distributed across an entire binding surface, even for some extensive protein–protein interfaces, and PPI antagonism can be realised by targeting small but critical 'hotspots'.[21–24] This appears to hold for some members of the two classes of lysine PTM reader proteins discussed here. In these cases it has been possible to identify key 'hotspots' that serve as anchors for small-molecule binding, then to grow out from these to gain affinity and selectivity. From a drug design perspective, a particularly attractive feature of these protein families is that these hotspots align with conserved structural features within the families, thereby enabling an efficient target-class approach to drug discovery.[25] In a similar fashion to targeting the ATP-binding hinge site for kinases, an acknowledged hotspot, targeting the Kac recognition site of bromodomains and Kme aromatic cage of methyl readers may yield similar success in finding chemical starting points for optimisation. A common observation amongst small-molecule PPI antagonists is that they often do not precisely mimic all the interactions made by their natural counterparts. Instead, their more diverse chemical scaffolds allow them to exploit a range of interactions besides the key subset known to be critical for affinity. The result is a variety of effective PPI inhibitors that bind the same target in distinct ways and are being pursued for a range of indications.[26]

The chemically diverse, potent small-molecule inhibitors of the BET (bromodomain and extraterminal domain) family of bromodomains are an example of this. These exhibit excellent selectivity profiles within the bromodomain family, and good pharmacokinetic properties with the potential to deliver profound *in vivo* pharmacology in complex diseases such as sepsis and cancer. Nanomolar BD inhibitors of BAZ2B and CREBBP have also been created using structure-based drug design and target-class approaches, demonstrating PPI chemical tractability exists outside the BET bromodomain subfamily. Progress has extended to inhibitors of methyl lysine readers, with the discovery of selective Kme reader inhibitors of L3MBTL3,[27] as well as of L3MBTL1,[28] WDR5[29,30] and JARID1A PHD.[31] This demonstrates a level of druggability amongst reader proteins that rivals many epigenetic enzyme classes.

8.3 Multivalency and Context

Isolated single histone modifications are rare in nature.[32] Mass spectrometry analysis has shown that multiple PTMs are often present on the same or adjacent histones. Histone PTMs and DNA methylation together provide highly specific recognition motifs to ensure that environmental stimuli trigger the correct downstream response pathways. The concerted action of multiple RDs is employed to decipher these motifs. Here, nature uses the power of a combinatorial approach, bringing together in a variety of ways a small number of discrete reader subunits, each of which senses a specific mark. Individual RDs may bind their cognate marks relatively weakly as single entities, but when covalently concatenated together within the same protein, the resulting multivalency of interaction confers both enhanced specificity and affinity. The same effect can be created when domains are brought together as part of different proteins within the megadalton protein–protein complexes often found on chromatin, rather than by formal covalent linkage.

Nature's elegant solution to providing fidelity in the interpretation of PTMs provides an additional challenge for small-molecule intervention, because successful targeting of epigenetic reader proteins requires the disruption of multivalent protein–protein complexes. The ease with which a RD inhibitor can disrupt the RD/histone interaction and the functional consequences of this antagonism is therefore highly situation dependent and difficult to predict given the network of interdependent associations that exist. A RD inhibitor therefore has the capacity to interfere with a variety of recruitment activities all linked to transcriptional control and do so in a context- or state-selective manner.

There is a tacit understanding that kinetic and binding parameters derived from purified proteins, especially when these are truncated or mutated, may for numerous reasons not be identical to those in a physiological context. Despite this, the use of purified protein assays serves an essential fit-for-purpose role to generate molecular SAR that guides compound optimisation. The use of isolated excised RDs can fulfil a similar role. However, it is important to remember that these small domains (often only ~100 amino acids) are designed to operate within large multi-domain proteins and complexes, so *in vitro* results may not straightforwardly translate to physiological observations.[33] It is therefore vital that appropriate thought is given to the choice of truncate or complex used within *in vitro* assays, and that extension beyond an individual reader domain may deliver more useful guidance to inhibitor discovery in some instances. This has been exemplified for the methyl reader L3MBTL3 family, where the use of a protein consisting of the first three MBT domains rather than just one was critical to the development of the nanomolar UNC1215 molecule.[27] Similarly, the tetrameric multi-protein core MLL1 complex was necessary to understand the ability of WDR5/MLL1 antagonists to modulate the demethylase activity of MLL1.[29,30]

8.4 Bromodomains – Druggable Acetyl Lysine Readers

8.4.1 What are Bromodomains?

Protein lysine acetylation (Kac) is an abundant and important PTM found to occur physiologically on over 1750 proteins.[34,35] Histone acetylation is most commonly associated with gene activation. Bromodomain-containing proteins (BCPs) recruited to sites of lysine acetylation function as scaffolds for the assembly of macromolecular complexes that alter chromatin accessibility and allow the recruitment of transcription factors or activation of RNA polymerases.

Bromodomains (BDs) are believed to be the main 'readers' of the histone acetylation mark.[36] They are small (~110 residue) evolutionary and structurally conserved modules present in a large number of chromatin-associated proteins and in nearly all nuclear histone acetyltransferases. Bromodomains are named after the *Drosophila* gene *brahma* (Hindu God of creation) when they were first identified, and soon afterwards homologues were located in humans and yeast. BDs are most commonly found as single copies within a polypeptide chain, frequently in combination with other epigenetic RDs such as PHD, PWWP and Tudor.[32] In instances where multiple BDs occur within a single protein, they are often contiguous (*e.g.* the BET family of proteins,[37] which contain two bromodomains each, or polybromodomain-1,[38] which contains six). However, BDs within the same protein may exhibit no greater similarity to each other than to BDs within other proteins. This suggests the multiple BD repeats are not the product of gene duplication and that the individual domains have distinct roles.[39,40] The human genome encodes at least 56 bromodomains within 42 different proteins as the phylogenetic tree of Figure 8.1 shows.

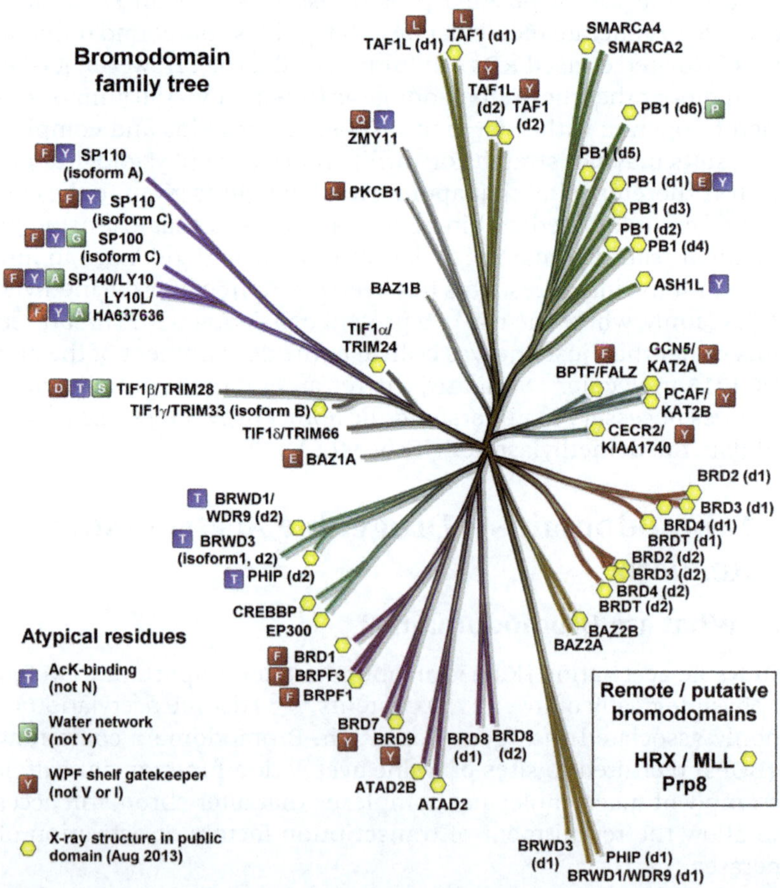

Figure 8.1 Bromodomain phylogenetic family tree.

Forty-three of the 56 BDs possess a conserved tyrosine and asparagine that form a key part of an acetyl lysine binding motif, as highlighted in the sequence alignment (Figure 8.2). The remaining BDs have atypical recognition signatures, whose functional consequences are not well understood. In BDs where the 'conserved' asparagine is replaced by other amino acids such as threonine or tyrosine residues (*e.g.* PB1-BD1), acetyl lysine binding capability is sometimes retained,[41] although whether this is the physiologically relevant recognition partner remains to be confirmed.

The first three-dimensional structure of a bromodomain, that of PCAF (p300/CREBBP-associated factor), was determined by NMR in 1999.[42] Since then three-dimensional structures of well over half of the bromodomain family have been determined by X-ray crystallography and/or NMR, revealing that they share a common fold consisting of four antiparallel α-helices (αZ–αA–αB–αC) arranged in a left-handed twist (Figure 8.3). The acetyl lysine

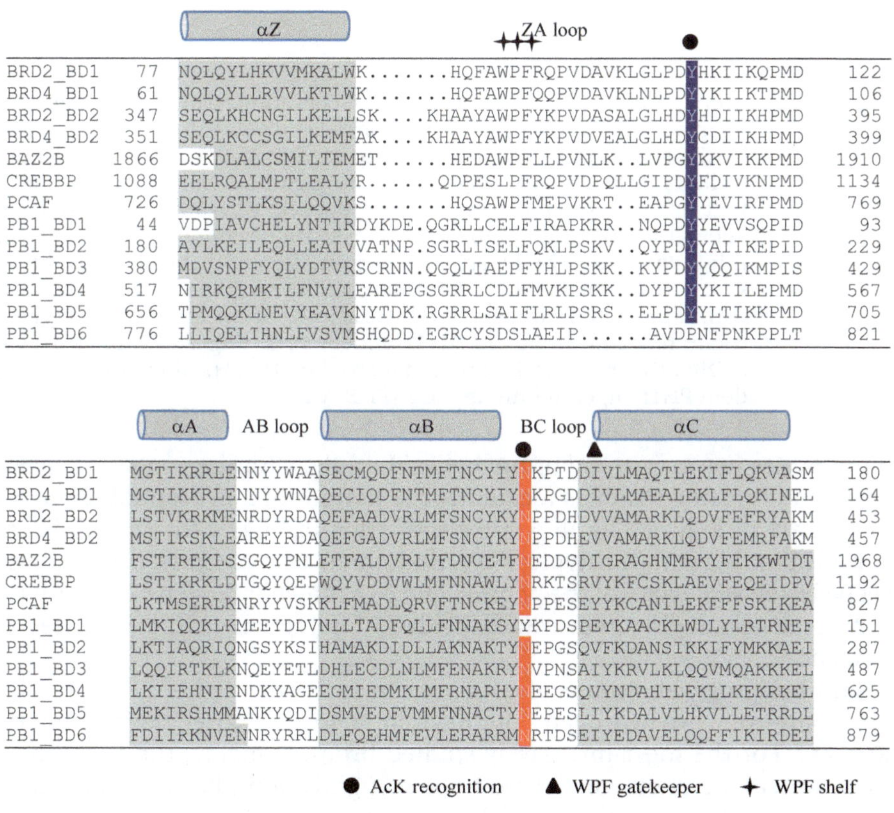

Figure 8.2 Sequence alignment of 13 bromodomains. The secondary structural elements (grey) and conserved recognition tyrosine (blue) and asparagine (red) are highlighted. The atypical bromodomains of the first and sixth bromodomains of human polybromodomain-1 are shown (PB1-BD1/6). The first has a tyrosine in place of the asparagine and the sixth replaces the conserved tyrosine with a proline residue.

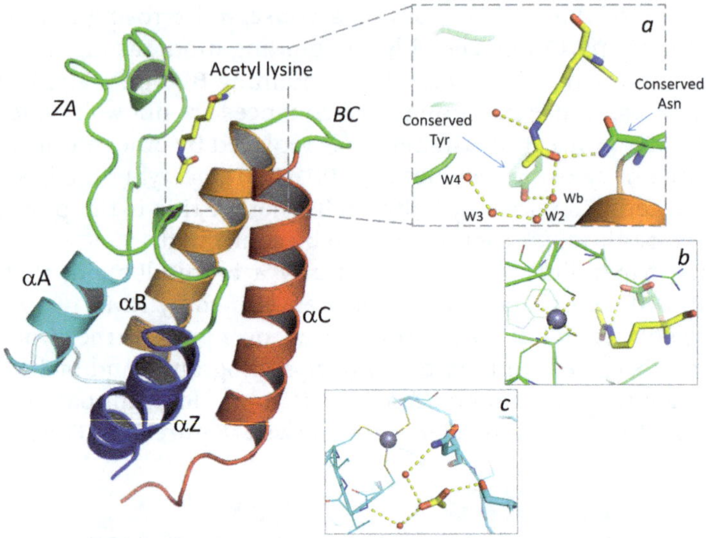

Figure 8.3 Bromodomain structure and AcK recognition. The archetypal four-helix bundle topology of bromodomains (αZ–αA–αB–αC) is illustrated by the structure of BRD4-BD1 complexed to the H4K5acK8ac peptide (yellow) (PDB: 3UVW). Highlighted in green are the ZA and BC loops that flank the acetyl lysine binding pocket. Variability in these loops allows the AcK residues to be recognised in the context of differing peptide sequences. The insert on the right shows (a) the conserved tyrosine and asparagine in stick format and the hydrogen bonding network within the AcK binding site to the carbonyl of the acetyl; (b) acetylated lysine recognition in the NMR tandem PHD finger of human DPF3b (PDB: 2KWJ); (c) binding of acetate to the first PHD domain of the tandem PHD finger of human MOZ (PDB: V43).

binding site lies at one end of the helical bundle, with the conserved asparagine and tyrosine residues located at the bottom of a hydrophobic pocket. The insert in Figure 8.3a illustrates the structural basis for acetyl-histone peptide binding within a 'typical' bromodomain. The carbonyl oxygen of the acetyl lysine side chain of the peptide makes a direct hydrogen bond with the conserved asparagine and a water-bridged interaction with the conserved tyrosine.

This bridging water molecule (referred to as Wb throughout) is part of an unusually extensive hydrogen-bonding water network buried deep within the Kac site. For the most intensively studied bromodomains, the BET family, these waters are often considered to be so tightly bound as to be effectively part of the protein structure. For others, this degree of structural conservation remains to be seen. The methyl group of the Kac sits in a small hydrophobic pocket. This pocket can accommodate larger substitutions, for example a propyl moiety. The biological relevance of this tolerance is unclear, but histones containing both propionylated and butyrylated lysine residues have been found in yeast and mammals.[43] The charge of an unmodified lysine would sit uncomfortably within this largely hydrophobic environment, explaining the preference for a neutralising modification. The often weak affinities of BDs for Kac peptides suggest that BDs could be considered as negative filters that disfavour binding to a charged native lysine peptide as much as positive readers that promote interactions with their acetylated form. This may suggest that the full spectrum of native modifications 'recognised' by BDs has not yet been fully discovered.

Despite similar mechanisms of Kac recognition, subtle differences within this pocket amongst BD family members may be exploited to give differential binding of small molecules. However, a more obvious area to derive small-molecule selectivity is the region immediately surrounding the Kac pocket. The long ZA and shorter BC loops that flank the Kac recognition site are the most variable regions between BD family members (Figure 8.2), and allow individual BDs to discriminate between Kac within the context of different peptide sequences.

Recently it has also been proposed that BDs may not be the sole readers of the acetyl lysine mark. A study of the tandem PHD finger of human DPF3b, which functions in association with the BAF remodelling complex to initiate gene transcription during heart and muscle development, was reported to bind H3 and H4 in an acetylation-dependent manner, although the differential affinity for the acetylated peptides was modest (~4-fold).[44] NMR and associated studies suggested a distinct mode of acetyl lysine recognition utilising a signature Asp-Phe motif (Figure 8.3b). An analogous Kac recognition site has also been proposed for the first PHD domain of the MOZ tandem PHD motif.[45] In this instance the binding of an acetate ion within a high-resolution crystal structure of MOZ (Figure 8.3c), as well as NMR chemical shift experiments were used to support this interaction model. The tractability of these new KAc-binding proteins to small-molecule inhibitor intervention is unexplored, but their relatively open binding sites may suggest a lower tractability than bromodomains, similar to that of other PHD domains (see Section 8.5).

8.4.2 Phenotypic Discovery of BET Bromodomain Inhibitors

Phenotypic screening[46] has been shown to be a powerful way of finding pharmacologically innovative compounds, which recent analysis has shown to lead to more approved first-in-class chemical entities (NCEs) than target-based approaches.[47] This is true both historically and since the change to target-focused drug discovery in the 1990s. However, subsequent identification of the molecular target and mechanism of action have been recognised as being very important in many instances. This knowledge facilitates a hypothesis-driven understanding of how first-in-class compounds may be advanced to best-in-class follower molecules with an improved safety profile or longer duration of action. It is notable that a number of epigenetic targets and inhibitors were initially discovered by phenotypic screening. For example, vorinostat (SAHA) was found from a screen measuring the ability of compounds to induce cell differentiation. This yielded a series of bishydroxamic acids, whose molecular mode of action (MMOA) was later discovered to be through HDAC inhibition (Figure 8.4).[48–50]

Therefore it may be of little surprise that the first potent bromodomain inhibitors disclosed, JQ1[51] (**1**) and I-BET762/819[52] (**2–3**) were derived from chemical starting points found by cellular assays focused on functional/phenotypic read-outs. These compounds specifically target the BET subfamily of tandem bromodomains comprising the ubiquitously expressed BRD2–4 and testis-specific BRDT proteins.[37] The GSK clinical candidate I-BET762, and other series, were derived from activators found in an Apo-A1 luciferase reporter assay. The molecular target for these compounds was deconvoluted using a combined chemoproteomic, biophysical and structural approach to reveal the compounds to be pan-BET bromodomain binders that antagonised acetyl lysine histone binding.[53] Equally serendipitously, anti-inflammatory screening by Mitsubishi Tanabe identified thienodiazepines (**4**) as potent inhibitors that were found to have BRD4 activity and anti-proliferative effects,[54,55] leading to the synthesis of JQ1. Most recently, RVX-208 (**5**), a clinical candidate for atherosclerosis, acute coronary syndrome and Alzheimer's disease, was also revealed to bind to the BET bromodomains.[56,57]

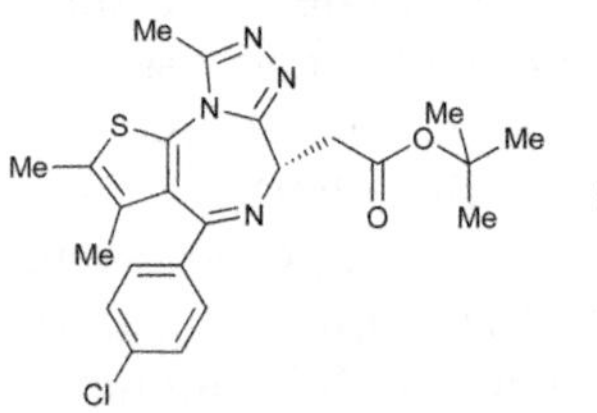
(**1**) (+)-JQ1

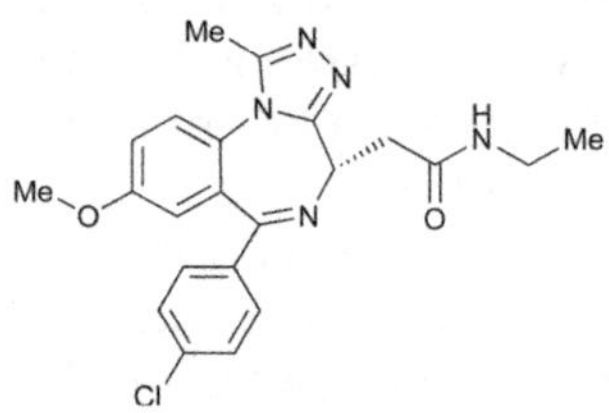
(**2**) I-BET762

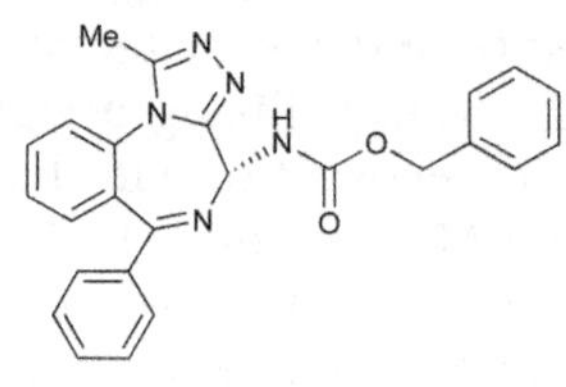
(**3**) I-BET819

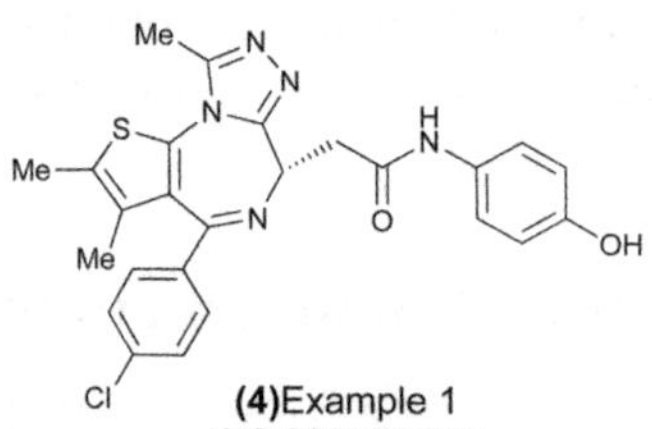
(**4**)Example 1
WO 2009/84693

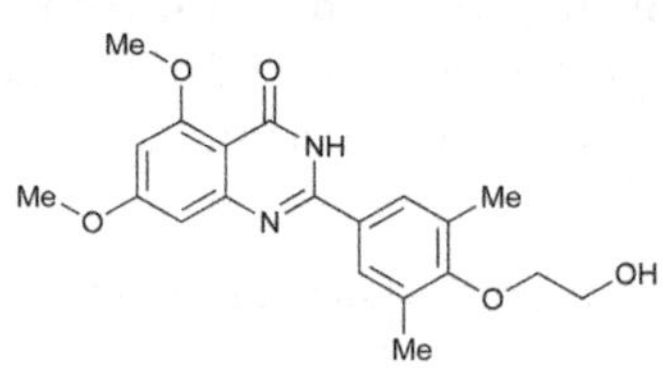
(**5**) RVX-208

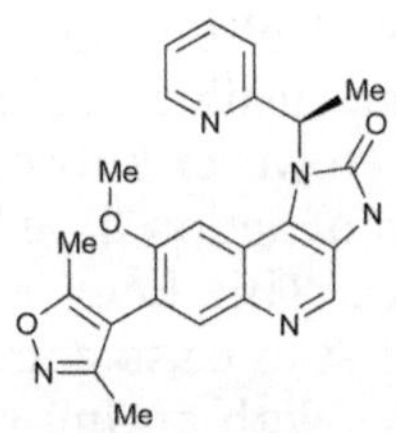
(**6**) I-BET151

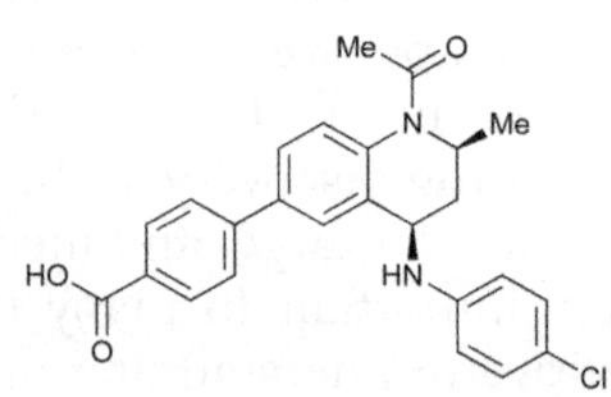
(**7**) I-BET726

Figure 8.4 BET bromodomain inhibitors.

This compound was progressed as a regulator of ApoA1 gene transcription without a MMOA before emerging BET literature prompted the testing of this compound for BET activity. The potential of phenotypic cellular assays to find epigenetic reader inhibitors in cases where the expected biology for inhibition is well understood is readily apparent. Unfortunately this degree of knowledge exists for only a very small minority of reader proteins.

Where this knowledge does exist, an appropriately configured assay in relevant cells can provide a direct read-out of the desired functional response that avoids target or mechanistic bias. This allows a greater breadth of access to targets within a regulatory pathway, both known and unknown, and an immediate indication of cellular activity. Compounds active in phenotypic screens may translate more effectively into therapeutic agents as they are often found and optimised based on their modulation of key cellular markers associated with disease. Potential drawbacks of this approach are that a compound with unknown MMOAs can limit the ability to predict a safety profile, and sometimes the ability to optimise the initial compounds found can be hindered, as structure–activity relationships from on- and off-target activities, cellular permeability and toxicity are compressed into a single read-out. Despite these caveats, a significant proportion of marketed drugs have been discovered using an appropriately configured 'black box' format. Unravelling the specific target(s) of a phenotypic compound can be difficult, especially if the compound has a low target affinity or is relatively non-specific. Two main chemoproteomic approaches have been successfully applied to achieve this goal: one based on compound affinity chromatography coupled with mass spectrometry[58] and the other based on stable isotope labelling by amino acids in cell culture (SILAC).[59] The discovery and subsequent molecular characterisation of the I-BET compounds serves as an illustrative case study of the considerations and techniques employed to successfully determine the MMOA of a phenotypic output.[53]

In the early 2000s GlaxoSmithKline began to investigate Apo-A1 up-regulation as a target for dyslipidaemia.[53] Whilst it was known that Apo-A1 up-regulation was associated with protection from atherosclerosis progression and with anti-inflammatory effects,[60] no mechanism by which up-regulation could be achieved with a small molecule was known.[61] Therefore, a stable human HepG2 cell line containing an Apo-A1 luciferase reporter was used to screen for activators without target bias. A variety of chemical templates was found which showed specific induction of the Apo-A1 reporter gene but had no effect on other similarly configured reporter systems. Compounds derived from these initial hits include benzodiazepine (BZD) compounds such as I-BET762 (**2**),[62] isoxazoles (IOXs) such as I-BET151 (**6**)[63,64] and tetrahydroquinolines (THQs) such as I-BET726 (**7**).[65,66]

The use of a cellular assay and the unknown MMOA were not a barrier in optimising several series to the point where they had good potency and reasonable pharmacokinetic properties. Despite the multiple chemotypes that were explored and extensive profiling amongst panels of known targets including GPCRs, ion channels, kinases, nuclear receptors, proteases, *etc.*, no molecular target(s) became apparent. A chemoproteomics approach, affinity chromatography followed by mass spectroscopy, was therefore pursued.

The known SAR within the BZD series allowed the design of a pair of active and inactive compounds with identical physical–chemical properties that could be linked to an affinity matrix. This formed an ideal platform to detect proteins from cell lysates that were preferentially retained on the active matrix over the inactive matrix. After elution, interacting proteins were identified by mass spectrometry. In multiple relevant cell types, the BET family of proteins consistently discriminated between the active and inactive matrices. Further insights into specificity were demonstrated by setting up a competition experiment. Adding active compounds to lysates prior to their application onto the active matrix prevented the retention of the BET proteins. This was true for active compounds from not only the BZD series but also from others, such as the isoxazoles, suggesting that the chemically diverse series of Apo-A1 up-regulators all bound to the BET proteins, possibly at a common site. Reassuringly, no competition was observed when inactive analogues from several series were used. To further localise a putative site of binding, pull-down experiments using transiently expressed N and C truncates of BRD2 identified the N-terminal tandem bromodomain region of these proteins as the targets for compound interaction.

A number of purified tandem and single bromodomain truncates of BRD2/3/4 were then produced to validate these observations in a series of biophysical experiments. Circular dichroism (CD) was first used to confirm the structural integrity of the expressed proteins, as stable and helical. Differential thermal shift analysis visualised by CD determined that the active compounds from a number of series enhanced the thermal stability of BRD2/3/4 bromodomains. Inactive members of the same series induced no stabilisation. This gave qualitative verification that the differential associations seen with endogenous proteins within cell lysates could be reproduced with these isolated proteins. To obtain a quantitative measure of binding affinity, the BET proteins were immobilised onto a surface plasmon resonance (SPR) chip over which compound solutions were injected. Analysis of these data suggested that compounds bound with submicromolar affinities. There was also an indication that the behaviour of the isolated N- and C-terminal bromodomains (BD1 and BD2, respectively) may be subtly different to those observed in the tandem domain constructs. To characterise the binding fully, isothermal titration calorimetry (ITC) was used. The K_d affinities of I-BET819 to the tandem domain constructs of BRD2/3/4 were calculated to be very similar: 52 nM, 19 nM and 24 nM, respectively. The thermodynamic profile was similar for all three proteins: the binding was found to be strongly enthalpically driven with a small unfavourable entropic component. Surprisingly, as the two domains are not identical, the compound was found to bind with a stoichiometry of two ligands per tandem bromodomain, or one compound to each bromodomain BD1 and BD2 of these proteins. The ability of compounds to bind to both BD1 and BD2 was confirmed using isolated BD1 and BD2 of BRD2 and BRD4. This binding of the molecules to both BDs of the BET proteins, whilst not interacting with BDs of other proteins, is a very fortunate selectivity pattern.

Elucidation of the binding characteristics of the I-BET compounds was completed using X-ray crystallography. These compounds bound at the Kac site of the bromodomains and are direct antagonists of the BD/Kac-histone PPI. The molecular details of the interactions are discussed in the structure-based discovery section of this chapter (8.4.3) and fully rationalise the SAR established from the Apo-A1 cellular activity. Interestingly, despite a wealth of SAR generated amongst diverse chemotypes, it was not apparent from these data that the native ligand contained an acetyl lysine moiety and knowledge of a wealth of inhibitor compounds did not help to identify the target, or even target class.

A MMOA can be established only when a causal link can be made between engagement of the putative molecular target and a biological response. Brd4 siRNA knockdown experiments showed a dose-dependent Apo-A1 up-regulation in HepG2 cells, thereby establishing a link between BRD4 knockdown and induction of Apo-A1. In addition, an excellent quantitative correlation was established between the Apo-A1 functional response and the *in vitro* binding affinities of compounds across multiple series to BRD2/3/4 in a binding assay utilising a fluorescently-tagged analogue of I-BET819.

In summary, the ability of cellular phenotypic assays to find BD inhibitors, and techniques to deconvolute their MMOA, has been comprehensively demonstrated by the discovery of chemical starting points that led to I-BET and JQ1 compounds. Phenotypic screening is a valuable way of revealing tractable targets and novel modes of disease intervention without preconceived bias. Once the MMOA has been found, target-focused assays coupled with molecular insights from computational and crystallographic studies can be an efficient mechanism for further drug discovery efforts. The remarkable ability of BET bromodomain inhibitors to perturb a vast array of biological systems[67] may reflect their pharmacological efficacy[68] and their mode of discovery.

8.4.3 Structure-Based Discovery of Bromodomain Inhibitors

Structure-based drug design (SBDD) or fragment-based drug design (FBDD) approaches have been particularly successful for protein systems where there is good structural understanding of the essential elements for recognition, affinity and selectivity. The kinase family is one example of such a system,[69] where ATP-site binding compounds normally contain motifs that bind to the 'hinge' hotspot. Selectivity is derived by accessing features beyond the purine pocket such as the back-pocket and DFG-out conformations. The strong conservation of a KAc-binding signature within the BD protein family suggests that Kac mimetics may serve as an analogous hotspot for BD inhibitors. Productive interactions outside of this conserved pocket have been best established for the BET bromodomains. The following section covers SBDD approaches to bromodomain inhibitor discovery in chronological order, starting with the pioneering work that first demonstrated the interaction of small molecules with the BDs of PCAF and CREBBP (Figure 8.5).

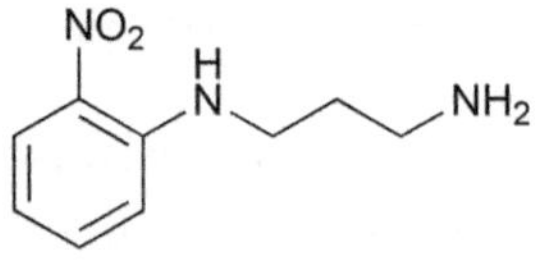

(8) PCAF inhibitor
IC50 5.1mM

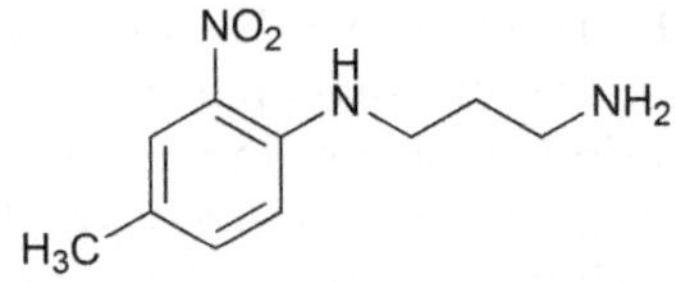

(9) NP1 PCAF inhibitor
IC50 1.6mM

Figure 8.5 PCAF bromodomain inhibitors.

8.4.3.1 NMR Screening of PCAF and CREBBP

Zhou and co-workers were the pioneers of BD inhibitor discovery. Following their solution by NMR of the structure of PCAF,[42] they elucidated an essential transcriptional role of this BD in the recognition of K50ac of HIV-1 Tat that suggests this may be a novel target for anti-HIV therapy.[70] To investigate this idea further, NMR screening was used to identify small molecules that bound to PCAF preferentially in the peptide binding groove rather than within the KAc-binding pocket, rationalising that such compounds may have greater selectivity over other BDs.[71] Several active molecules were found, including an *N*1-aryl-propane-1,3-diamine (**8**), which had comparable affinity to the Tat peptide. The relative affinities of this compound and analogues were determined by ELISA using an immobilised biotinylated Tat-K50ac peptide and GST-PCAF. The most active compound (**9**), a 4-methyl analogue compound **8**, had a modest IC_{50} of 1.6 μM. Further studies undertaken to optimise this series did not yield a significant increase in affinity.[72] However, members of this class of *N*-aryl-diamine compounds were shown to be inactive against the BDs of CREBBP and TIF1β, demonstrating that selectivity had been achieved by these small molecules.

Building on this success, Zhou *et al.*[73] applied a slightly modified NMR screening methodology to look for acetyl lysine mimetic inhibitors of CREBBP that compete for its interaction with p53-K382ac. Computational methods were used to select a focused set of 200 compounds that consisted of one aromatic ring connected by various linkers to a $-N(R)COCH_3$ group. Nitrogen-15 HSQC screening yielded an impressive hit rate of 7%. Despite the set selection targeting the conserved Kac site, the CREBBP hits exhibited some selectivity over the bromodomain of PCAF, reiterating the potential to differentiate between bromodomains even for compounds binding within the Kac recognition pocket. This high confirmed hit rate also indicated the amenability of bromodomains to a structure and fragment-based approach, although again, in this example only modest micromolar inhibitors could be developed. Examples of these compounds (**10–13**) are shown in Figure 8.6. The most potent of these, MS7972 (**12**) and MS2126 (**13**), were able to almost completely block the interaction between CREBBP BRD and the p53-Kac382 peptide at 50 μM and 100 μM in an ELISA assay and an equilibrium binding constant of 19.6 μM of MS7972 (**13**) was determined by using tryptophan fluorescence consistent with the ELISA observation.

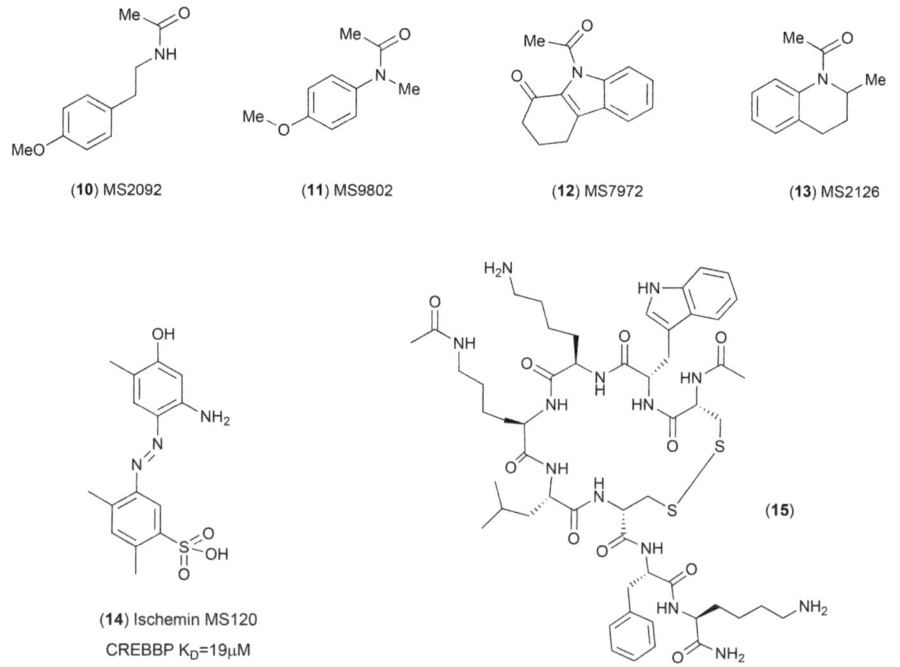

Figure 8.6 CREBBP bromodomain inhibitors.

This group has also published the discovery of azobenzene compounds, such as MS120 (**14**) with affinity in the tens to hundreds of micromolar range for CREBBP,[74] and cyclic peptide mimetics of p53 (**15**) of moderate micromolar affinity for CREBBP.[75]

Whilst it was encouraging that these molecules of relatively affinity could be found, this left unanswered the question of whether it would be possible to obtain high-affinity drug-like BD antagonists with therapeutic potential. The discovery of BET inhibitors such as JQ1 (**1**) and I-BET762 transformed the perception of bromodomains and epigenetic readers in general as druggable targets. These molecules proved that drug-like compounds were able to deliver profound and distinct pharmacology effects *in vitro* and *in vivo* in multiple disease models. Since their disclosure it has become apparent that these PPI targets may also be amenable to structure-guided methods of drug discovery such SBDD and FBDD. Structural understanding of the SAR around the BET inhibitors, exemplified in the next section by the clinical candidate I-BET762 (**2**), has driven the development of novel compounds for these and other bromodomains.

8.4.3.2 BET Bromodomain Structure–Activity Relationship

Many of the potent and selective BET ligands, such as JQ1 (**1**) and I-BET762 (**2**), exploit three key interaction regions within the BET bromodomains. These are the KAc-binding pocket, the WPF shelf and the ZA channel (Figure 8.7).

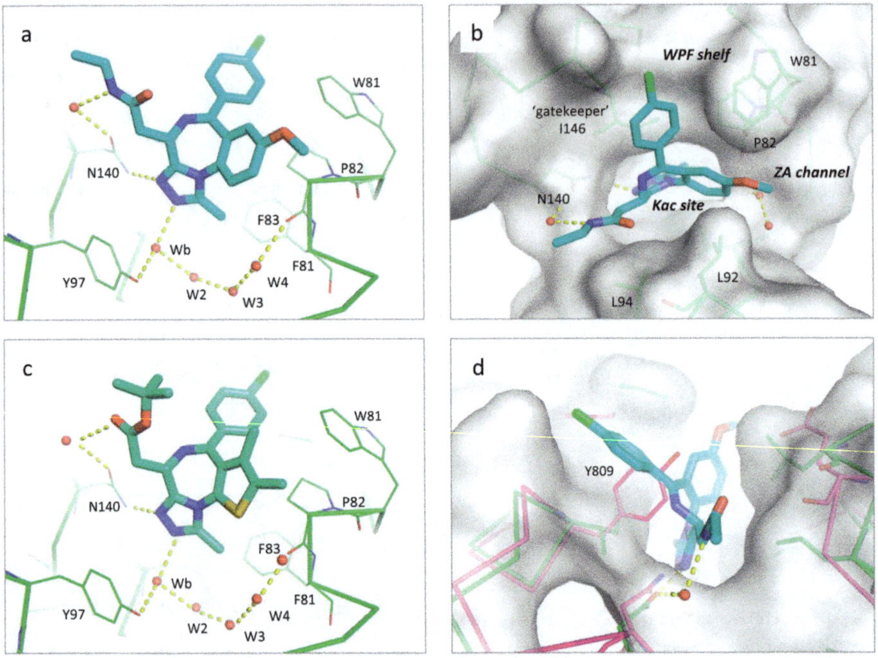

Figure 8.7 (a) I-BET726 binding in BRD4-BD1 (PDB: 3P5O). (b) I-BET726 binding in BRD4-BD1 with protein surface depicted. (c) JQ1 binding in BRD4-BD1 (PDB: 3MXF). (d) Overlay of (b) with PCAF protein (PDB: 3GGF3). The larger Y809 gatekeeper of PCAF labelled.

JQ1 (**1**) and I-BET762 (**2**) are both based upon a triazolo-diazepine (TZD) scaffold that mimics the Kac interactions observed in BET/peptide complexes. The methyl group of the heterocycle binds into the small hydrophobic pocket that recognises the methyl group of the acetyl head group of the acetyl lysine. This is formed in large part by the side chain of Phe83 in BRD4-BD1, which is perfectly conserved in the other BET bromodomains (Figure 8.7a/c). The two hydrogen bond acceptor nitrogen atoms of the 1,2,4-triazolyl ring together mimic the carboxyl group of the acetyl lysine head group. One ring nitrogen (N1) accepts a hydrogen bond from the water molecule that bridges to the hydroxyl group of the side chain of the conserved tyrosine residue (Tyr97 in BRD4-BD1). The second nitrogen (N2) accepts a hydrogen bond from the side chain NH_2 group of the conserved asparagine (Asn140 in BRD4-BD1). The ability of a triazolyl ring to mimic the acetyl group in this way was unexpected but important, as it suggested diverse chemotypes can act as antagonists of the bromodomain/Kac peptide interaction. The fused phenyl ring of I-BET762 and thiophene of JQ1 extend beyond the region occupied by the histone peptide, binding into a lipophilic pocket termed the ZA channel (Figure 8.7b), sandwiched between side chains of the long ZA loop (notably Pro82 and Leu92 in BRD4-BD1). The curvature of the *R* enantiomer of the diazepine template provides good shape complementarity to the protein, explaining the high potency of these compounds and the preference for this enantiomer. The pendant chlorophenyl group interacts with another hydrophobic region of the BC loop, named the WPF shelf, because it includes the conserved WPF motif present in all of the BET family bromodomains (Figure 8.1). In structures of BRD4-BD1 and BRDT-BD1 bound to diacetylated histone peptides,[40,41] whilst the first Kac mark lies within the established Kac pocket, the second Kac sits across the WPF shelf region, with the Cg and Ce atoms of the lysine side chain overlaying closely with the positions of the pendant chlorophenyl ring of the TZD compounds. This reinforces the hypothesis that occupation of the WPF shelf and surrounding regions of the site are important for both binding and governing selectivity.

BET family specificity is achieved by making distinct contacts outside the acetyl lysine cavity with the BC and ZA loops, which differ significantly between BCPs in both composition and length (Figure 8.1). Differences in these loop regions between the BET family domains are relatively minor so the TZD compounds bind unselectively to all of them. For example, a 'gatekeeper' residue (Ile146 in the N-terminal bromodomain of BRD4) is located at the entrance to the WPF shelf. The size of this residue governs access to this shelf, but in the BET bromodomains it is either isoleucine or valine (Figure 8.1). This is not the case for all bromodomains. In the PCAF bromodomain (Figure 8.1) the gatekeeper, a much larger tyrosine residue, would be expected to hinder access to the WPF shelf and thus clash with the binding of the pendant chlorophenyl of the TZD compounds (Figures 8.2 and 8.7d). Similar arguments can be used to explain the high degree of selectivity of the TZDs for other bromodomains.

The 1,4-benzodiazepine and thiazolodiazepine skeletons of the TZDs are among medicinal chemistry's most widely used scaffolds. First discovered in the 1960s, there are many marketed pharmaceutical agents that contain this motif principally as GABA receptor agonists for CNS clinical indications, such as the anxiolytic alprazolam and etizolam. Differential SAR between the BET bromodomain and GABA allows selectivity between these proteins to be achieved. For example the marketed drugs are invariably achiral, as the GABA receptor shows little enantiomeric preference. Analogues with substituents at the *meta* or *para* positions of the pendant chlorophenyl ring are well tolerated in BET but lack activity at the GABA receptors.

8.4.3.3 Exploring the BET Key Interaction Regions in Turn

8.4.3.3.1 Focusing on the Kac Pocket. All reported BET bromodomain inhibitors reported occupy the Kac pocket, emphasising its importance as an anchor for subsequent interactions. Several fragment-based approaches, supported by crystallography, encapsulate the hierarchy of importance of the interactions that can be made within this site. The first study used a small fragment set of a few thousand compounds chosen to contain the two features required for faithful Kac mimicry: a hydrogen-bonding functionality and a small alkyl substituent.[76] The analysis of 40 crystal structures of BRD2-BD1 with diverse fragments from this effort gives insights into preferences of these two aspects. If fragments are not able to make hydrogen-bonding interactions with both the conserved asparagine and the conserved bridging water Wb then invariably the latter is found to be maintained, suggesting interactions with Wb are more critical. Even when both hydrogen-bonding interactions are found, the fragment is almost always closer to the bridging water than to the asparagine, again implying that the H-bond interaction with this water dominates that with the asparagine, but that both influence ligand positioning. This was explicitly exemplified by the crystal structures of BRD2-BD1 with fragments **16–20** shown in Figure 8.8.

This finding was not predicted but appears consistent with observations made since, both with more elaborated molecules and additional fragment efforts. For example, the sedative midazolam, a diazepine template that contains an imidazole rather than a triazole ring, is able to weakly bind to BRD4-BD1.[77] It is unable to form a hydrogen bond to the conserved Asn residue, but its ability to do so with the bridging Wb water is adequate to yield sufficient affinity for generation of a well-resolved X-ray co-crystal structure (PDB: 3U5K). Although it has been remarked that this structure lacks two of the 'conserved' four water molecules at the bottom of the Kac site, re-examination of the electron density map suggests that these molecules are probably present, albeit at lower occupancy than normally observed. Three (**23–25**) out of four fragments (**22–25**) reported by Zhao *et al.*,[78] found by high concentration X-ray co-crystallisation of compounds with BRD4-BD1, confirm that both interactions with the Wb and Asn156 are preferred, and that the hydrogen bond to Wb tends to be shorter and stronger. The fourth

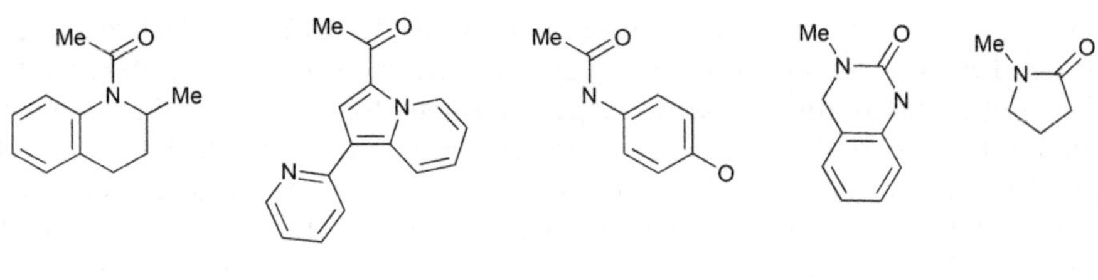

(16) **(17)** **(18)** **(19)** **(20)**

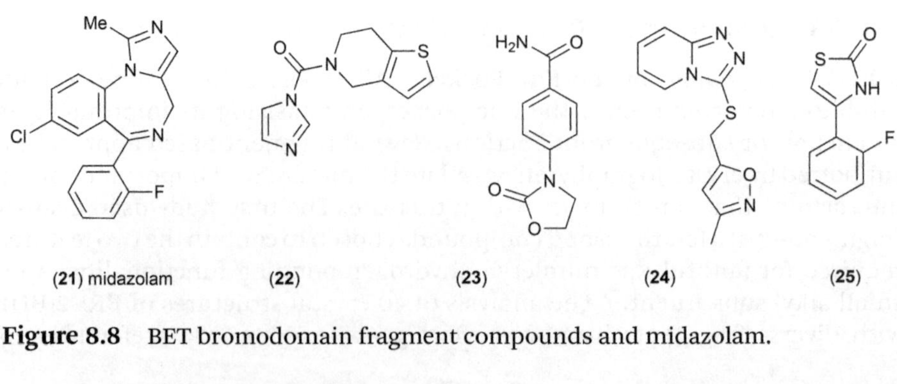

(21) midazolam **(22)** **(23)** **(24)** **(25)**

Figure 8.8 BET bromodomain fragment compounds and midazolam.

fragment revealed (**22**) was the most difficult to position within the electron density seen in the active site. The authors' favoured placement put the thiophene ring deeper in the Kac site than is normally seen. In this mode the bridging water, Wb, is displaced by the sulphur atom of the thiophene ring. Additionally, since this compound makes no interactions with either Tyr113 or Asn156, perhaps unsurprisingly, this fragment was essentially inactive when tested at 100 μM in a binding assay, as were fragments **23** and **24**. Only the thiazolidinone (**25**), in which the placement of atoms most resembles the native Kac, gave significant inhibition at 100 μM (~43%), illustrating that whilst many chemotypes can be tolerated within the Kac pocket and are seen crystallographically, not all make effective interactions.

One heterocycle that does bind efficiently and has been identified by several groups is the methyl or dimethyl isoxazole.[79–81] Its interactions very closely resemble those made by the methyltriazoles of the TZD compounds, and this prompted one group to elaborate this fragment to create a hybrid isoxazole azepine scaffold (**28**).[81]

8.4.3.3.2 Outside the Kac Pocket. Guided by SAR around the TZD series the hybrid isoxazole azepine and other fragment starting points have been optimised by exploiting the WPF shelf and/or ZA regions known to deliver affinity and selectivity in the BET proteins. Figure 8.9 summarises some of these efforts, highlighting representative starting fragments and optimised compounds, where submicromolar inhibitors have been delivered.[78–82] All of these start from Kac warheads and then extend in a variety of ways out onto the WPF shelf and/or ZA channel. The diversity of chemotypes and optimisation strategies provides us with knowledge of these peripheral pockets.

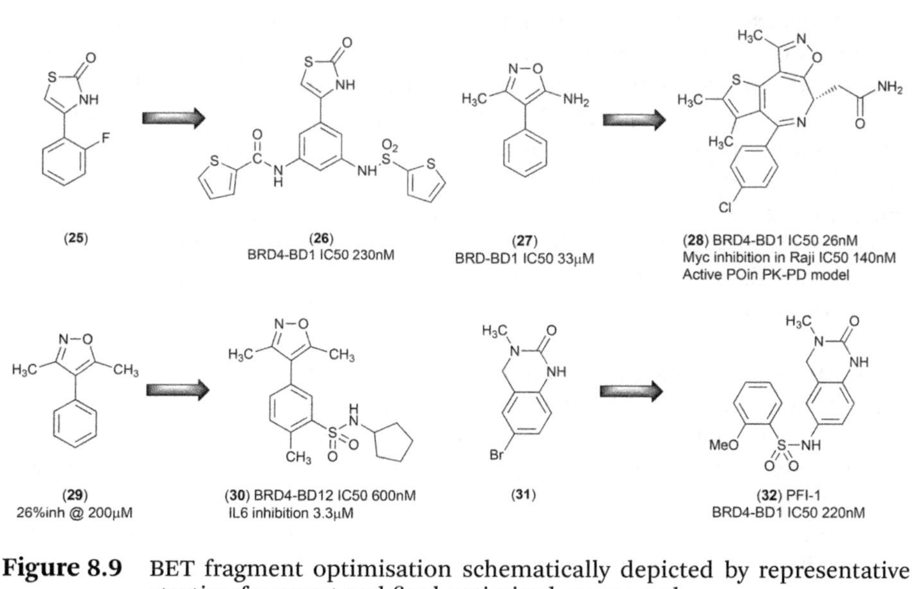

Figure 8.9 BET fragment optimisation schematically depicted by representative starting fragment and final optimised compound.

The WPF shelf is predominantly lipophilic, and can tolerate saturated and non-saturated hydrophobic groups. Occupation of this shelf can enhance affinity, but the optimal substituent depends on how this substituent is projected onto the shelf and therefore on the nature of the linker and point of linkage from the Kac core template. Additionally, where the WPF shelf group accepts a ring system, whether the ring sits flat on the shelf or at an alternative angle can give rise to differences in preferred ring size that is also trajectory dependent. For example, Bamborough *et al.*[80] explored a variety of substituents directed onto the WPF shelf from the phenylsulphonamide isoxazoles, and found a saturated cyclopentyl group to be optimal (**30**) and superior to aromatic phenyl rings. In contrast, Gehling *et al.*,[81] working on the hybrid isoxazole azepine template (**28**), found that in addition to the favoured chlorophenyl of the TZD, other aromatic rings, such as the 2-pyridine or aminopyridine were also tolerated on the WPF shelf, but replacement with a saturated or polar saturated system such as a propyl or tetrahydropyran moiety was not well tolerated. As with the TZD compounds, *ortho*-substitution of the pendant aromatic ring was detrimental, as this induces a rotation in the inter-ring torsion angle, leading to a decrease in the surface area of the pendant phenyl ring in contact with the hydrophobic surface of Ile146 on the WPF shelf. Where the core Kac template lacks the ability to gradually 'curve' onto the WPF shelf, as with the TZDs, a pronounced kink must be induced to access this lipophilic region. This has been achieved with a sulphonamide or reverse sulphonamide as exemplified by the isoxazoles (**30**) and 3,4-dihydro-3-methyl-2(1*H*)-quinazolines (**32**).[82] It is notable that the vector from such sulphonamide linkages from the bicyclic quinazoline core differs from that of the IOX and TZD compounds. This alters the substitution preferences so *ortho*-substituted phenyl rings are not only tolerated but yield enhanced affinity, as exemplified by the chemical probe PFI-1 (**32**) (Figure 8.10).

PFI-1 does not occupy the ZA channel, but this region was explored by Zhao *et al.* during their optimisation of the thiazolidinones.[78] Using a sulphonamide linker (**33**) and (**34**) were made, intending to access the WPF shelf. Both compounds were found to have 4.1 μM affinity in a fluorescence anisotropy assay, but surprisingly the crystal structure of **33** revealed that the longer linker placed the aromatic ring in the ZA channel and not on the WPF shelf as intended. In this binding mode the sulphonamide oxygens of **34** point towards the protein to make interactions with often conserved waters that bridge to the backbone carbonyl groups of P82 and Q85 and the side chain amide of Q85 in BRD4-BD1. This projects the pendant benzyl group of **33** through the ZA channel and beyond. Combining elements that could simultaneously access the WPF and ZA channel gave submicromolar compounds, the most potent of which is **26** (230 nM).

The relatively open nature of the Kac pocket allows solvent-exposed groups to be attached at several positions in these molecules to modulate physicochemical and pharmacokinetic properties, such as the placement of a solubilising morpholino group.[80] It has also been exploited to add chemical diversity to pre-existing core structures (**35**).[83] Two other templates for

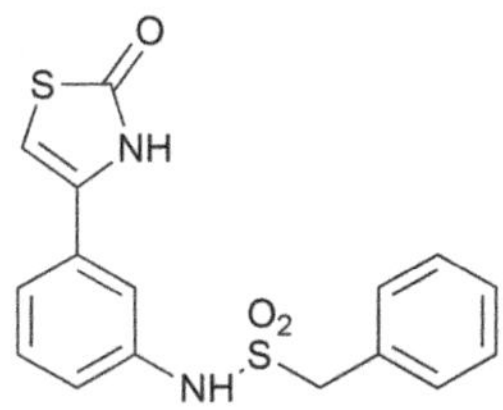

(**33**)

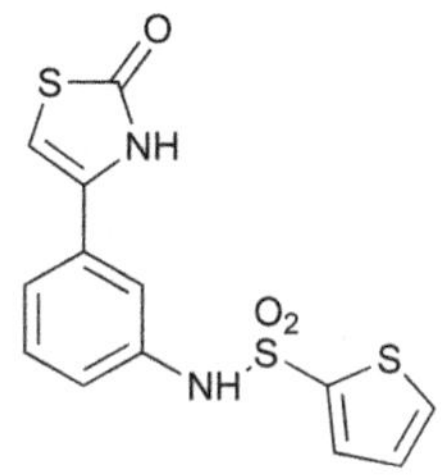

(**34**)

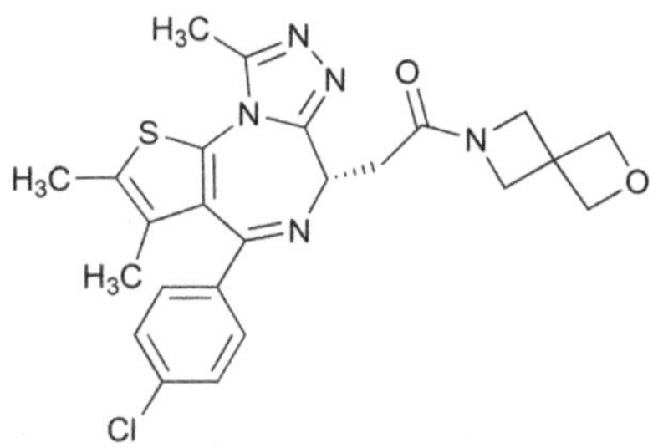

(**35**)

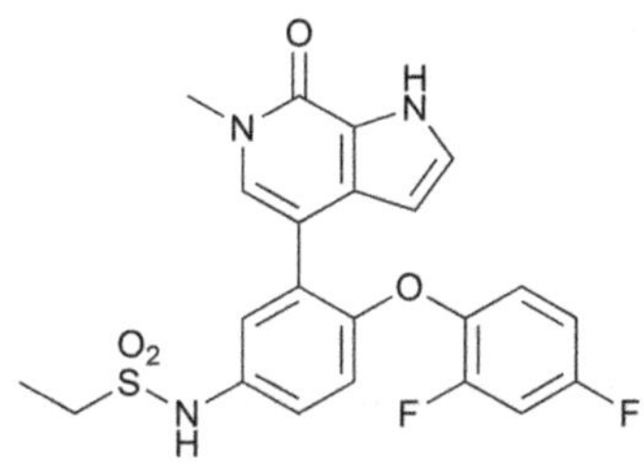

(**36**)

Example 36 (W021013097601 A1)
BRD4-BD1 (Ki 3.5nM)
BRD4_BD2 (Ki 1.4nM)
MX-1 profileration EC50 17.4nM

Figure 8.10 BET bromodomain compounds.

which nanomolar *in vivo* active BET compounds have been reported but not discussed in detail are the GSK tetrahydroquinolines (*e.g.* I-BET726[62] (**7**)) and the Abbvie 6-methyl-1*H*-pyrrolo[2,3-*c*]pyridine-7(6*H*)-one series (**36**).[84] Details of their discovery are yet to be published, but it is likely that, in both instances, the halogen substituted aromatic rings sit on the WPF shelf and that the benzoic acid and reverse ethyl sulphonamide, respectively, lie within the ZA channel.

In summary, extraordinary progress has been made towards the design, synthesis and biological characterisation of potent and selective BET bromodomain ligands. A diverse range of chemotypes have emerged as capable of targeting the Kac pocket of these proteins. There are heterocyclic acetyl lysine mimics such as the triazolobenzodiazepines, dimethylisoxazoles and pyridones, and there are templates that use the acetyl group itself attached to a number of scaffolds such as the tetrahydroquinolines. All the reported ligands thus far are essentially pan-BET compounds that show similar activities against both domains of all the BRD2/BRD3/BRD4/BRDT isoforms. A key challenge that remains is the development of potent and isoform/domain-selective BET inhibitors. Given the high sequence homology between these protein domains, the lack of selectivity is unsurprising and based on the primary sequence, domain selectivity may be easier to obtain than isoform selectivity. Indeed, a hint of this lies in the observation that some compounds do display a degree of differential BD1/BD2 activity, with 10-fold differences accessible in many templates. Whether this gap can be increased to give true domain-selective inhibitors remains to be seen. The pharmacology of pan, domain and isoform selective inhibitors will also be of great interest.

8.4.3.4 Prospects for Bromodomain Inhibitors Beyond BET

There is considerable interest in the use of experimental[85,86] and computational methods[87,88] to assess the chemical tractability of putative targets for small-molecule drug discovery as it is estimated that 60% of projects fail to identify tractable starting points of appropriate quality.[89] Numerous *in silico* methods have been proposed, with their major attraction being the ability to evaluate a large portfolio of proteins with relatively little resource. The accuracy with which this estimated tractability is predictive is likely to depend on a number of factors, including the available information associated with the target.

The 3D structure of a protein provides a rich source of information that improves these predictions as it reveals the nature, accessibility and size of likely interaction sites. SiteMap[88] is one *in silico* method that makes use of crystal structures. Its evaluation of the ligandability of 24 bromodomains outside the BET family alongside the BET proteins showed high variability across the family.[90] Of the three classifications used (difficult, intermediate and druggable), approximately half of those studied (including EP300 and BAZ2B) were classed as difficult. The rest were almost equally divided amongst the druggable (*e.g.* PCAF) and intermediate (*e.g.* CREBBP) categories.

Although research into bromodomain inhibitors within and outside the BET family is intense, few structures of potent, selective non-BET inhibitors have been disclosed. The two examples below (GSK2801,[91] I-CBP112[92]) have both been released as part of the SGC (Structural Genomics Consortium) epigenetic chemical probe initiative in partnership with GSK and demonstrate the value of structure-based design for this target class.

GSK2801 is an inhibitor of the BAZ2 subfamily of bromodomains (BAZ2A/B), deemed a 'difficult' target by SiteMap.[88] These BCPs are characterised by a carboxy-terminal bromodomain adjacent to a PHD finger and a WACZ motif. In addition, four other conserved motifs are found in the N-terminus of BAZ family members, namely the LH motif (a leucine-rich helical domain), the ZB2 motif and the BAZ1 and BAZ2 motifs. Less is known about their biological function. It has been suggested that BAZ2B regulates nucleosome mobilisation by the ATP-dependent chromatin remodelling factor ISWI mediated by the BAZ1 motif.[93] The related BAZ2A is thought to be essential for heterochromatin formation leading to transcriptional silencing of certain rRNAs[94] and plays a central role in the NoRC (nucleolar remodelling complex).

A BAZ2B high-throughput screen was run at the NCGC (National Institutes of Health Chemical Genomics Center, http://www.ncgc.nih.gov/) on behalf of the SGC, using an AlphaScreen peptide displacement format to find bromodomain chemical tools to further explore the biology of this protein. As yet no potent confirmed hits have emerged from this effort.

An alternative chemical starting point was the indolizine compound **17**, which was originally discovered as a BET inhibitor.[76] During cross-screening against other bromodomains, the group at GSK found it to have comparable affinity against BAZ2B, and carried out an optimisation to improve the selectivity over the BET proteins. This was accomplished by exploiting the shorter ZA loop of BAZ2B that gives rise to a wider ZA channel in the region of the benzoic acid of 7 (Figure 8.2). Placing a large *ortho*-substituent onto this ring induces a less planar inter-ring torsional preference of approximately 80°, which is well accommodated in BAZ2B but which would clash with the Leu94 residue located on the ZA loop of BRD4-BD1. This and other factors contributed to the chemical probe GSK2801, a 110 nM compound with 500-fold selectivity *versus* BRD4 (http://www.thesgc.org/scientists/chemical_probes/GSK2801). The full details of the optimisation of this probe will be published in due course.

Focused screening of Kac pharmacophore targeted libraries and cross-screening from BET compounds have delivered chemical probes and starting points from a number of distinct templates for another bromodomain-containing protein, CREBBP, a target classed as 'intermediate' tractability. The most potent of these compounds, I-CBP112 (**38**) and SGC-CBP30, are dual CREBBP and EP300 inhibitors with similar activity against these phylogenetically adjacent bromodomains (Figures 8.1 and 8.11). The structure of SGC-CBP30 has not been released at the time of writing. Analogous to other potent BD compounds, I-CBP112 exploits regions outside the Kac pocket for selectivity and affinity. The WPF motif of the BET family has the sequence LPF in CREBBP and EP300, where the leucine residue can be observed to sit in conformations that leave one face of the proline solvent exposed. As a result the LPF shelf of CREBBP is considerably more spacious than the WPF shelf of BET, allowing larger groups to fit on this expanded lipophilic area, such as the dimethyoxyphenyl of I-CBP112.

In summary, the evidence so far demonstrates that finding BD inhibitors outside of the BET family is achievable, whether *in silico* predictions deem

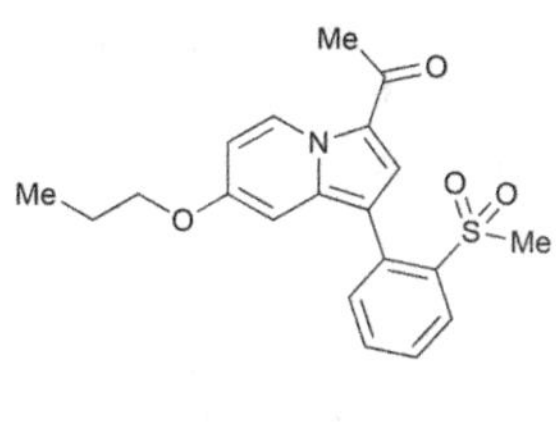

(**37**) GSK2801

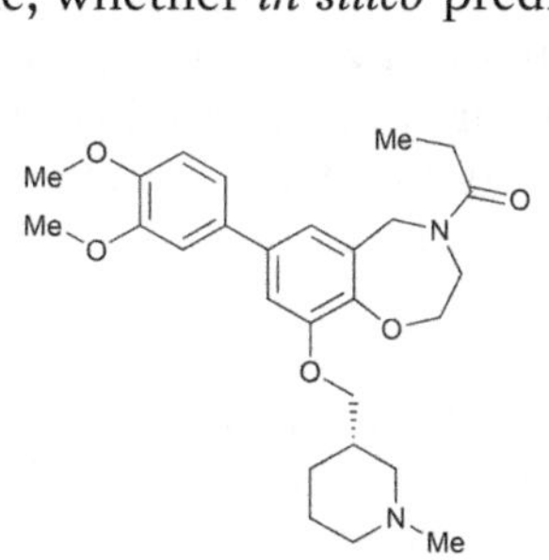

(**38**) I-CBP112

Figure 8.11 BAZ2B chemical probe (GSK2801) and CREBBP bromodomain chemical probe (I-CBP112).

the bromodomains to be difficult, intermediate or druggable. However, as the optimisation and SAR around these non-BET inhibitors have not been published, it is not possible to say whether *in silico* methods have correctly forecast the ease with which success was obtained.

8.5 Progress in Methyl Lysine Reader Inhibitors

8.5.1 Histone Methylation and Methyl Reader Domains

Histone methylation is a very common and versatile PTM found on basic residues such as lysine and arginine.[95] These amino acids can exist in multiple methylation states: mono- (me1), di- (me2), tri- (me3) and even unmethylated. Arginine dimethylation can also occur in a symmetric and asymmetric manner. This array of methyl marks is 'read' by >200 methyl reader domains thought to be contained in the human genome.[96,97] At least four families of reader domains recognise methylated lysine: the zinc finger plant homeodomains (PHDs), the WD40 repeat proteins (WD40rs), the ankyrin repeats and the large Royal superfamily consisting of the Tudor, Agenet, chromo, Pro-Trp-Trp-Pro (PWWP) and malignant brain tumour (MBT) domains.[25,97]

Unlike acetylation, methylation produces no change in the overall charge of the modified histone residues and all methylated forms of lysine and arginine are anticipated to be positively charged at physiological pH. Incremental addition of methyl groups does increase the hydrophobicity and cation radius of the lysine methylammonium and arginine guanindino groups, and decreases their hydrogen bond donor potential. Whilst methyl lysine readers can adopt a number of protein folds, convergent evolution has developed a common mechanism for the recognition of methylated lysine states, maybe highlighting the difficulty of designing protein structural motifs able to both accommodate and discriminate the unusual mix of formal positive charge and hydrophobicity within these PTMs. Higher methylated states (Kme2–3) are invariably recognised by aromatic cages usually formed by two to four aromatic residues (typically Tyr, Trp and Phe), which interact with the methylammonium ions by cation–π interactions (Figure 8.12 left).[25] One or more acidic residues may be present in the binding pocket, and these help to differentiate between methylation states by interacting with any non-methylated H-bond donors on the cation. To facilitate binding of these larger cations, the histone peptides often bind in a rather shallow solvent-exposed site, participating in a so-called 'surface recognition' mode (Figure 8.12 middle). Lower methylation states can also make use of this aromatic cage, with an increased number of supplementary acid residues to accommodate the greater potential H-bonding valency of the cation. Additionally, the more compact nature of lower methylation states allows their recognition within more enclosed pockets that sterically exclude higher methylated states, where the aromatic cage may lie at the bottom of this pocket. This mode of recognition has been termed 'cavity insertion' (Figure 8.12 right).

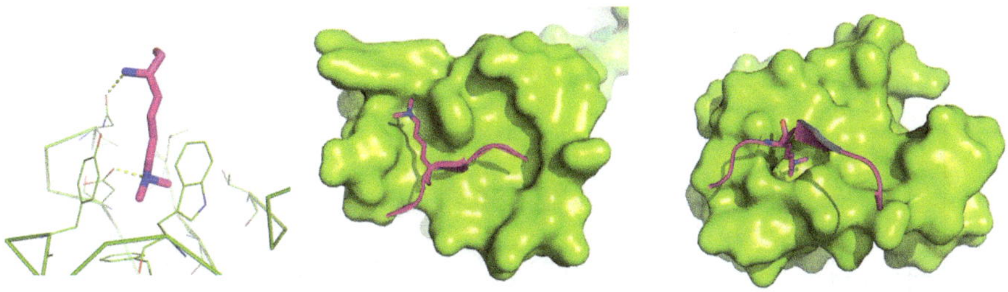

Figure 8.12 Methyl reader recognition motifs. (Left) Aromatic cage second MBT domain of L3MBTL1 (lime green) with histone H4K20me2 (12–30) (magenta) binding into the (PDB: 2RJF). (Middle) Binding of H3(1–15) K4Me3 (magenta) to PHD finger of BPTF (lime green) (PDB: 2F6J). (Right) L3MBTL1–H4K20me2 as in (Left) with solvent accessible surface of the protein shown (PDB: 2RJF).

In a similar study to that discussed for bromodomains, Santiago *et al.*[98] used computational algorithms to predict the chemical tractability of 38 non-redundant methyl readers for which there are X-ray structures. The diversity of protein families and methyl reader pockets gave a highly variable picture, but some trends were seen. Folds containing deep pockets, for example MBT and PWWP proteins, were more likely to be tractable, whereas those with rather open shallow sites, such as PHD domains, were not. For methyl readers of the WDR and Chromo families the situation was slightly more complicated. Tractability was found, but located at sites other than the methylamino acid pocket. For WDR, this was the proximal non-methylated arginine pocket, and for CBX chromoproteins it was the groove occupied by the backbone of the histone tail.

These predictions seem consistent with subjective judgment, but a limitation is the focus on rigid snapshots of structures of isolated domains and fragments of full-length proteins. Neglecting the importance of the context of a domain and of physiological complexes has consequences, both positive and negative, for inhibitor discovery, which are highlighted in the following case histories.

8.5.2 Inhibitors of MBT Domains

MBT domains are motifs of ~100 AA that are found in multiple repeats within proteins. To date, nine proteins containing a total of 27 different MBT domains have been identified (Figure 8.13). Individual MBTs recognise lower methylation states, Kme1 or 2, within peptides *via* a cavity-insertion mode where the methyl lysine side chain is fully surrounded by a deep and narrow cavity. This aspect makes them relatively attractive for small-molecule inhibitor discovery in comparison to the shallow solvent exposure Kme pockets seen within some other Kme readers. Many of the MBT proteins have been

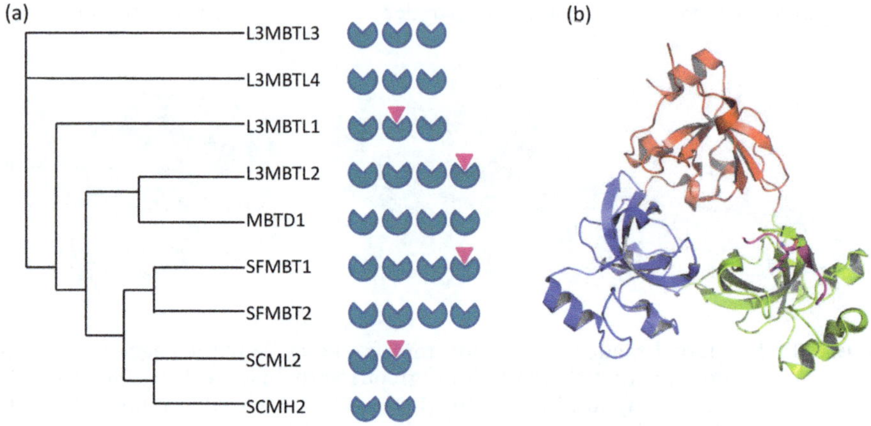

Figure 8.13 (a) Schematic illustration of number of MBT within each member of the MBT family of proteins; the pink triangle highlights MBT domains demonstrated to bind methyl lysine peptides. (b) Crystal structure of the first three MBT domains (blue, yellow and red, respectively) of L3MBTL1 with histone H4K20me2 (12–30) shown in magenta binding into the second MBT domain (PDB: 2RJF).

crystallised to reveal that sequential MBTs can form compact structures such as the tri-propeller of L3MBTL1 shown in Figure 8.13. Despite the potential for multiple MBT repeats within each MBT to interact in a polyvalent manner with partner proteins, there is no evidence to suggest this is relevant to histone peptide recognition.[99,100]

In contrast to this wealth of structural knowledge, the biological function of these domains is poorly understood. The best-studied member of this family, L3MBTL1, is known to act as a 'chromatin lock'[101] to repress expression of E2F-regulated genes such as the growth-related and oncogenic gene c-myc. It has also been shown to bind to the tumour suppressor p53 *via* methylation-dependent recognition of K382.[102] Depletion of L3MBTL1 has been linked to DNA breaks,[103] DNA replication and genomic instability.[104] The second MBT domain of L3MBTL1 has been shown to bind methyl lysine, but the role of the other MBT domains remains a mystery, although it has been speculated that they may be involved in recognition of multiple nucleosomes or residues other than methyl lysine.[102]

Chemical probes to each of the MBT domains would greatly enhance the ability to dissect the function of these proteins. Attempting a target-class approach, the Frye group at UNC has employed a number of screening methodologies against a panel of MBTs (and other Kme reader domains). Initially a virtual screening strategy was employed with the aim of identifying non-peptidic, cell penetrant probes for MBT-containing proteins.[105] Using a database of commercially available compounds, 2D and 3D *in silico* methods were used to select a set of 51 compounds. These were experimentally tested in AlphaScreen assays[106] *versus* truncates of L3MBTL1/3/4 and MBTD1. Nineteen compounds from this set were found to give dose-dependent effects within the tens of micromolar range. These included direct Kme1/2 analogues such as **39–41** as well as examples where the methylated amine is replaced by the pyrrolidine (**42**).

Taking an orthogonal approach to explore alternative methyl lysine mimetics, the group also explored a peptidomimetic avenue. A small array of compounds based on a 'tripeptide' motif were made and tested in an AlphaScreen assay. The first two amino acids of each peptide were a histidine, intended to interact with Thr385 on the protein surface, followed by a phenylalanine in place of arginine for ease of purification. The third 'amino acid' was varied, from the different states of methylated lysine (Kme0–3) to their methylethylamine (**43**) and pyrrolidine (**44**) equivalents, as shown in Figure 8.14.

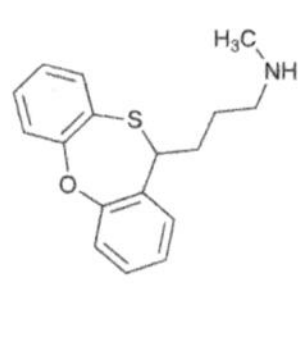

(39)
L3MBTL1 14μM
L3MBTL3 N.D.
L3MBTL4 > 100μM
MBTD1 >100μM

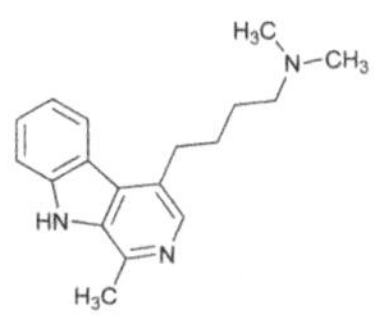

(40)
L3MBTL1 21μM
L3MBTL3 5.7μM
L3MBTL4 > 100μM
MBTD1 73μM

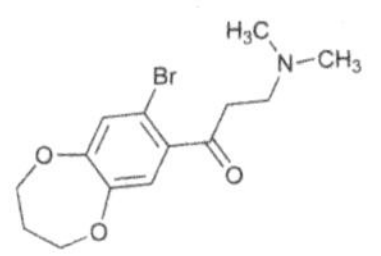

(41)
L3MBTL1 >100μM
L3MBTL3 >100μM
L3MBTL4 12μM
MBTD1 > 100μM

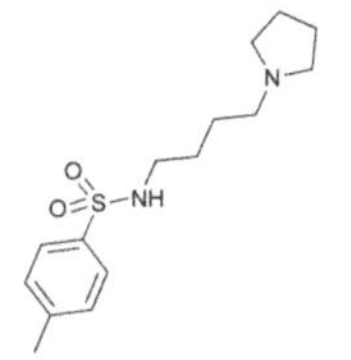

(42)
L3MBTL1 17μM
L3MBTL3 57μM
L3MBTL4 >100μM
MBTD1 59μM

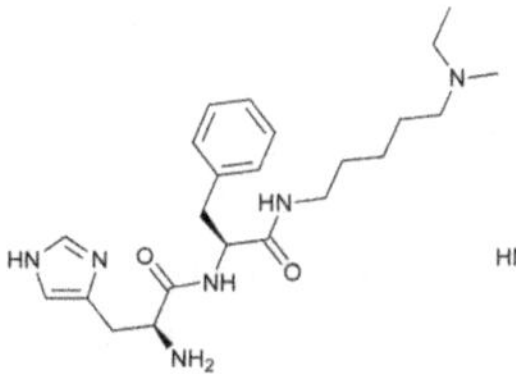

(43)
L3MBTL1 78μM

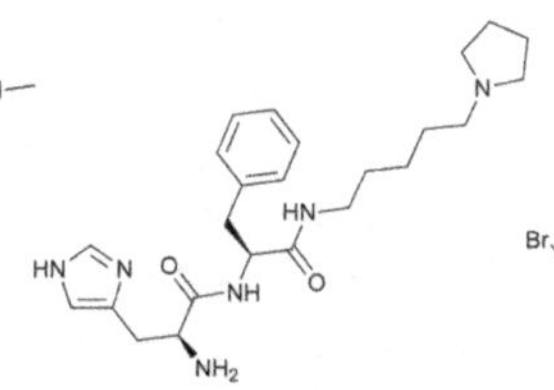

(44)
L3MBTL1 37μM

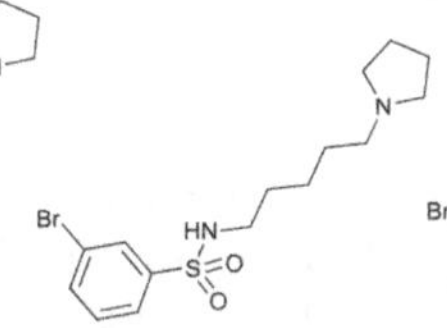

(45)
L3MBTL1 25μM

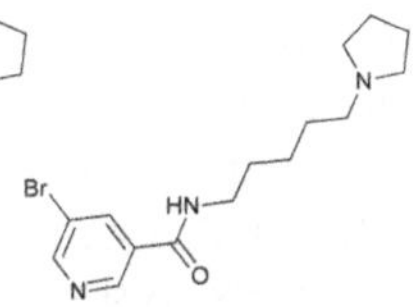

(46)
L3MBTL1 46μM
L3MBTL3 28μM
L3MBTL4 inactive
SFMBT inactive
MBTD1 inactive
PHF13 48μM
CBX inactive

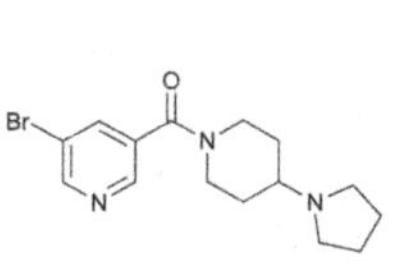

(47)
UNC669 IC50 (AlphaScreen)
L3MBT1 7.9μM
L3MBT3 4.8μM
L3MBTL4 >30μM (41%)
MBTD1 >30μM (23%)
SFMBT1 > 30μM
CBX7 >30μM(9%)

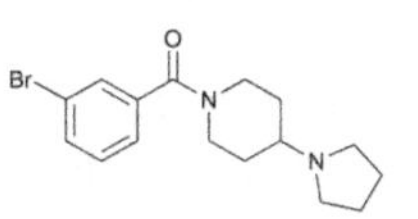

(48)
UNC926 IC50 (AlphaScreen)
L3MBT1 3.9μM
L3MBT3 3.2μM
L3MBTL4 15.6μM
MBTD1 >30μM (44%)
SFMBT1 > 30μM (12%)
CBX7 >30μM(11%)

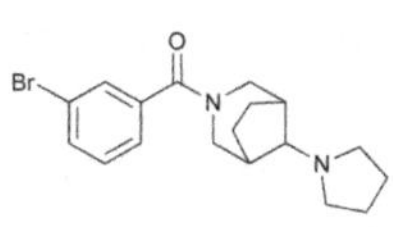

(49)
IC50 (AlphaScreen)
L3MBT1 23.9μM
L3MBT3 3.6μM
L3MBTL4 >30μM (16%)
MBTD1 >30μM (31%)
SFMBT1 > 30μM (n/a)
CBX7 >30μM (n/a)

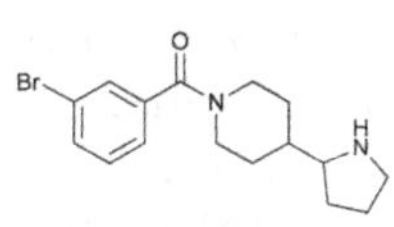

(50)
IC50 (AlphaScreen)
L3MBT1 >30μM (6%)
L3MBT3 >30μM (52%)
L3MBTL4 >30μM (n/a)
MBTD1 9.2μM
SFMBT1 > 30μM (n/a)
CBX7 >30μM (n/a)

Figure 8.14 L3MBTL methyl lysine inhibitors.

Both these latter compounds were found to have higher affinity than any of the methyl lysine tripeptides for L3MBTL1, but were still relatively weak given the size of these molecules.

As there seemed little advantage in having the first two amino acids in the tripeptide scaffold compared to the simple aromatic anchors found by virtual screening, the group used both the aryl sulphonamide and nicotinic scaffolds (**45, 46**) as the basis of future optimisations.[107] More than 350 compounds were synthesised to explore the SAR within this methyl lysine pocket and beyond. Disappointingly the SAR outside of the methyl lysine pocket appeared rather flat, which may in part reflect the fact that the affinities of these compounds lie in the range where weaker compounds are difficult to quantify. The SAR amongst the more active compounds reinforced this hypothesis in the absolute requirement for a basic centre within the Kme pocket; for example, replacement of the pyrrolidine with morpholine or cyclopentyl resulted in inactive compounds. There also appeared to be an optimal size of substituents within this pocket, because swapping the pyrrolidine with smaller mono- or dimethylamines resulted in decreased activity, as did replacing it with a larger piperidine or methylation of the pyrrolidine at the 2- or 3-positions. Efforts to make affinity improvements by changing the nature of the linker between the aromatic anchor and the Kme mimetic, or moving or replacing the bromine, did not give improvement over UNC669 (**47**) and UNC926 (**48**) for L3MBTL1. However, interestingly, as some compounds were also tested against a panel of Kme proteins (the MBT domains of L3MBTL3/4, SFMBTL1 and MBTD1, the PHD domain of PHF13-PHD and the chromodomain CBX7), there emerged indications of differential SAR between proteins within this panel. For example, replacement of the bromine with an acid at this *meta*-position on UNC926 (**48**) was not tolerated in L3MBTL1 or L3MBTL4, but produced only a 2-fold decrease in affinity for L3MBLT3 to 9.9 μM, showing signs that divergent SAR may be possible outside the Kme site. Additionally, some differences could be detected within the methyl lysine site itself as L3MBTL3 seems to be more tolerant of bulky linkers as in **49**. Also, movement of the attachment point of the pyrrolidine was not tolerated except for **50**, which showed selective inhibition of MBTD1.

The predicted binding modes of these compounds were confirmed with the co-crystal structure of UNC669 (**47**) in the first three MBT domains of L3MBTL1 (PDB: 3P8H). As expected this inhibitor binds only to the second MBT domain of this protein. The pyrrolidine interacts with Asp355 within the aromatic cage as expected, but is positioned slightly higher in the pocket than the native Kme peptides. UNC669 (**47**) exhibited 5-fold greater affinity than the cognate H4K20me1 peptide for L3MBTL1, with improved ligand efficiency, and represents a significant advance in small-molecule inhibitors of the MBT domains.

Building on this success, the group designed a series of compounds that incorporated two Kme mimetics into the same molecule, including UNC1021 (**51**). Although there was no evidence of polyvalency as a mechanism for increasing potency for native peptides for MBT domains, they were inspired by Kme binders such as 53BP1,[103,108] a mammalian DNA repair factor containing tandem Tudor domains, which was known to recognise an adjacent unmodified arginine in addition to the H4K20me2 modification. These dibasic compounds were tested across a panel of methyl lysine readers in the group's target-class cross-screening approach. UNC1021 (**51**) showed modest micromolar activity for 53BP1 (IC_{50} 17 μM) and similar potencies for L3MBTL1 and L3MBTL4, but surprisingly high affinity (IC_{50} 40 nM) for L3MBTL3. UNC1215 (**52**) was made to improve the cellular activity of this starting point and displayed a similar affinity and selectivity profile to UNC1021. UNC1079 (**53**) was also synthesised as a negative control with knowledge of the strict size requirement of the Kme pocket within the L3MBTL family, resulting in a decrease of >1000-fold in activity *versus* L3MBTL3. The co-crystal structure of UNC1215–L3MBTL3 explains the basis for this compound's selectivity and affinity. It binds as a unique 2 : 2 complex, as shown in Figure 8.15. The compound bridges between two L3MBTL3 protein molecules to induce dimer formation. One pyrrolidine interacts with the aromatic cage of domain 2, consisting of Phe387, Phe405, Trp408 and Tyr412, and a key hydrogen bond with Asp381 of one protein monomer. Simultaneously, its other basic pyrrolidine engages the methyl lysine binding site of domain 1 of the second protein monomer. The interactions with domain 1 are less extensive, as the molecule sits more shallowly within the aromatic cage. Given this unusual binding mode, gel filtration experiments were conducted to confirm the ability of the compound to induce protein dimerisation of the first three MBT domains of L3MBTL3 and thus bind to this protein in solution. FRAP experiments using a transiently transfected GFP fusion protein of the three MBT domains of L3MBTL3 showed that UNC1215, but not its negative control UNC1079, was able to alter chromatin mobility with a half-maximum effective concentration of 50–100 nM, confirming cell penetration and target engagement of this chemical probe. UNC1215 was also found to be able to disrupt nuclei foci formation of this truncated protein in transfected U2OS and HEK293 cells, but not when full-length L3MBTL3 was used. This led the authors to propose that other domains in the full-length protein may act cooperatively with the MBT domains to facilitate foci formation, and that in their presence chemical probes such as UNC1215 have insufficient potency for efficacy due to the avidity of these increased interactions.

This example highlights both the opportunities and challenges of the multivalent interactions that are so prevalent within the physiological environment of a reader protein. By engaging MBT domains on separate proteins, UNC1215 achieves high affinity, yet its ability to inhibit these domains in a cellular context may be limited.

8.5.3 Inhibitors of WDR5

MLL (mixed lineage leukaemia) proteins are histone H3 lysine 4 (H3K4) methyltransferases (HMTs) that catalyse mono-, di- and trimethylation through their evolutionary conserved SET domains. The most studied of these

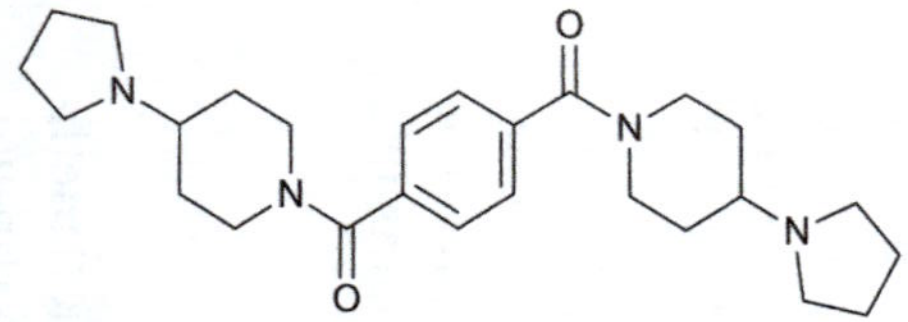

(**51**) UNC 1021

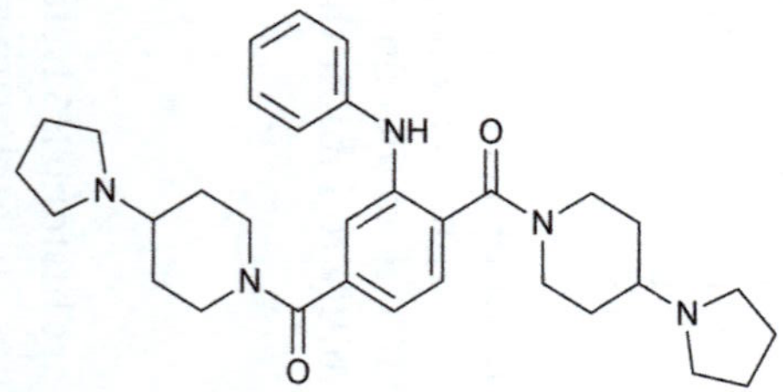

(**52**) UNC 1215

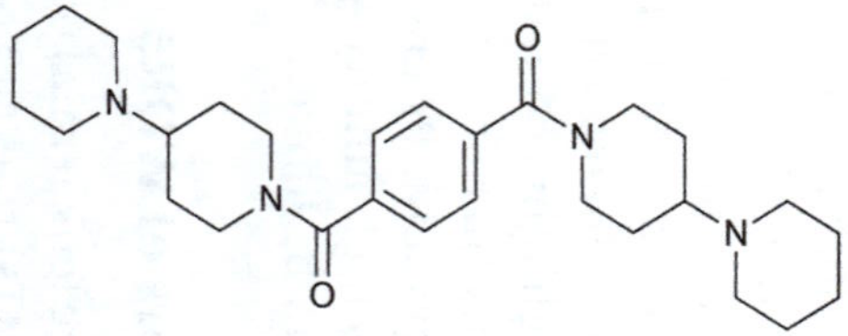

(**53**) UNC1079
('negative' control)

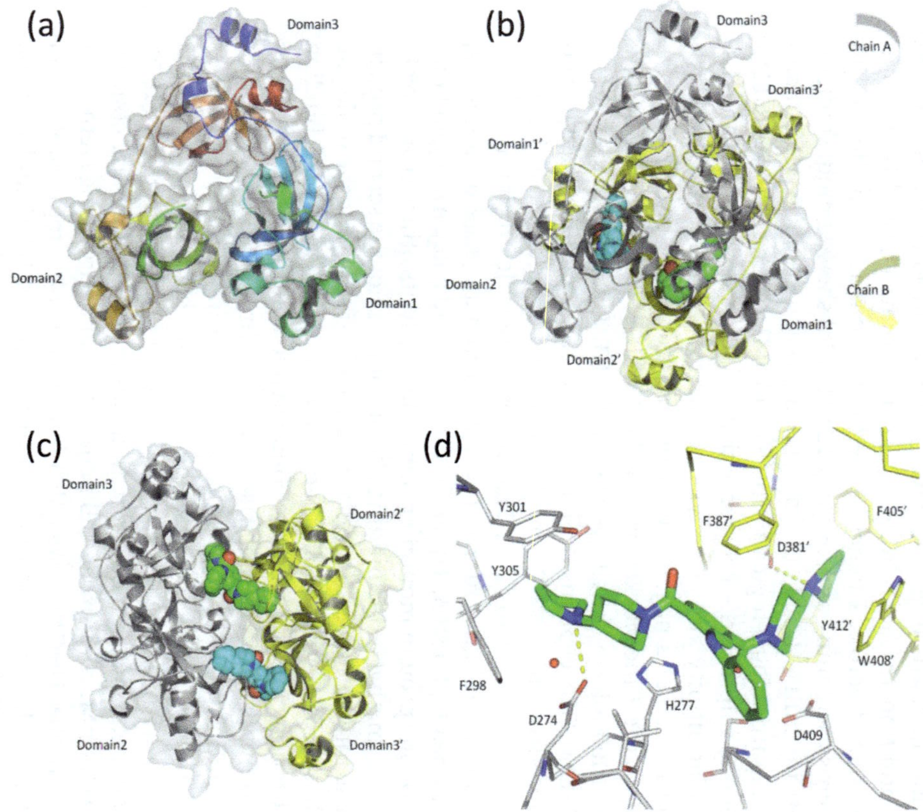

Figure 8.15 L3MBTL3 tool compounds UNC1021, UNC1215, UNC1079. (a) First three MBT domains of L3MBTL3 in ribbon format (truncated coloured using a spectrum starting with blue at N-terminus through to red at the C-terminus). The anticlockwise organisation of the domains 1, 2, 3 are labelled accordingly (PDB: 4FL6). (b) 2:2 L3MBTL3–UNC1215 complex. One protomer of the dimer shown in grey, the other in yellow. Atoms of UNC1215 are depicted as spheres coloured according to atom type (oxygen in red, nitrogen in dark blue), with one molecule having carbons in cyan, the other in green (PDB: 4FL6). (c) Rotated view of (b) (PDB: 4FL6). (d) Details of UNC1215 pocket within L3MBTL3 dimer (PDB: 4FL6).

proteins, MLL1, is often misregulated in lymphoblastic and acute myeloid leukaemias (AML)[109] and genetic abnormality of this protein accounts for 5–10% of AML in adults and 70% in infants.[110] Wild-type MLL1 activity is tightly controlled by the formation of a core complex composed of MLL1, WDR5 (WD40-repeat protein 5), RbBP5 (retinoblastoma-binding protein 5) and ASH2L (absent small homeotic discs-2-like), where all four components are required for optimal activity.[111] The MLL1/WDR5 interaction is mediated by the WDR5 interacting (WIN) motif of approximately 12 amino acids and is required to maintain the integrity of not only the MLL1 complex, but also those of MLL2, MLL3 and MLL4, whose expression can be altered in other forms of cancer.[112,113]

Antagonism of the MLL/WDR5 subunit interaction has been proposed as an alternative cancer strategy to active site inhibition of MLL1. Two groups have pursued this goal using different approaches, both guided by structural insights. The first essentially followed a small-molecule approach from the start. Screening a diverse library of 16 000 small molecules in a WDR5 peptide–MLL1 fluorescence displacement assay at 50 μM, 17 compounds were found to give a greater than 40% inhibition. Only one of these, WDR5-0101 (**54**, K_d 5.5 μM) had K_d less than 60 μM.[30] Analogues included WDR5-0102 (**55**, K_d 4 μM) and WDR5-103 (**56**, K_d 450 nM) for which co-crystal structures (PDB: 3SMR, 3UR4) were obtained. These confirmed binding to the deep central cavity of the doughnut-shaped WD40-repeat fold. This site is the arginine pocket identified by the SiteMap analysis of Santiago *et al.*[98] SAR exploration around three points of diversity around WDR5-0102 has also been described.[114] By far the most fruitful area of optimisation was changing substituents that occupy the shallow lipophilic shelf at the entrance to the arginine binding site, yielding WDR5-0103 (**57**, K_d 300 nM, PDB: 4AI9) as the most potent compound. Compound WDR5-0103 showed no activity *versus* a panel of human HMTs and SET domains, but when tested in an *in vitro* MLL1 catalytic assay that requires the presence of the tetrameric core MLL complex, was found to have inhibition in the hundreds of micromolar range. A complex multi-protein equilibrium modulates MLL cellular activity and the effectiveness of WDR5–MLL inhibitors to change this balance will depend on the local concentration of the protein components and the nature of the complexes formed. Therefore whilst these initial PPI inhibitors of the WDR5–MLL demonstrate inhibition in simple competition assays it is likely that inhibitors of far greater potency are necessary to assess the full potential of this mode of MLL inhibition. This has been achieved by following an alternative peptidomimetic approach (Figure 8.16).[29]

Co-crystal structures of WDR5 with a MLL1 peptide[115,116] suggested that not all 12 amino acids of the WIN motif were critical for binding. Truncations revealed that the CO-Ala-Arg-Ala-NH tripeptide corresponding to MLL1 (3764–3766) was both necessary and sufficient to bind WDR5, and that the Ac-Ala-Arg-Ala-NH2 tripeptide had a 150 nM affinity for WDR5.[117] Exploration of peptidomimetics from this starting point led to MM-102, which is >100 times more potent than the tripeptide. MM-102 has subnanomolar affinity in a

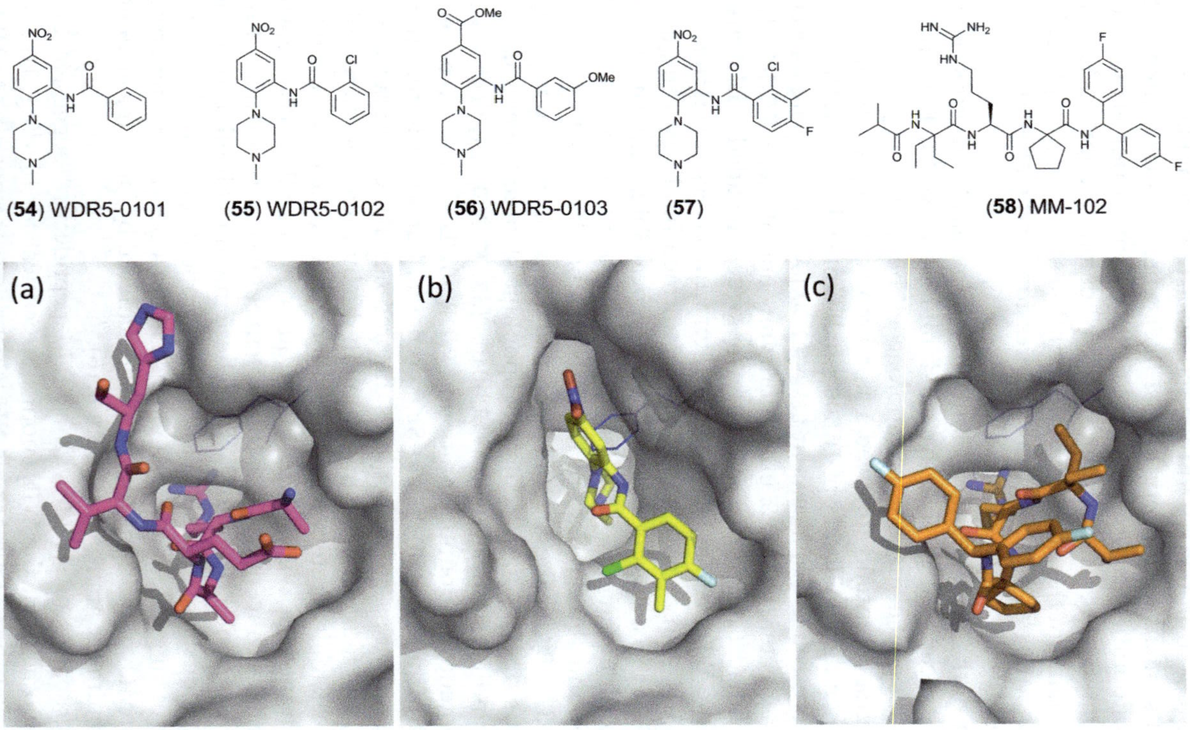

Figure 8.16 WDR5 inhibitors (54–58). WD40 domain of WDR5 in complex with (a) unmodified H3(1–12) (magenta) (PDB: 3MEH). Position of Phe133 in this complex highlighted in blue. (b) Compound 57 (PDB: 3SMR). Phe133 movement shown by blue line depicting its original position, as in (a), and current position in the ligand complex in stick format beneath the protein surface. (c) Peptidomimetic MM-102 (PDB: 4GM8). Phe133 position in this complex is shown as in blue and resembles that in peptide-bound structures as in (a).

displacement assay and the ability to inhibit MLL1 core complex methyltransferase activity in the single-digit nanomolar range.[29] Treatment of MLL1-AF9 transduced murine myeloblasts with MM-102 reduced mRNA levels of two critical MLL1 target genes, *HoxA9* and *Meis-1* required for leukemogenesis, without altering those of housekeeping genes such as *GAPDH*. Approximately 70% inhibition of HoxA9 and 40% Meis-1 mRNA levels was observed with 50 μM MM-102. This successful proof of concept study suggests that potent small-molecule inhibitors of the WDR5–MLL1 interaction can effectively inhibit MLL1-mediated gene transcription in cancer caused by misregulation of MLL1. This represents a novel therapeutic strategy for acute leukaemia.[118]

8.6 Conclusions

Phenotypic screening has been, and will continue to be, a valuable way of finding new targets and pathways for novel medicines, and has delivered the first therapeutically viable inhibitors of bromodomains. Rational structure- and computational-based approaches then provide an efficient platform to expand the chemical knowledge around these novel targets that can generate compounds of increased safety and efficacy. Where these targets belong to protein families, a target-class approach, for example by cross-screening of compound libraries designed to exploit common key recognition features within the family, can be used to find useful tools for chemical biology as well as new starting points for lead optimisation efforts. Structural insights from protein–partner interactions can also be a source of inspiration for chemical tools as the WDR40 peptidomimetic illustrates, allowing innovative mechanisms to be validated. In summary, success in discovery of reader domain inhibitors has been achieved by a variety of diverse approaches that have provided valuable chemical tools to dissect the underlying biology. Their key challenge is realisation of their potential to translate from compounds with unique pharmacology into clinically efficacious and differentiated medicines.

References

1. M. A. Dawson, T. Kouzarides and B. J. Huntly, *N. Engl. J. Med.*, 2012, **367**, 647.
2. C. H. Arrowsmith, C. Bountra, P. V. Fish, K. Lee and M. Schapira, *Nat. Rev. Drug Discovery*, 2012, **11**, 384.
3. C. W. Chung and D. F. Tough, *Drug Discovery Today: Ther. Strategies*, 2012, **9**, e111–e120.
4. D. R. Owen and J. D. Trzupek, *Drug Discovery Today: Technol.*, 2012, e29–e34.
5. A. L. Hopkins and C. R. Groom, *Nat. Rev. Drug Discovery*, 2002, **1**, 727.
6. A. P. Russ and S. Lampel, *Drug Discovery Today*, 2005, **10**, 1607.
7. J. P. Overington, B. Al-Lazikani and A. L. Hopkins, *Nat. Rev. Drug Discovery*, 2006, **5**, 993.

8. A. D. Keefe, S. Pai and A. Ellington, *Nat. Rev. Drug Discovery*, 2010, **9**, 537.
9. P. Vlieghe, V. Lisowski, J. Martinez and M. Khrestchatisky, *Drug Discovery Today*, 2010, **15**, 40.
10. T. Yokota, S. Takeda, Q. L. Lu, T. A. Partridge, A. Nakamura and E. P. Hoffman, *Arch. Neurol.*, 2009, **66**, 32.
11. M. P. Stumpf, T. Thorne, E. de Silva, R. Stewart, H. J. An, M. Lappe and C. Wiuf, *Proc. Natl. Acad. Sci. U. S. A.*, 2008, **105**, 6959.
12. K. Venkatesan, J. F. Rual, A. Vazquez, U. Stelzl, I. Lemmens, T. Hirozane-Kishikawa, T. Hao, M. Zenkner, X. Xin and K. I. Goh, *Nat. Methods*, 2008, **6**, 83.
13. I. M. Nooren and J. M. Thornton, *EMBO J.*, 2003, **22**, 3486.
14. A. Mullard, *Nat. Rev. Drug Discovery*, 2012, **11**, 173.
15. A. R. Horswill and S. J. Benkovic, *Curr. Protoc. Protein Sci.*, 2006, ch. 19.
16. G. L. Verdine and G. J. Hilinski, *Methods Enzymol.*, 2012, **503**, 3.
17. E. Miranda, I. K. Nordgren, A. L. Male, C. E. Lawrence, F. Hoakwie, F. Cuda, W. Court, K. R. Fox, P. A. Townsend, G. K. Packham, S. A. Eccles and A. Tavassoli, *J. Am. Chem. Soc.*, 2013, **135**, 10418.
18. C. A. Lipinski, F. Lombardo, B. W. Dominy and P. J. Feeney, *Adv. Drug Delivery Rev.*, 1997, **23**, 3.
19. L. Pagliaro, J. Felding, K. Audouze, S. J. Nielsen, R. B. Terry, C. Krog-Jensen and S. Butcher, *Curr. Opin. Chem. Biol.*, 2004, **8**, 442.
20. P. J. Hajduk, W. R. Galloway and D. R. Spring, *Nature*, 2011, **470**, 42.
21. M. R. Arkin and J. A. Wells, *Nat. Rev. Drug Discovery*, 2004, **3**, 301.
22. J. A. Wells and C. L. McClendon, *Nature*, 2007, **450**, 1001.
23. A. Domling, *Curr. Opin. Chem. Biol.*, 2008, **12**, 281.
24. X. Morelli and T. Hupp, *EMBO Rep.*, 2012, **13**, 877.
25. S. D. Taverna, H. Li, A. J. Ruthenburg, C. D. Allis and D. J. Patel, *Nat. Struct. Mol. Biol.*, 2007, **14**, 1025.
26. X. Morelli, R. Bourgeas and P. Roche, *Curr. Opin. Chem. Biol.*, 2011, **15**, 475.
27. L. I. James, D. Barsyte-Lovejoy, N. Zhong, L. Krichevsky, V. K. Korboukh, J. M. Herold, C. J. MacNevin, J. L. Norris, C. A. Sagum, W. Tempel, E. Marcon, H. Guo, C. Gao, X. P. Huang, S. Duan, A. Emili, J. F. Greenblatt, D. B. Kireev, J. Jin, W. P. Janzen, P. J. Brown, M. T. Bedford, C. H. Arrowsmith and S. V. Frye, *Nat. Chem. Biol.*, 2013, **9**, 184.
28. J. M. Herold, L. A. Ingerman, C. Gao and S. V. Frye, *Curr. Chem. Genomics*, 2011, **5**, 51.
29. H. Karatas, E. C. Townsend, F. Cao, Y. Chen, D. Bernard, L. Liu, M. Lei, Y. Dou and S. Wang, *J. Am. Chem. Soc.*, 2013, **135**, 669.
30. G. Senisterra, H. Wu, A. Allali-Hassani, G. A. Wasney, D. Barsyte-Lovejoy, L. Dombrovski, A. Dong, K. T. Nguyen, D. Smil, Y. Bolshan, T. Hajian, H. He, A. Seitova, I. Chau, F. Li, G. Poda, J. F. Couture, P. J. Brown, R. Al-Awar, M. Schapira, C. H. Arrowsmith and M. Vedadi, *Biochem. J.*, 2013, **449**, 151.
31. E. K. Wagner, N. Nath, R. Flemming, J. B. Feltenberger and J. M. Denu, *Biochemistry*, 2012, **51**, 8293.

32. A. J. Ruthenburg, H. Li, D. J. Patel and C. D. Allis, *Nat. Rev. Mol. Cell Biol.*, 2007, **8**, 983.
33. M. Bantscheff, C. Hopf, M. M. Savitski, A. Dittmann, P. Grandi, A. M. Michon, J. Schlegl, Y. Abraham, I. Becher, G. Bergamini, M. Boesche, M. Delling, B. Dumpelfeld, D. Eberhard, C. Huthmacher, T. Mathieson, D. Poeckel, V. Reader, K. Strunk, G. Sweetman, U. Kruse, G. Neubauer, N. G. Ramsden and G. Drewes, *Nat. Biotechnol.*, 2011, **29**, 255.
34. T. Kouzarides, *EMBO J.*, 2000, **19**, 1176.
35. K. T. Smith and J. L. Workman, *Nat. Biotechnol.*, 2009, **27**, 917.
36. S. Mujtaba, L. Zeng and M. M. Zhou, *Oncogene*, 2007, **26**, 5521.
37. B. Florence and D. V. Faller, *Front. Biosci.*, 2001, **6**, D1008–D1018.
38. M. Thompson, *Biochimie*, 2009, **91**, 309.
39. E. Shang, H. D. Nickerson, D. Wen, X. Wang and D. J. Wolgemuth, *Development*, 2007, **134**, 3507.
40. J. Moriniere, S. Rousseaux, U. Steuerwald, M. Soler-Lopez, S. Curtet, A. L. Vitte, J. Govin, J. Gaucher, K. Sadoul, D. J. Hart, J. Krijgsveld, S. Khochbin, C. W. Muller and C. Petosa, *Nature*, 2009, **461**, 664.
41. P. Filippakopoulos, S. Picaud, M. Mangos, T. Keates, J. P. Lambert, D. Barsyte-Lovejoy, I. Felletar, R. Volkmer, S. Muller, T. Pawson, A. C. Gingras, C. H. Arrowsmith and S. Knapp, *Cell*, 2012, **149**, 214.
42. C. Dhalluin, J. E. Carlson, L. Zeng, C. He, A. K. Aggarwal and M. M. Zhou, *Nature*, 1999, **399**, 491.
43. F. Vollmuth and M. Geyer, *Angew. Chem., Int. Ed.*, 2010, **49**, 6768.
44. L. Zeng, Q. Zhang, S. Li, A. N. Plotnikov, M. J. Walsh and M. M. Zhou, *Nature*, 2010, **466**, 258.
45. M. Ali, K. Yan, M. E. Lalonde, C. Degerny, S. B. Rothbart, B. D. Strahl, J. Cote, X. J. Yang and T. G. Kutateladze, *J. Mol. Biol.*, 2012, **424**, 328.
46. J. A. Lee, M. T. Uhlik, C. M. Moxham, D. Tomandl and D. J. Sall, *J. Med. Chem.*, 2012, **55**, 4527.
47. D. C. Swinney and J. Anthony, *Nat. Rev. Drug Discovery*, 2011, **10**, 507.
48. M. Duvic and J. Vu, *Expert Opin. Invest. Drugs*, 2007, **16**, 1111.
49. V. M. Richon, Y. Webb, R. Merger, T. Sheppard, B. Jursic, L. Ngo, F. Civoli, R. Breslow, R. A. Rifkind and P. A. Marks, *Proc. Natl. Acad. Sci. U. S. A.*, 1996, **93**, 5705.
50. V. M. Richon, S. Emiliani, E. Verdin, Y. Webb, R. Breslow, R. A. Rifkind and P. A. Marks, *Proc. Natl. Acad. Sci. U. S. A.*, 1998, **95**, 3003.
51. P. Filippakopoulos, J. Qi, S. Picaud, Y. Shen, W. B. Smith, O. Fedorov, E. M. Morse, T. Keates, T. T. Hickman, I. Felletar, M. Philpott, S. Munro, M. R. McKeown, Y. Wang, A. L. Christie, N. West, M. J. Cameron, B. Schwartz, T. D. Heightman, T. N. La, C. A. French, O. Wiest, A. L. Kung, S. Knapp and J. E. Bradner, *Nature*, 2010, **468**, 1067.
52. E. Nicodeme, K. L. Jeffrey, U. Schaefer, S. Beinke, S. Dewell, C. W. Chung, R. Chandwani, I. Marazzi, P. Wilson, H. Coste, J. White, J. Kirilovsky, C. M. Rice, J. M. Lora, R. K. Prinjha, K. Lee and A. Tarakhovsky, *Nature*, 2010, **468**, 1119.
53. C. Chung, H. Coste, J. H. White, O. Mirguet, J. Wilde, R. L. Gosmini, C. Delves, S. M. Magny, R. Woodward, S. A. Hughes, E. V. Boursier,

H. Flynn, A. M. Bouillot, P. Bamborough, J. M. Brusq, F. J. Gellibert, E. J. Jones, A. M. Riou, P. Homes, S. L. Martin, I. J. Uings, J. Toum, C. A. Clement, A. B. Boullay, R. L. Grimley, F. M. Blandel, R. K. Prinjha, K. Lee, J. Kirilovsky and E. Nicodeme, *J. Med. Chem.*, 2011, **54**, 3827.
54. K. Adachi, H. Hikawa, M. Hamada, J.-I. Endoh, S. Ishibuchi, N. Fujie, M. Tanaka, K. Sugahara, K. Oshita and M. Murata, Preparation of thienotriazolodiazepine compounds having an action of inhibiting the CD28 costimulatory signal in T cell, WO 2006-JP310709, 2011, 2006 (Mitsubishi Pharma Corporation, Japan).
55. S. Miyoshi, S. Ooike, K. Iwata, H. Hikawa and K. Sugahara, Antitumor Agent, WO 2008-JP73864 (2008) [WO 2009084693 (2009)] (Mitsubishi Tanabe Pharma Corporation, Japan).
56. E. McNeill, *Curr. Opin. Invest. Drugs*, 2010, **11**, 357.
57. P. R. Young, E. Johnson, E. Campeau, S. Attwell, L. Tsujikawa, S. Wasiak, D. Gilham, H. Hansen, G. Wagner, N. C. W. Wong, J. Johansson, S. J. Nicholls, A. Gordon, S. Nissen and K. G. McLure, *RVX-208 is a BET Inhibitor Currently in Phase II Human Clinical Trials for the Treatment of Atherosclerosis, 2nd Epigenetics in Drug Discovery*, Boston, MA, 2013.
58. G. C. Terstappen, C. Schlupen, R. Raggiaschi and G. Gaviraghi, *Nat. Rev. Drug Discovery*, 2007, **6**, 891.
59. M. Mann, *Nat. Rev. Mol. Cell Biol.*, 2006, **7**, 952.
60. P. J. Barter, S. Nicholls, K. A. Rye, G. M. Anantharamaiah, M. Navab and A. M. Fogelman, *Circ. Res.*, 2004, **95**, 764.
61. N. C. Wong, *Curr. Opin. Invest. Drugs*, 2007, **8**, 718.
62. O. Mirguet, R. Gosmini, J. Toum, C. A. Clément, M. Barnathan, J. M. Brusq, J. E. Mordaunt, R. M. Grimes, M. Crowe, O. Pineau, M. Ajakane, A. Daugan, P. Jeffrey, L. Cutler, A. C. Haynes, N. N. Smithers, C. W. Chung, P. Bamborough, I. J. Uings, A. Lewis, J. Witherington, N. Parr, R. K. Prinjha and E. Nicodème, *J. Med. Chem.*, 2013, **56**, 7501–7515.
63. J. Seal, Y. Lamotte, F. Donche, A. Bouillot, O. Mirguet, F. Gellibert, E. Nicodeme, G. Krysa, J. Kirilovsky, S. Beinke, S. McCleary, I. Rioja, P. Bamborough, C. W. Chung, L. Gordon, T. Lewis, A. L. Walker, L. Cutler, D. Lugo, D. M. Wilson, J. Witherington, K. Lee and R. K. Prinjha, *Bioorg. Med. Chem. Lett.*, 2012, **22**, 2968.
64. O. Mirguet, Y. Lamotte, F. Donche, J. Toum, F. Gellibert, A. Bouillot, R. Gosmini, V. L. Nguyen, D. Delannee, J. Seal, F. Blandel, A. B. Boullay, E. Boursier, S. Martin, J. M. Brusq, G. Krysa, A. Riou, R. Tellier, A. Costaz, P. Huet, Y. Dudit, L. Trottet, J. Kirilovsky and E. Nicodeme, *Bioorg. Med. Chem. Lett.*, 2012, **22**, 2963.
65. A. Wyce, G. Ganji, K. C. Smitheman, C. w. Chung, S. Korenchuk, Y. Bai, O. Barbash, B. Le, P. D. Craggs and M. T. McCabe, *PLoS One*, 2013, **8**, e72967.
66. E. H. Demont and R. L. M. Gosmini, Tetrahydroquinoline derivatives as bromodomain inhibitors and their preparation and use in the treatment of cancer, Autoimmune and Inflammatory Diseases, WO 2010-EP66701 (2009) [WO 2011054848 (2011)].
67. BET inhibitor biology review, 2013.

68. D. C. Swinney, *Nat. Rev. Drug Discovery*, 2004, **3**, 801.
69. P. Bamborough, *Expert Opin. Drug Discovery*, 2012, **7**, 1053.
70. S. Mujtaba, Y. He, L. Zeng, A. Farooq, J. E. Carlson, M. Ott, E. Verdin and M. M. Zhou, *Mol. Cell*, 2002, **9**, 575.
71. L. Zeng, J. Li, M. Muller, S. Yan, S. Mujtaba, C. Pan, Z. Wang and M. M. Zhou, *J. Am. Chem. Soc.*, 2005, **127**, 2376.
72. C. Pan, M. Mezei, S. Mujtaba, M. Muller, L. Zeng, J. Li, Z. Wang and M. M. Zhou, *J. Med. Chem.*, 2007, **50**, 2285.
73. Sachchidanand, L. Resnick-Silverman, S. Yan, S. Mutjaba, W. J. Liu, L. Zeng, J. J. Manfredi and M. M. Zhou, *Chem. Biol.*, 2006, **13**, 81.
74. J. C. Borah, S. Mujtaba, I. Karakikes, L. Zeng, M. Muller, J. Patel, N. Moshkina, K. Morohashi, W. Zhang, G. Gerona-Navarro, R. J. Hajjar and M. M. Zhou, *Chem. Biol.*, 2011, **18**, 531.
75. G. Gerona-Navarro, R. Yoel, S. Mujtaba, A. Frasca, J. Patel, L. Zeng, A. N. Plotnikov, R. Osman and M. M. Zhou, *J. Am. Chem. Soc.*, 2011, **133**, 2040.
76. C. W. Chung, A. W. Dean, J. M. Woolven and P. Bamborough, *J. Med. Chem.*, 2012, **55**, 576.
77. P. Filippakopoulos, S. Picaud, O. Fedorov, M. Keller, M. Wrobel, O. Morgenstern, F. Bracher and S. Knapp, *Bioorg. Med. Chem.*, 2012, **20**, 1878.
78. L. Zhao, D. Cao, T. Chen, Y. Wang, Z. Miao, Y. Xu, W. Chen, X. Wang, Y. Li, Z. Du, B. Xiong, J. Li, C. Xu, N. Zhang, J. He and J. Shen, *J. Med. Chem.*, 2013, **56**, 3833.
79. D. S. Hewings, M. Wang, M. Philpott, O. Fedorov, S. Uttarkar, P. Filippakopoulos, S. Picaud, C. Vuppusetty, B. Marsden, S. Knapp, S. Conway and T. D. Heightman, *J. Med. Chem.*, 2011, **54**, 6761.
80. P. Bamborough, H. Diallo, J. D. Goodacre, L. Gordon, A. Lewis, J. T. Seal, D. M. Wilson, M. D. Woodrow and C. W. Chung, *J. Med. Chem.*, 2012, **55**, 587.
81. V. S. Gehling, M. C. Hewitt, R. G. Vaswani, Y. Leblanc, A. Côté, C. G. Nasveschuk, A. M. Taylor, J. C. Harmange, J. E. Audia, E. Pardo, S. Joshi, P. Sandy, J. A. Mertz, R. J. Sims, L. Bergeron, B. M. Bryant, S. Bellon, F. Poy, H. Jayaram, R. Sankaranarayanan, S. Yellapantula, N. Bangalore Srinivasamurthy, S. Birudukota and B. K. Albrecht, *ACS Med. Chem. Lett.*, 2013, **4**, 835–840.
82. P. V. Fish, P. Filippakopoulos, G. Bish, P. E. Brennan, M. E. Bunnage, A. S. Cook, O. Federov, B. S. Gerstenberger, H. Jones, S. Knapp, B. Marsden, K. Nocka, D. R. Owen, M. Philpott, S. Picaud, M. J. Primiano, M. J. Ralph, N. Sciammetta and J. D. Trzupek, *J. Med. Chem.*, 2012, **55**, 9831.
83. N. Schmees, J. Kuhnke, B. Haendler, P. Lienau, A. E. Fernandez-Montalvan, P. Lejeune, S. Siegel and W. Scott, 6h-thieno [3, 2-f][1, 2, 4] triazolo [4, 3-a][1, 4] diazepines, WO Pat. App. PCT/EP2012/066,600, 2012.
84. L. Wang, J. K. Pratt, K. F. Mcdaniel, Y. Dai, S. D. Fidanze, L. Hasvold, J. H. Holms, W. M. Kati, D. Liu and R. A. Mantei, Bromodomain inhibitors, WO Pat. App. PCT/CN2012/086,357, 2012.
85. P. J. Hajduk, J. R. Huth and S. W. Fesik, *J. Med. Chem.*, 2005, **48**, 2518.

86. F. N. Edfeldt, R. H. Folmer and A. L. Breeze, *Drug Discov. Today*, 2011, **16**, 284.
87. S. Henrich, O. M. H. Salo-Ahen, B. Huang, F. F. Rippmann, G. Cruciani and R. C. Wade, *J. Mol. Recognit.*, 2010, **23**, 209.
88. T. A. Halgren, *J. Chem. Inf. Model.*, 2009, **49**, 377.
89. D. Brown and G. Superti-Furga, *Drug Discovery Today*, 2003, **8**, 1067.
90. L. R. Vidler, N. Brown, S. Knapp and S. Hoelder, *J. Med. Chem.*, 2012, **55**, 7346.
91. GSK801, http://www.thesgc.org/scientists/chemical_probes/GSK2801, 2013.
92. I-CBP112, http://www.thesgc.org/scientists/chemical_probes/ICBP112, 2013.
93. M. H. Jones, N. Hamana, J. I. Nezu and M. Shimane, *Genomics*, 2000, **63**, 40.
94. C. Guetg, P. Lienemann, V. Sirri, I. Grummt, D. L. Hernandez-Verdun, M. O. Hottiger, M. Fussenegger and R. Santoro, *EMBO J.*, 2010, **29**, 2135.
95. T. J. Wigle and R. A. Copeland, *Curr. Opin. Chem. Biol.*, 2013, **17**, 369.
96. J. M. Herold, T. J. Wigle, J. L. Norris, R. Lam, V. K. Korboukh, C. Gao, L. A. Ingerman, D. B. Kireev, G. Senisterra, M. Vedadi, A. Tripathy, P. J. Brown, C. H. Arrowsmith, J. Jin, W. P. Janzen and S. V. Frye, *J. Med. Chem.*, 2011, **54**, 2504.
97. M. A. Adams-Cioaba and J. Min, *Biochem. Cell Biol.*, 2009, **87**, 93.
98. C. Santiago, K. Nguyen and M. Schapira, *J. Comput.-Aided Mol. Des.*, 2011, **25**, 1171.
99. R. Bonasio, E. Lecona and D. Reinberg, MBT domain proteins in development and disease, *Semin. Cell Dev. Biol.*, 2010, **21**(2), 221.
100. N. Nady, L. Krichevsky, N. Zhong, S. Duan, W. Tempel, M. F. Amaya, M. Ravichandran and C. H. Arrowsmith, *J. Mol. Biol.*, 2012, **423**, 702–718.
101. P. Trojer, G. Li, R. J. Sims III, A. Vaquero, N. Kalakonda, P. Boccuni, D. Lee, H. Erdjument-Bromage, P. Tempst and S. D. Nimer, *Cell*, 2007, **129**, 915.
102. L. E. West, S. Roy, K. Lachmi-Weiner, R. Hayashi, X. Shi, E. Appella, T. G. Kutateladze and O. Gozani, *J. Biol. Chem.*, 2010, **285**, 37725.
103. K. Acs, M. S. Luijsterburg, L. Ackermann, F. A. Salomons, T. Hoppe and N. P. Dantuma, *Nat. Struct. Mol. Biol.*, 2011, **18**, 1345.
104. N. Gurvich, F. Perna, A. Farina, F. Voza, S. Menendez, J. Hurwitz and S. D. Nimer, *Proc. Natl. Acad. Sci. U. S. A.*, 2010, **107**, 22552.
105. D. Kireev, T. J. Wigle, J. Norris-Drouin, J. M. Herold, W. P. Janzen and S. V. Frye, *J. Med. Chem.*, 2010, **53**, 7625.
106. T. J. Wigle, J. M. Herold, G. A. Senisterra, M. Vedadi, D. B. Kireev, C. H. Arrowsmith, S. V. Frye and W. P. Janzen, *J. Biomol. Screening*, 2010, **15**, 62.
107. J. M. Herold, L. I. James, V. K. Korboukh, C. Gao, K. E. Coil, D. J. Bua, J. L. Norris, D. B. Kireev, P. J. Brown and J. Jin, *MedChemComm*, 2012, **3**, 45.
108. M. V. Botuyan, J. Lee, I. M. Ward, J. E. Kim, J. R. Thompson, J. Chen and G. Mer, *Cell*, 2006, **127**, 1361.

109. A. V. Krivtsov and S. A. Armstrong, *Nat. Rev. Cancer*, 2007, **7**, 823.
110. P. M. Ayton and M. L. Cleary, *Oncogene*, 2001, **20**, 5695.
111. J. Wysocka, T. Swigut, T. A. Milne, Y. Dou, X. Zhang, A. L. Burlingame, R. G. Roeder, A. H. Brivanlou and C. D. Allis, *Cell*, 2005, **121**, 859.
112. D. G. Huntsman, S. F. Chin, M. Muleris, S. J. Batley, V. P. Collins, L. M. Wiedemann, S. Aparicio and C. Caldas, *Oncogene*, 1999, **18**, 7975.
113. Y. W. Cho, T. Hong, S. Hong, H. Guo, H. Yu, D. Kim, T. Guszczynski, G. R. Dressler, T. D. Copeland and M. Kalkum, *J. Biol. Chem.*, 2007, **282**, 20395.
114. Y. Bolshan, M. Getlik, E. Kuznetsova, G. A. Wasney, T. Hajian, G. Poda, K. T. Nguyen, H. Wu, L. Dombrovski, A. Dong, G. Senisterra, M. Schapira, C. H. Arrowsmith, P. J. Brown, R. Al-awar, M. Vedadi and D. Smil, *ACS Med. Chem. Lett.*, 2013, **4**, 353.
115. J. J. Song and R. E. Kingston, *J. Biol. Chem.*, 2008, **283**, 35258.
116. A. Patel, V. Dharmarajan and M. S. Cosgrove, *J. Biol. Chem.*, 2008, **283**, 32158.
117. H. Karatas, E. C. Townsend, D. Bernard, Y. Dou and S. Wang, *J. Med. Chem.*, 2010, **53**, 5179.
118. S. Wang, H. Karatas, Y. Dou, E. Townsend, D. Bernard and F. Cao, Cyclized Peptidomimetic small molecule inhibitors of the WDR5 and MLL1 interaction, US Pat. 20,130,150,309, 2013.

CHAPTER 9

Who will Benefit from Epigenetic Drugs?

DOMINIQUE VERHELLE*[a] AND DAVID J. SHIELDS[a]

[a]Pfizer Inc, Oncology Research Unit, 10724 Science Center Drive, San Diego, CA 92121, USA
*E-mail: dominique.verhelle@pfizer.com

9.1 Introduction

Targeted drug discovery efforts in oncology have traditionally focused on genetic lesions such as gene mutations, amplifications or translocations that directly modulate tumour cell proliferation, survival, migration or invasion. Such genetic events operate in a tumour cell autonomous fashion or function at the interface between neoplastic cells and the microenvironment by modulating angiogenesis, the immune response or heterotypic tumour/stroma signalling. More recently, genetic lesions have also been shown to confer a selective advantage on tumour cells through epigenetic mechanisms. Dysregulation of epigenetic control can impact longer-term reprogramming mechanisms, which calibrate the level of gene expression without eliciting structural changes in the genome. Both DNA and histones may be covalently modified and the resultant changes in epigenetic marks contribute to the malignant phenotype. Notably, the altered mark profiles are reversible and consequently, novel therapies have been developed to target the aberrant enzymatic activities and restore epigenetic control. However, despite increased research and discovery efforts on epigenetic targets, only

RSC Drug Discovery Series No. 48
Epigenetics for Drug Discovery
Edited by Nessa Carey

Published by the Royal Society of Chemistry, www.rsc.org

six drugs have been approved to date for the treatment of cancer patients (Vidaza, Dacogen, Istodax, Zolinza, Faridaq and Beleodaq). Of note, the biggest clinical impact of these epigenetic modifiers in oncology has been among haematological malignancies. While the clinical efficacy and subsequent approval of these agents attests to the importance of epigenomic changes, an increased understanding of epigenetic mechanisms of action (MOA) is required to develop responder strategies, expand the target indications and maximise return on drug discovery investment.

9.2 Epigenetic Modifications

Two layers of chemical modifications edit the genome: DNA undergoes methylation, while histones are subject to covalent modification with many different chemical groups.

9.2.1 DNA Methylation in Cancer

DNA methylation leads to the addition of one methyl group to carbon 5 of the cytosine base. DNA methylation occurs mainly on CG-rich regions of the human genome found in CpG islands[1] or CpG shores[2] of gene promoters, or downstream in intra- and intergenic regions.[3] DNMT3A and DNMT3B catalyse *de novo* DNA methylation, while DNMT1 maintains the DNA methylation status. In contrast, DNA demethylation is not an active mechanism, and can arise when the maintenance of DNA methylation is inhibited during cell replication or as a result of hydroxymethylation catalysed by the ten-eleven translocation (TET) family of enzymes.[4] In the neoplastic state, both hypomethylation and hypermethylation of CpG islands have been identified. Hypomethylation, which occurs mainly at DNA-repetitive regions, activates genes with tumour-promoting functions and engenders a loss of genome stability and imprinting that contributes to the development of malignancies such as colorectal carcinoma and Wilms' tumour.[5–7] Hypermethylation most commonly marks promoter CpG islands, causing transcriptional silencing of tumour suppressor genes such as BRCA1 and RAD51C in breast and ovarian cancers[8,9] or hMLH1 in colorectal cancer.[10–13]

9.2.1.1 Aberrant DNA Methylation in Tumorigenesis

Aberrant CpG island (CGI) hypermethylation is a hallmark of cancer characterised by the tumour-specific hypermethylation of several hundred CGIs.[14] In 1999, two distinct subgroups of colorectal cancers were defined based on the degree of tumour-specific methylation (low or high), and the latter group were described to exhibit a 'CpG island methylator phenotype' (CIMP).[15] Subsequently, CIMP was reported to be positively associated with KRAS mutations, but negatively associated with TP53 mutations in colorectal cancers.[16] CIMP has also been described in gastric, lung, liver and ovarian malignancies, as well as leukaemia.[17,18] The CIMP concept was further refined when the distribution of methylation indices was recognised to be a gradual rather than bimodal distribution, which led to the notion of CIMP-I (intermediate) in addition to CIMP high and CIMP low.[19,20]

Dysregulation of DNA methylation in cancer can arise following genetic alterations of DNMT enzymes. For example DNMT3A mutations have been described in approximately 20% of acute myeloid leukaemia (AML) patients, with a particular enrichment in those with an intermediate risk cytogenetic profile. Such patients have a poor clinical outcome despite a lack of modifications at the 5-methylcytosine. Rather, mutant DNMT3A enzymes have an impaired ability to bind transcription factors, leading to DNA methylation at aberrant genomic locations.[21]

Hypermethylation can also result from the dysregulation of the TET family proteins.[22] The family consists of three members (TET1, TET2 and TET3), which share a high level of homology in their C-terminal catalytic domains. TET1 and TET2 were recently shown to play an important role in DNA demethylation through the oxidation of 5-mC (methylcytosine) to 5-hmC (hydroxymethylcytosine), 5-fC (formylcytosine) and 5-caC (carboxylcytosine). Such 5-hmC modifications are known to be prevalent in cancers, but the impact of altered 5-hmC levels on tumour development remains to be elucidated.[23,24]

Atypical DNA methylation profiles can also arise following mutations in metabolic genes such as isocitrate dehydrogenase (IDH) 1 and 2. IDH1/2 mutations have been reported in 30% of AML patients and have been linked to bone marrow environment changes and impaired cellular differentiation.[25,26] Epigenetic profiling shows that IDH1/2 mutated AML patients display global hypermethylation and furthermore, unique hypermethylation signatures.[27]

9.2.1.2 Clinical Use of Hypomethylating Agents

Two of the most studied epigenetic agents are the hypomethylating agents, 5-azacytidine and 5-aza-2-deoxycytidine (decitabine), with the latter having deoxyribose as its sugar base. Both compounds have the same mode of action when integrated into DNA. Once inside the cell, decitabine is activated to 5-aza-2′-deoxycytidine-5′-triphosphate by deoxycytidine kinase and incorporated *in lieu* of cytidine into DNA. In contrast, the chemically related 5-azacytidine contains a ribose sugar ring and is incorporated primarily into RNA, and at a lower frequency into DNA. The product of the incorporation interacts covalently with DNA methyltransferase leading to degradation of the enzyme. Both drugs are approved for MDS and low blast count AML,[28] based on the survival benefit to patients. Oral 5-azacytidine (CC-486), developed by Celgene, is currently in trials as a single-agent therapy in MDS, AML and head and neck cancer, while combinations with carboplatin are being tested in bladder and ovarian cancers and with abraxane in non-small cell lung cancer (NSCLC), pancreatic cancer and melanoma.

Moreover, oral Vidaza is being combined with the HDAC inhibitor entinostat for several solid tumours, while SGI-110, a pro-drug of 5-aza-2-deoxycytidine is in phase I and phase II trials as a single agent for MDS, AML and hepatocellular carcinoma. Combinations with carboplatin or irinotecan are also being assessed in phase 1 trials for ovarian cancer and colorectal cancer, respectively.

9.2.1.3 *Possible Explanation for Superior Efficacy in Haematological Malignancies*

While hypomethylating agents have demonstrated efficacy in haematological malignancies, clinical trials in solid tumours have not yielded positive results to date. Dose-limiting toxicities included grade 4 neutropenia and myelosuppression when hypomethylating agents were combined with interleukin-2 and carboplatin, respectively.[29,30] Several explanations could be advanced for the unremarkable single-agent activity of hypomethylating drugs in solid tumours. A combination of slowly proliferating cells coupled to drugs with a short half-life may lead to limited drug incorporation. Accordingly, hypomethylating agents that function independently of cell replication time may have greater efficacy. However, the observed adverse events remain prohibitive and could be attributed to a number of different phenomena, particularly (i) reactivation of undesirable genes may induce deleterious effects, and thereby decrease the therapeutic index, and (ii) use of high-dose hypomethylating agents may increase off-target effects. Notably, in epithelial malignancies, both decitabine and 5-azacytidine are prescribed at higher doses than in haematological cancers and indeed, high concentrations of these agents have been shown to induce DNA damage.[31] Importantly, recent reports support the notion that low-dose treatment with hypomethylating agents may restore sensitivity to standard of care therapy, opening the door for combination opportunities. Further underlining the merits of this paradigm, pretreatment with azacytidine in platinum-resistant ovarian cancer restored the sensitivity to carboplatin and significantly improved clinical outcome.[32]

9.2.2 Histone Modifications in Cancer

Histones are critical regulators of gene expression and in turn are subject to tight control *via* multiple post-translational modifications such as acetylation, methylation, phosphorylation, ubiquitylation and sumoylation. Modifications of the histone tail serve to alter chromatin structure and/or recruit transcription machinery complexes. Three classes of histone modifiers have been described so far: writers add epigenetic groups to the histone tails, readers recognise the appended groups and recruit additional proteins to the DNA, while erasers remove these modifications.[33]

9.2.2.1 *Examples of Histone Modifiers as Potential Players in Cancer*

Histone acetylation is regulated by two families of enzymes, histone acetyltransferases (HATs) and histone deacetylases (HDACs). HATs catalyse the transfer of an acetyl group to lysine residues on the histone tail, which promotes a more relaxed form of chromatin and allows recruitment of transcription factors. HDACs reverse the acetylation to induce a condensed chromatin structure, and thereby inhibit transcription.[34] Histone acetylation can also promote transcriptional regulation independent of structural changes in chromatin – certain proteins that modulate gene activation or repression contain a bromodomain, which recognises acetylated lysines and thereby enables protein–protein interactions and recruitment to the region of interest.[35]

Histones can also be methylated on lysine or arginine residues. Lysine methyl transferases (KMTs) catalyse mono-, di- and trimethylation of lysine residues, while protein arginine methyl transferases (PRMTs) catalyse mono- or dimethylation of arginine residues. As with histone acetylation, methylation can be reversed enzymatically. While many lysine demethylases have been identified, the demethylation of arginine occurs through an unknown mechanism. Analogous to bromodomain recognition of acetylated lysines, certain methyl-binding domains recognise different levels of lysine or arginine methylation, which enables calibration of transcriptional activation or repression based on the degree of lysine/arginine methylation.[36]

9.2.2.2 *Clinical Uses of Histone Modifier Inhibitors*

9.2.2.2.1 Histone Acetylation. While acetylation can be modulated by either HATs or HDACs, inhibitors of HDACs (HDACi) have been prioritised over the last decade. HDACis are divided into different classes based on their chemical properties, including hydroxamic acids, such as suberoylanilide hydroxamic acid (SAHA), trichostatin A (TSA), LBH589 (panobinostat) and PXD101 (belinostat); short-chain fatty acids, such as sodium butyrate (NaBu), 4-phenylbutyrate (4-PB) and valproic acid; cyclic tetrapeptides, such as trapoxin, apicidin and depsipeptide (also known as FK228 or romidepsin); benzamides, such as MS-275, CI-994, MGCD0103, and a variety of other chemical compounds and synthetic inhibitors. Vorinostat (SAHA), Istodax (romidepsin), Beleodaq (belinostat) and Faridak (panobinostat) are currently approved for the treatment of cutaneous T-cell lymphoma (CTCL) and peripheral T-cell lymphoma (PTCL), but have not yet yielded compelling data in other cancer types. Second-generation inhibitors in early clinical development such as entinostat and pivanex are more selective for specific classes of HDAC and have been tested against NSCLC in combination with standard of care therapy. While erlotinib (trade name Tarceva) combined with entinostat did not improve progression-free or overall survival, when compared with erlotinib monotherapy, high cadherin expression levels at the time of diagnosis were indicative of an increased sensitivity to HDACi/EGFR–TKI inhibition

(9.4 *vs.* 5.4 months; overall survival), providing the basis for a biomarker-driven validation study.[37] Additionally, panobinostat (LBH589) has shown single-agent activity in patients with Hodgkin's lymphoma who relapsed or were refractory to autologous transplantation.[38] Although the MOA of HATs is understood, few inhibitors have been described and none are currently in clinical development. The chemical and clinical tractability of this class of enzymes remains to be determined.

Histone acetylation also promotes transcription by providing a binding site for proteins involved in gene activation such as the bromodomain-containing proteins (BCPs). The dysregulation of the bromo and extraterminal (BET) family of BCPs, which includes BRD2, BRD3, BRD4 and a testis-specific protein BRDT, has been implicated in the development of cancers such as nut midline carcinoma and AML.[39–42] The most compelling demonstration that bromodomain targeting may be relevant to oncology therapy is the use of the BET inhibitors, iBet151 and JQ1. Each inhibitor showed strong efficacy in AML models as well as a nut midline carcinoma model in which the BRD4 gene is translocated to the NUT gene, leading to abnormal gene expression.[43,44] Currently, four BET inhibitors are in clinical trials: GSK525762, OTX015 and CPI-0610 are under evaluation in various carcinomas and haematological malignancies, while RVX-208 is being tested in atherosclerosis and type 2 diabetes.

9.2.2.2.2 Histone Methylation. Tight regulation of the expression, activity and recruitment of histone methyltransferases (HMTs) and histone demethylases (HDMs) is essential as a loss of control may shift the transcriptional balance, leading to inappropriate gene expression programmes that impact cell fate and the genesis of disease. Among the arginine and lysine HMTs encoded by the human genome, at least half have been implicated in the development of cancer or other diseases.[45,46] To date, the most studied HMTs in the oncology field are EZH2, DOT1L and NSD2.

EZH2 is over-expressed in many different types of cancer and elevated levels have been proposed as a molecular marker of prostate cancer progression and metastasis, as well as a driver of breast cancer aggressiveness and bladder cancer progression.[45] Yet, it remains to be determined whether EZH2 over-expression is a driver of disease progression in these solid tumour indications or a molecular marker of a particular tumour trait such as increased cell cycle progression. By contrast, it is established that gain-of-function mutations in the EZH2 catalytic domain (*e.g.* Tyr641Asn) are sufficient to drive tumorigenesis in particular subclasses of non-Hodgkin's lymphoma such as diffuse large B-cell lymphoma (DLBCL) and follicular lymphoma (FL). Moreover, these active site mutants confer sensitivity to EZH2 inhibitors, underlining the dependency of certain DLBCL and FL tumours on aberrant EZH2 activity. Currently, the two most advanced EZH2 inhibitors, EPZ-6438 and GSK2816126, are in phase I/II trials of non-Hodgkin lymphoma patients. Both compounds are potent, selective and *S*-adenosyl-methionine-competitive. GSK126 is an intravenous agent while EPZ-6438 is an oral agent. Recently, CPI-1205 a third inhibitor entered phase I trials.[47,48]

DOT1L is a histone methyltransferase that modifies nucleosomal histone H3 on lysine 79. In contrast to other lysine histone methyltransferases, DOT1L does not contain a SET domain that represents the catalytic activity of these enzymes. DOT1L has an established driver role in AML patients who harbour translocations comprised of MLL and a fusion partner such as AF4, AF9, ENL, AF10 or AF6.[49] In these scenarios, DOT1L is aberrantly targeted to DNA and its catalytic activity on lysine 79 of histone H3 contributes to up-regulation of transforming genes.[45,50] The DOT1L inhibitor EPZ 005676 is currently under evaluation in phase I trials of acute leukaemia patients with rearrangement of the *MLL* gene. EPZ 5676 is a potent *S*-adenosyl methionine competitor, selective for DOT1L.[51] A third example of an HMT relevant to oncology is NSD2 – its expression is correlated with tumour aggressiveness and poor prognosis in several types of cancer,[52] but to date, direct evidence supporting a driver role for NSD2 in tumorigenesis has only been established in specific types of multiple myeloma harbouring a t(4;14) translocation.[53]

On the opposite side of the methylation reaction, both flavin-dependent monoamine oxidase and Jumonji domain-containing histone demethylase families have been linked to tumorigenesis. Pharmacological as well as genetic evidence support the role of LSD1 (lysine-specific demethylase-1) as a driver in AML and ALL. LSD1, also known as KDM1, demethylates histone H3 through an FAD-dependent oxidative reaction. Indeed, gene silencing approaches as well as treatment with LSD1 inhibitors (single agent or in combination with ATRA) have each showed demonstrable effects on myeloid cell differentiation. Our understanding of the underlying MOA will continue to be refined as more selective LSD1 inhibitors (reversible and irreversible) enter clinical development such as those from Oryzon and GSK. Beyond LSD1, the JMJD2 and JARID families of demethylases are also the focus of drug development in epigenetic therapy, but these efforts are still in the preclinical phase.[54,55]

9.2.2.2.3 Possible Explanation for Superior Efficacy of HDACis in Haematological Malignancies. As with hypomethylating agents, the efficacy of HDAC inhibitors in solid tumours has been underwhelming, as exemplified by data from trials of panobinostat combined with bevacizumab and everolimus in advanced solid tumours,[56] as well as a phase II study of panobinostat combined with bortezomib in pancreas cancer patients who had progressed on standard of care gemcitabine-based therapy.[57] Of potential relevance, HDACs not only deacetylate histones, but also other non-histone proteins such as transcription factors, signal transducers, oncoproteins and tumour suppressor proteins.[58] Modulation of the acetylation status of these non-histone substrates may be a contributory factor in the toxicities observed in solid tumours. Key to achieving efficacy in solid tumours will not only be the identification of tolerable and efficacious doses, but also the selection and stratification of patients according to a dependency on particular acetylation, methylation or other epigenetic profiles that arise from dysregulation of the key regulatory enzymes.

9.3 Epigenetic Programmes in Development and as Predictors of Response

In this section, we will take a more in-depth look at patient stratification approaches and discuss some epigenetic programmes in development that may feature distinct responder strategies. As discussed earlier, aberrant epigenetic profiles may arise from genetic lesions such as somatic mutations, copy number/gene expression changes or genetic rearrangements that directly impact a key regulatory node. Such changes are particularly amenable to precision medicine approaches and the development of molecularly guided responder strategies. However, in some cases, a sole genetic driver may not be operative and instead, a shift in the balance of activities such as KMT/KDM may confer sensitivity to a therapeutic approach. In each of these targeted therapy scenarios, it is imperative to elucidate the underlying biology and MOA of a particular epigenetic target. In particular, non-catalytic scaffolding or recruitment functions of an epigenetic target can confound the interpretation of data, underlining the importance of tool compounds or other modulators of activity in the development of responder strategies.

9.3.1 DOT1L Inhibitor

The lead compound in this class of therapeutic agents, EPZ-5676, is currently in clinical trials for patients with MLL-rearranged AML.[50,59] MLL-rearranged leukaemia is universally associated with a chromosomal translocation impacting the MLL gene at chromosome 11q23.[40,60–62] The resultant fusion proteins invariably encompass a binding domain that recruits DOT1L to target genes and induces methylation of lysine 79 on histone H3. H3K79 methylation is an active epigenetic mark that leads to aberrant transcriptional activation of leukemogenic genes such as HOXA5 and HOXA9.[63] While DOT1L inhibitors block global H3K79 methylation in leukaemia cells, irrespective of MLL gene integrity, the inhibitor selectively kills cells with MLL translocations *in vitro*. Moreover, a DOT1L inhibitor has been shown to provide a dose-dependent survival advantage over vehicle-treated controls in preclinical *in vivo* studies.[51] EPZ-5676 recently advanced to the clinic and the emerging efficacy data are intriguing. After 5 days of dosing, a striking 90% reduction in circulating blasts was observed and this correlated with a 60% decrease in the H3K79me2 marks from circulating blasts. While the treatment had to be discontinued for CNS-related complications, these first-in-patient data yielded proof of efficacy. Furthermore, this programme represents an example of a chromosomal translocation that does not encode a new oncogenic driver, but rather confers a secondary dependency on a druggable node, by recruitment of DOT1L to key gene loci and modulation of the H3K79 methylation status. Studies are ongoing to understand the MOA of the DOT1L inhibitor and expand the target populations for such assets.

9.3.2 EZH2 Inhibitor

An EZH2 inhibitor from Epizyme recently entered phase I trials for patients with subtypes of non-Hodgkin's lymphoma (NHL) such as diffuse large B-cell lymphoma (DLBCL) or follicular lymphoma (FL). EZH2 is the SET domain KMT that represents the catalytic subunit of the multi-protein Polycomb repressive complex 2 (PRC2). This complex modulates trimethylation of lysine 27 on histone H3, which is associated with transcriptional silencing of tumour suppressors such as p16-INK4A.[64] Several mechanisms have been described that lead to hyper-trimethylation of H3K27, which is associated with tumorigenesis in a number of haematological and solid tumour indications. Causal mechanisms include over-expression or amplification of EZH2, somatic mutations in the EZH2 catalytic domain, and loss-of-function mutations in the cognate H3K27 demethylase, UTX (KDM6a). Of the various genetic events that give rise to increased H3K27 methylation, only two have proved predictive of response in preclinical studies; EZH2 somatic mutations in certain NHL subtypes and SMARCB1-deficiency in malignant rhabdoid tumour cells.[65] Interestingly several somatic EZH2 mutations, including Tyr641Asn, have also been identified in melanoma, but the pathophysiological consequences and impact on response to EZH2 inhibition remains unknown. Historically, much of the data supporting EZH2 as a pharmacologically tractable target centred on the many solid tumour malignancies that feature over-expression of EZH2. In fact, genetic ablation of EZH2 had significant effects on cellular proliferation in many of those solid tumour indications. However, elevated expression has not yet proved predictive of response to EZH2 inhibition. To date, five different EZH2 somatic mutations (Y641F, Y641N, Y641S, Y641H and A677G) have been reported in DLBCL or FL.[66] While these mutations are heterozygous, high levels of trimethylation are apparent in patient specimens. However, it is also noteworthy that heightened levels of H3K27me3 are not globally predictive of response to EZH2 inhibition. More recently, responder gene signatures have been sought, but contextual EZH2 activity appears important, because the putative signatures identified to date are not globally applicable across indications or even multiple cell lines within the same indication harbouring similar genetic events. Thus, while EZH2 remains the focus of significant drug discovery interest, the molecular determinants of response and the associated basis for indication expansion and patient selection remain elusive.

9.3.3 NSD2 Inhibitor

Development of NSD2 (WHSC1, MMSET) inhibitors is still in its infancy, but specific NSD2-centric genetic events may demarcate patient populations that will prove responsive to agents targeting this KMT. Despite a burgeoning number of cancer subtypes that show elevated NSD2 gene expression, only a subset of haematological malignancies have exhibited a dependency on this KMT to date. Specifically, a subclass of multiple myeloma cancer cells harbouring the t(4;14) chromosomal translocation showed a dependency on the NSD2 fusion derivative in RNA interference studies *in vitro*. Of note, the 15% of multiple myeloma cases that are positive for the translocation represent one of the worst prognostic subgroups. The t(4;14) translocation results in massive over-expression of NSD2, which leads to increased dimethylation of lysine 36 on histone H3. Accordingly, genetic knockdown of the NSD2 fusion attenuated proliferation and this was associated with a decrease in the H3K36me2 mark – this histone methylation mark is associated with regions that are transcriptionally active.[67] Recently, gain-of-function mutants of NSD2 (E1099K) with increased catalytic activity were identified in acute lymphoblastic leukaemia (ALL).[68] While a driver role for such mutants might be predicted, the ramifications for disease progression remain to be determined. Similarly, the exact consequences of NSD2 over-expression will remain unclear until a tool compound is available, but data from genetic studies would suggest that over-expression alone does not suffice as either a driver of tumorigenesis or a marker of responsiveness in the indications studied to date.

9.3.4 Bromodomain Inhibitor

Several BET inhibitors are currently in phase I clinical trials for oncology indications such as NUT midline carcinoma (I-Bet762) and haematological malignancies (OTX015 and CP-0610). These programmes are yielding proof of concept validation for epigenetic readers as therapeutic targets in oncology, despite the absence of a catalytic domain. Agents such as I-Bet762 and OTX-015 show highly specific binding to the bromodomains of BRD2, BRD3 and BRD4, abrogating the engagement of the bromodomain with acetyl lysine residues.[69,70] So far, the use of bromodomain inhibitors has been invoked in instances where tumours harbour fusions that include a bromodomain-containing protein or fusions that complex with a BET protein. As an example of the former, the translocation of the NUT gene with BRD4 causes mislocalisation of NUT to chromatin and this is a proposed driver event in NUT midline carcinoma. Administration of a BET inhibitor displaces the fusion protein from chromatin, induces rapid differentiation and growth arrest *in vitro*, and promotes tumour differentiation, regression and increased survival in preclinical *in vivo* studies.[43] In an illustrative example of the latter, a bromodomain inhibitor also demonstrated preclinical efficacy *in vitro* and *in vivo* against acute myeloid leukaemia cells that harboured MLL-AF4 or MLL-AF9 rearranged genes.[44] Notably, BET inhibitors may gain purchase beyond AML subtypes with MLL rearrangements, because several recent studies yielded compelling evidence that BET inhibitors may prove efficacious in neuroblastoma, glioblastoma and lung cancer, independent of effects on c-MYC transcriptional regulation.[71–73]

9.4 In Search of the Next Wave of Epigenetic Targets and Development of Responder Strategies

As highlighted in the previous examples, tumour subtypes that may prove responsive to targeted epigenetic inhibition may not necessarily feature a genetic driver event in the epigenetic target *per se*. Therefore, direct interrogation of the mutational status of an epigenetic target may not suffice to define drivers of disease and predictors of response. Accordingly, a multi-tiered profiling strategy may be required to not only define the mechanistic basis of disease progression but also identify molecular markers that might predict response.

9.4.1 Profiling Targets of Interest

While the precise details and stage gates of a profiling cascade will differ based on target, indication, MOA and epigenetic subclass, certain strategies may have general applicability. Take the example of EZH2 to illustrate how profiling approaches can help to drive understanding of the target, develop a responder strategy and expand the indication. EZH2 regulates gene transcription by methylating histone H3 in nucleosomes – increased methylation at H3K27 is associated with transcriptional repression. EZH2 mutations in the SET domain are sufficient to driver cancer cell proliferation, repress many important cell cycle regulators and confer sensitivity to EZH2 inhibition. However, such somatic mutations have only been identified in DLBCL, FL and at low frequency in melanoma. Several companies, including Epizyme/Eisai, GSK and Constellation, are developing EZH2 inhibitors following the Tyr641 mutation paradigm with the anticipation that additional applications will emerge as the knowledge base matures. In order to identify and expand the patient populations, several hypotheses for driving malignancy can be put forward and evaluated, *e.g.* mutation of EZH2 is required; EZH2 protein is over-expressed due to gene translocation or amplification in cancer cells; the level of EZH2 substrate H3K27me3 is higher in cancer cells; EZH2 alone or in complex with PRC2 components is aberrantly recruited to promoter regions. To test these and other hypotheses, a plethora of profiling approaches is available to interrogate each purported driver event.

Next-generation sequencing, whole exome sequencing, RNA sequencing and SNP arrays can provide meaningful data on somatic mutations, gene expression, copy number variations, structural events such as insertions/deletions, translocations, gene isoform changes, *etc.* If targeted expression analysis of epigenetic network genes is preferred or necessary due to budgetary or other constraints, platforms such as nanostring can provide expression data on up to 800 genes. Similarly, targeted sequencing platforms can rapidly provide data on known somatic mutations. In the realm of small RNAs, miRNA Seq will provide expression data on the known miRnome, while targeted array-based approaches can provide miRNA read-outs pertinent to subsets of disease or specific biologies.

Analysis of the methylation status of a target residue such as H3K27 is often performed by antibody-based approaches such as immunoblotting, immunohistochemistry or immunofluorescence analysis. However, in the absence of selective antibodies, proteomic-based approaches such as single reaction monitoring or the multiplex variant thereof, multiple reaction monitoring, can not only provide insight into the status of a particular mark but provide a systems level overview of the status of hundreds of different post-translational modifications. These data may prove insightful as to the status of directly proximal pharmacodynamic read-outs as well as compensatory mechanisms that are operative.

While EZH2 Tyr641 mutation induces a global increase in H3K27me3, in some instances determination of substrate modification may require the use of chromatin immunoprecipitation (ChIP)-based analysis that is more specific to targeted genes, such as RT-PCR ChIP or other variants of the technology (ChIP-Seq, rev-ChIP, *etc.*). An understanding of the mechanism by which the drug target is impacting cancer cell fate may greatly facilitate the discovery of new responder populations or biomarkers of same. The subject of our case study, EZH2, has been shown to methylate histone H3 and to repress gene transcription. As with other epigenetic protein families, non-histone proteins may also be modified by EZH2. For example, studies have suggested that EZH2 methylates the transcription factor GATA4 as well as the nuclear receptors AR and ROR.[74–76] Such data can provide guidance towards indications in which AR and ROR may play an important role such as prostate cancer and breast cancer, respectively.[75,76] Identification of additional substrates can be facilitated by proteomic-based labelling approaches such as SILAC[77] or the use of orthogonal tagging strategies that involve the generation of analogue-specific derivatives of the target enzyme.[78] The discovery of new substrates can illuminate the underlying MOA, expand target indications and identify combination therapy opportunities.

9.4.2 Substrate for Target Profiling

Substrate selection can also be governed by particular protein–protein interactions. EZH2 catalytic activity is associated with the PRC2 complex, but reports have suggested that EZH2 can act through different protein complexes or in a solo capacity and change its substrate preference accordingly.[75] Understanding the relevant substrates and context of EZH2 activity in a particular neoplastic setting may be key to the design of a relevant and efficacious inhibitor. Just as an agent may be developed with a focus on the Tyr641 mutated derivative of EZH2 rather than the wild-type enzyme, distinct chemical matter may be a better fit for EZH2 that resides in a complex with a particular composition of interacting proteins. In fact, selectivity for EZH2 in a specific complex may not only ensure correct contextual efficacy but also constrain the knock-on effects from substrates that may have safety liabilities and which could ultimately impact the therapeutic index. As a case in point, a peptide competitor has been described that inhibits the interaction of EZH2 and EED, and thereby destabilises the PRC2 complex. However, EZH2 activity is not inhibited, so this mode of intervention may be advantageous in certain non-neoplastic settings where EZH2 activity is required.[79]

As with the selection of profiling methodology, the substrate for profiling can impact the outcome of the profiling campaign and the subsequent direction of the drug discovery programme or patient stratification work-stream. Preclinical *ex vivo* models such as clonal cell lines, primary-derived lines, thick slice cultures and tumour/stroma co-culture systems are readily available options for profiling substrate. In some instances, the underlying biology of a target may mandate the growth of cells in specific conditions such as 3D, spheroid, mini-organ or in media conditions that maintain a particular state of differentiation or stemness. Conversely, a simple 2D *in vitro* culture system may suffice to enable the expected anti-proliferative or cell differentiation phenotype elicited by the therapeutic agent. Multiple preclinical *in vivo* model systems can also provide substrate for profiling such as cell-based xenografts, tumour/stroma co-implantations, patient-derived xenografts, genetically engineered mouse models, humanised models or other model organisms such as zebrafish and *Drosophila*. Beyond laboratory models, profiling patient specimens can be particularly instructive for the relevant biology – the menu of starting materials for profiling could include primary tumours, metastatic lesions, circulating or secreted fluids containing circulating tumour cells (CTCs), nucleic acids (CNAs) or microRNAs. Approaches can be indication-specific, centred on histology or subtype, or based on a particular molecular spectrum such as driver mutations in KRAS or PIK3CA, genetic events in chromatin remodelling machinery, *etc.* Moreover, comparative analyses between pretreated and treated tumours, responders and relapsed refractory, sensitive *vs.* resistant patients, may prove particularly informative for the development of responder strategies. While the inherent heterogeneity of patient specimens or higher end preclinical models such as PDXs (patient-derived xenografts) can be a positive means to expand the scope of a target identification, validation or patient stratification campaign, data interpretation also requires more stringent criteria. Analysis of patient specimens can be confounded by intratumoural heterogeneity, so multi-point sampling from a given tumour may be required to identify representative molecular traits. For example, specific somatic mutations may not be detected if profiling is conducted on a single region of the tumour. Conversely, identification of somatic mutations or other genetic events does not alone confirm driver status so a validation cascade is required to distinguish between driver and passenger events. Notwithstanding these caveats, the increasing availability of patient specimens, more complex and representative preclinical models and an expanded toolbox of profiling options will greatly facilitate the discovery of new epigenetic targets and the opportunities for parallel development of responder hypotheses.

In conclusion, precision medicine approaches are essential to ensure that quality chemical matter has a home and optimal preclinical settings are identified to maximise the probability of enabling efficacy. Moreover, while biotech companies may purposely focus on small indications that represent an unmet medical need and thereby, a fast track to FDA approval *via* orphan drug designation, pharmaceutical companies can be reluctant to deploy drug discovery resources for such niche patient populations. Accordingly, translational medicine efforts are being instigated at ever earlier stages of drug development to ensure maximal coverage of single-agent and combination therapy opportunities. The emerging preclinical and clinical data on epigenetic targets indicate great promise and the alignment with precision medicine approaches is timely given the substantial growth in the number of molecularly guided trials, which are set to expand in tandem with the rapid growth of epigenetic drug discovery programmes.

References

1. M. Esteller, *N. Engl. J. Med.*, 2008, **358**, 1148–1159.
2. K. D. Hansen, W. Timp, H. C. Bravo, S. Sabunciyan, B. Langmead, O. G. McDonald, B. Wen, H. Wu, Y. Liu, D. Diep, E. Briem, K. Zhang, R. A. Irizarry and A. P. Feinberg, *Nat. Genet.*, 2011, **43**, 768–775.
3. A. K. Maunakea, R. P. Nagarajan, M. Bilenky, T. J. Ballinger, C. D'Souza, S. D. Fouse, B. E. Johnson, C. Hong, C. Nielsen, Y. Zhao, G. Turecki, A. Delaney, R. Varhol, N. Thiessen, K. Shchors, V. M. Heine, D. H. Rowitch, X. Xing, C. Fiore, M. Schillebeeckx, S. J. Jones, D. Haussler, M. A. Marra, M. Hirst, T. Wang and J. F. Costello, *Nature*, 2010, **466**, 253–257.
4. M. Tahiliani, K. P. Koh, Y. Shen, W. A. Pastor, H. Bandukwala, Y. Brudno, S. Agarwal, L. M. Iyer, D. R. Liu, L. Aravind and A. Rao, *Science*, 2009, **324**, 930–935.
5. A. Eden, F. Gaudet, A. Waghmare and R. Jaenisch, *Science*, 2003, **300**, 455.
6. A. P. Feinberg, *Cancer Res.*, 1999, **59**, 1743s–1746s.
7. A. Kaneda and A. P. Feinberg, *Cancer Res.*, 2005, **65**, 11236–11240.
8. O. A. Stefansson, J. G. Jonasson, K. Olafsdottir, H. Hilmarsdottir, G. Olafsdottir, M. Esteller, O. T. Johannsson and J. E. Eyfjord, *Epigenetics*, 2011, **6**, 638–649.
9. T. Hansmann, G. Pliushch, M. Leubner, P. Kroll, D. Endt, A. Gehrig, S. Preisler-Adams, P. Wieacker and T. Haaf, *Hum. Mol. Genet.*, 2012, **21**, 4669–4679.
10. T. L. Chan, S. T. Yuen, C. K. Kong, Y. W. Chan, A. S. Chan, W. F. Ng, W. Y. Tsui, M. W. Lo, W. Y. Tam, V. S. Li and S. Y. Leung, *Nat. Genet.*, 2006, **38**, 1178–1183.
11. A. Dobrovic and D. Simpfendorfer, *Cancer Res.*, 1997, **57**, 3347–3350.
12. G. Deng, A. Chen, J. Hong, H. S. Chae and Y. S. Kim, *Cancer Res.*, 1999, **59**, 2029–2033.
13. E. Hatzimichael, G. Dranitsaris, A. Dasoula, L. Benetatos, J. Stebbing, T. Crook and K. L. Bourantas, *Clin. Lymphoma Myeloma*, 2009, **9**, 239–242.
14. J. F. Costello, M. C. Fruhwald, D. J. Smiraglia, L. J. Rush, G. P. Robertson, X. Gao, F. A. Wright, J. D. Feramisco, P. Peltomaki, J. C. Lang, D. E. Schuller, L. Yu, C. D. Bloomfield, M. A. Caligiuri, A. Yates, R. Nishikawa,

H. Su Huang, N. J. Petrelli, X. Zhang, M. S. O'Dorisio, W. A. Held, W. K. Cavenee and C. Plass, *Nat. Genet.*, 2000, **24**, 132–138.
15. M. Toyota, N. Ahuja, M. Ohe-Toyota, J. G. Herman, S. B. Baylin and J. P. Issa, *Proc. Natl. Acad. Sci. U. S. A.*, 1999, **96**, 8681–8686.
16. M. Toyota, M. Ohe-Toyota, N. Ahuja and J. P. Issa, *Proc. Natl. Acad. Sci. U. S. A.*, 2000, **97**, 710–715.
17. C. J. Marsit, E. A. Houseman, B. C. Christensen, K. Eddy, R. Bueno, D. J. Sugarbaker, H. H. Nelson, M. R. Karagas and K. T. Kelsey, *Cancer Res.*, 2006, **66**, 10621–10629.
18. M. Toyota, N. Ahuja, H. Suzuki, F. Itoh, M. Ohe-Toyota, K. Imai, S. B. Baylin and J. P. Issa, *Cancer Res.*, 1999, **59**, 5438–5442.
19. M. Toyota, C. Ho, N. Ahuja, K. W. Jair, Q. Li, M. Ohe-Toyota, S. B. Baylin and J. P. Issa, *Cancer Res.*, 1999, **59**, 2307–2312.
20. A. Rashid, L. Shen, J. S. Morris, J. P. Issa and S. R. Hamilton, *Am. J. Pathol.*, 2001, **159**, 1129–1135.
21. T. J. Ley, L. Ding, M. J. Walter, M. D. McLellan, T. Lamprecht, D. E. Larson, C. Kandoth, J. E. Payton, J. Baty, J. Welch, C. C. Harris, C. F. Lichti, R. R. Townsend, R. S. Fulton, D. J. Dooling, D. C. Koboldt, H. Schmidt, Q. Zhang, J. R. Osborne, L. Lin, M. O'Laughlin, J. F. McMichael, K. D. Delehaunty, S. D. McGrath, L. A. Fulton, V. J. Magrini, T. L. Vickery, J. Hundal, L. L. Cook, J. J. Conyers, G. W. Swift, J. P. Reed, P. A. Alldredge, T. Wylie, J. Walker, J. Kalicki, M. A. Watson, S. Heath, W. D. Shannon, N. Varghese, R. Nagarajan, P. Westervelt, M. H. Tomasson, D. C. Link, T. A. Graubert, J. F. DiPersio, E. R. Mardis and R. K. Wilson, *N. Engl. J. Med.*, 2010, **363**, 2424–2433.
22. C. Ye and L. Li, *Cancer Biol. Ther.*, 2014, **15**, 10–15.
23. C. G. Lian, Y. Xu, C. Ceol, F. Wu, A. Larson, K. Dresser, W. Xu, L. Tan, Y. Hu, Q. Zhan, C. W. Lee, D. Hu, B. Q. Lian, S. Kleffel, Y. Yang, J. Neiswender, A. J. Khorasani, R. Fang, C. Lezcano, L. M. Duncan, R. A. Scolyer, J. F. Thompson, H. Kakavand, Y. Houvras, L. I. Zon, M. C. Mihm Jr, U. B. Kaiser, T. Schatton, B. A. Woda, G. F. Murphy and Y. G. Shi, *Cell*, 2012, **150**, 1135–1146.
24. B. A. Orr, M. C. Haffner, W. G. Nelson, S. Yegnasubramanian and C. G. Eberhart, *PloS one*, 2012, **7**, e41036.
25. P. S. Ward, J. Patel, D. R. Wise, O. Abdel-Wahab, B. D. Bennett, H. A. Coller, J. R. Cross, V. R. Fantin, C. V. Hedvat, A. E. Perl, J. D. Rabinowitz, M. Carroll, S. M. Su, K. A. Sharp, R. L. Levine and C. B. Thompson, *Cancer Cell*, 2010, **17**, 225–234.
26. E. R. Mardis, L. Ding, D. J. Dooling, D. E. Larson, M. D. McLellan, K. Chen, D. C. Koboldt, R. S. Fulton, K. D. Delehaunty, S. D. McGrath, L. A. Fulton, D. P. Locke, V. J. Magrini, R. M. Abbott, T. L. Vickery, J. S. Reed, J. S. Robinson, T. Wylie, S. M. Smith, L. Carmichael, J. M. Eldred, C. C. Harris, J. Walker, J. B. Peck, F. Du, A. F. Dukes, G. E. Sanderson, A. M. Brummett, E. Clark, J. F. McMichael, R. J. Meyer, J. K. Schindler, C. S. Pohl, J. W. Wallis, X. Shi, L. Lin, H. Schmidt, Y. Tang, C. Haipek, M. E. Wiechert, J. V. Ivy, J. Kalicki, G. Elliott, R. E. Ries, J. E. Payton, P. Westervelt, M. H. Tomasson,

M. A. Watson, J. Baty, S. Heath, W. D. Shannon, R. Nagarajan, D. C. Link, M. J. Walter, T. A. Graubert, J. F. DiPersio, R. K. Wilson and T. J. Ley, *N. Engl. J. Med.*, 2009, **361**, 1058–1066.
27. M. E. Figueroa, O. Abdel-Wahab, C. Lu, P. S. Ward, J. Patel, A. Shih, Y. Li, N. Bhagwat, A. Vasanthakumar, H. F. Fernandez, M. S. Tallman, Z. Sun, K. Wolniak, J. K. Peeters, W. Liu, S. E. Choe, V. R. Fantin, E. Paietta, B. Lowenberg, J. D. Licht, L. A. Godley, R. Delwel, P. J. Valk, C. B. Thompson, R. L. Levine and A. Melnick, *Cancer Cell*, 2010, **18**, 553–567.
28. P. Fenaux, G. J. Mufti, E. Hellstrom-Lindberg, V. Santini, C. Finelli, A. Giagounidis, R. Schoch, N. Gattermann, G. Sanz, A. List, S. D. Gore, J. F. Seymour, J. M. Bennett, J. Byrd, J. Backstrom, L. Zimmerman, D. McKenzie, C. Beach, L. R. Silverman and M. D. S. S. S. G. International Vidaza High-Risk, *Lancet Oncol.*, 2009, **10**, 223–232.
29. J. A. Gollob and C. J. Sciambi, *Clin. Cancer Res.*, 2007, **13**, 5219–5225.
30. K. Appleton, H. J. Mackay, I. Judson, J. A. Plumb, C. McCormick, G. Strathdee, C. Lee, S. Barrett, S. Reade, D. Jadayel, A. Tang, K. Bellenger, L. Mackay, A. Setanoians, A. Schatzlein, C. Twelves, S. B. Kaye and R. Brown, *Am. J. Clin. Oncol.*, 2007, **25**, 4603–4609.
31. J. Liu, Y. S. Xie, F. L. Wang, L. J. Zhang, Y. Zhang and H. S. Luo, *Biomed. Pharmacother.*, 2013, **67**, 78–87.
32. D. Matei, F. Fang, C. Shen, J. Schilder, A. Arnold, Y. Zeng, W. A. Berry, T. Huang and K. P. Nephew, *Cancer Res.*, 2012, **72**, 2197–2205.
33. C. H. Arrowsmith, C. Bountra, P. V. Fish, K. Lee and M. Schapira, *Nat. Rev. Drug Discovery*, 2012, **11**, 384–400.
34. M. D. Shahbazian and M. Grunstein, *Annu. Rev. Biochem.*, 2007, **76**, 75–100.
35. P. Filippakopoulos, S. Picaud, M. Mangos, T. Keates, J. P. Lambert, D. Barsyte-Lovejoy, I. Felletar, R. Volkmer, S. Muller, T. Pawson, A. C. Gingras, C. H. Arrowsmith and S. Knapp, *Cell*, 2012, **149**, 214–231.
36. T. Kouzarides, *Cell*, 2007, **128**, 693–705.
37. S. E. Witta, R. M. Jotte, K. Konduri, M. A. Neubauer, A. I. Spira, R. L. Ruxer, M. Varella-Garcia, P. A. Bunn Jr. and F. R. Hirsch, *Am. J. Clin. Oncol.*, 2012, **30**, 2248–2255.
38. A. Younes, A. Sureda, D. Ben-Yehuda, P. L. Zinzani, T. C. Ong, H. M. Prince, S. J. Harrison, M. Kirschbaum, P. Johnston, J. Gallagher, C. Le Corre, A. Shen and A. Engert, *Am. J. Clin. Oncol.*, 2012, **30**, 2197–2203.
39. C. A. French, I. Miyoshi, J. C. Aster, I. Kubonishi, T. G. Kroll, P. Dal Cin, S. O. Vargas, A. R. Perez-Atayde and J. A. Fletcher, *Am. J. Pathol.*, 2001, **159**, 1987–1992.
40. C. A. French, I. Miyoshi, I. Kubonishi, H. E. Grier, A. R. Perez-Atayde and J. A. Fletcher, *Cancer Res.*, 2003, **63**, 304–307.
41. J. Zuber, J. Shi, E. Wang, A. R. Rappaport, H. Herrmann, E. A. Sison, D. Magoon, J. Qi, K. Blatt, M. Wunderlich, M. J. Taylor, C. Johns, A. Chicas, J. C. Mulloy, S. C. Kogan, P. Brown, P. Valent, J. E. Bradner, S. W. Lowe and C. R. Vakoc, *Nature*, 2011, **478**, 524–528.
42. B. Huang, X. D. Yang, M. M. Zhou, K. Ozato and L. F. Chen, *Mol. Cell. Biol.*, 2009, **29**, 1375–1387.

43. P. Filippakopoulos, J. Qi, S. Picaud, Y. Shen, W. B. Smith, O. Fedorov, E. M. Morse, T. Keates, T. T. Hickman, I. Felletar, M. Philpott, S. Munro, M. R. McKeown, Y. Wang, A. L. Christie, N. West, M. J. Cameron, B. Schwartz, T. D. Heightman, N. La Thangue, C. A. French, O. Wiest, A. L. Kung, S. Knapp and J. E. Bradner, *Nature*, 2010, **468**, 1067–1073.
44. M. A. Dawson, R. K. Prinjha, A. Dittmann, G. Giotopoulos, M. Bantscheff, W. I. Chan, S. C. Robson, C. W. Chung, C. Hopf, M. M. Savitski, C. Huthmacher, E. Gudgin, D. Lugo, S. Beinke, T. D. Chapman, E. J. Roberts, P. E. Soden, K. R. Auger, O. Mirguet, K. Doehner, R. Delwel, A. K. Burnett, P. Jeffrey, G. Drewes, K. Lee, B. J. Huntly and T. Kouzarides, *Nature*, 2011, **478**, 529–533.
45. M. Albert and K. Helin, *Semin. Cell Dev. Biol.*, 2010, **21**, 209–220.
46. S. Pal, R. A. Baiocchi, J. C. Byrd, M. R. Grever, S. T. Jacob and S. Sif, *EMBO J.*, 2007, **26**, 3558–3569.
47. S. K. Knutson, S. Kawano, Y. Minoshima, N. M. Warholic, K. C. Huang, Y. Xiao, T. Kadowaki, M. Uesugi, G. Kuznetsov, N. Kumar, T. J. Wigle, C. R. Klaus, C. J. Allain, A. Raimondi, N. J. Waters, J. J. Smith, M. Porter-Scott, R. Chesworth, M. P. Moyer, R. A. Copeland, V. M. Richon, T. Uenaka, R. M. Pollock, K. W. Kuntz, A. Yokoi and H. Keilhack, *Mol. Cancer Ther.*, 2014, **13**, 842–854.
48. M. T. McCabe, H. M. Ott, G. Ganji, S. Korenchuk, C. Thompson, G. S. Van Aller, Y. Liu, A. P. Graves, A. Della Pietra, 3rd, E. Diaz, L. V. LaFrance, M. Mellinger, C. Duquenne, X. Tian, R. G. Kruger, C. F. McHugh, M. Brandt, W. H. Miller, D. Dhanak, S. K. Verma, P. J. Tummino and C. L. Creasy, *Nature*, 2012, **492**, 108–112.
49. Y. Okada, Q. Feng, Y. Lin, Q. Jiang, Y. Li, V. M. Coffield, L. Su, G. Xu and Y. Zhang, *Cell*, 2005, **121**, 167–178.
50. S. R. Daigle, E. J. Olhava, C. A. Therkelsen, C. R. Majer, C. J. Sneeringer, J. Song, L. D. Johnston, M. P. Scott, J. J. Smith, Y. Xiao, L. Jin, K. W. Kuntz, R. Chesworth, M. P. Moyer, K. M. Bernt, J. C. Tseng, A. L. Kung, S. A. Armstrong, R. A. Copeland, V. M. Richon and R. M. Pollock, *Cancer Cell*, 2011, **20**, 53–65.
51. S. R. Daigle, E. J. Olhava, C. A. Therkelsen, A. Basavapathruni, L. Jin, P. A. Boriack-Sjodin, C. J. Allain, C. R. Klaus, A. Raimondi, M. P. Scott, N. J. Waters, R. Chesworth, M. P. Moyer, R. A. Copeland, V. M. Richon and R. M. Pollock, *Blood*, 2013, **122**, 1017–1025.
52. M. Morishita and E. di Luccio, *Biochim. Biophys. Acta*, 2011, **1816**, 158–163.
53. M. Chesi, E. Nardini, R. S. Lim, K. D. Smith, W. M. Kuehl and P. L. Bergsagel, *Blood*, 1998, **92**, 3025–3034.
54. Z. Q. Yang, I. Imoto, Y. Fukuda, A. Pimkhaokham, Y. Shimada, M. Imamura, S. Sugano, Y. Nakamura and J. Inazawa, *Cancer Res.*, 2000, **60**, 4735–4739.
55. G. Liu, A. Bollig-Fischer, B. Kreike, M. J. van de Vijver, J. Abrams, S. P. Ethier and Z. Q. Yang, *Oncogene*, 2009, **28**, 4491–4500.
56. J. H. Strickler, A. N. Starodub, J. Jia, K. L. Meadows, A. B. Nixon, A. Dellinger, M. A. Morse, H. E. Uronis, P. K. Marcom, S. Y. Zafar, S. T. Haley and H. I. Hurwitz, *Cancer Chemother. Pharmacol.*, 2012, **70**, 251–258.
57. H. Wang, Q. Cao and A. Z. Dudek, *Anticancer Res.*, 2012, **32**, 1027–1031.

58. B. N. Singh, G. Zhang, Y. L. Hwa, J. Li, S. C. Dowdy and S. W. Jiang, *Expert Rev. Anticancer Ther.*, 2010, **10**, 935–954.
59. K. M. Bernt, N. Zhu, A. U. Sinha, S. Vempati, J. Faber, A. V. Krivtsov, Z. Feng, N. Punt, A. Daigle, L. Bullinger, R. M. Pollock, V. M. Richon, A. L. Kung and S. A. Armstrong, *Cancer Cell*, 2011, **20**, 66–78.
60. J. L. Hess and B. A. Hug, *Proc. Natl. Acad. Sci. U. S. A.*, 2004, **101**, 16985–16986.
61. A. V. Krivtsov and S. A. Armstrong, *Nat. Rev. Cancer*, 2007, **7**, 823–833.
62. J. Yan, J. Diaz, J. Jiao, R. Wang and J. You, *J. Biol. Chem.*, 2011, **286**, 27663–27675.
63. A. V. Krivtsov, Z. Feng, M. E. Lemieux, J. Faber, S. Vempati, A. U. Sinha, X. Xia, J. Jesneck, A. P. Bracken, L. B. Silverman, J. L. Kutok, A. L. Kung and S. A. Armstrong, *Cancer Cell*, 2008, **14**, 355–368.
64. A. P. Bracken, D. Kleine-Kohlbrecher, N. Dietrich, D. Pasini, G. Gargiulo, C. Beekman, K. Theilgaard-Monch, S. Minucci, B. T. Porse, J. C. Marine, K. H. Hansen and K. Helin, *Genes Dev.*, 2007, **21**, 525–530.
65. S. K. Knutson, N. M. Warholic, T. J. Wigle, C. R. Klaus, C. J. Allain, A. Raimondi, M. Porter Scott, R. Chesworth, M. P. Moyer, R. A. Copeland, V. M. Richon, R. M. Pollock, K. W. Kuntz and H. Keilhack, *Proc. Natl. Acad. Sci. U. S. A.*, 2013, **110**, 7922–7927.
66. R. D. Morin, N. A. Johnson, T. M. Severson, A. J. Mungall, J. An, R. Goya, J. E. Paul, M. Boyle, B. W. Woolcock, F. Kuchenbauer, D. Yap, R. K. Humphries, O. L. Griffith, S. Shah, H. Zhu, M. Kimbara, P. Shashkin, J. F. Charlot, M. Tcherpakov, R. Corbett, A. Tam, R. Varhol, D. Smailus, M. Moksa, Y. Zhao, A. Delaney, H. Qian, I. Birol, J. Schein, R. Moore, R. Holt, D. E. Horsman, J. M. Connors, S. Jones, S. Aparicio, M. Hirst, R. D. Gascoyne and M. A. Marra, *Nat. Genet.*, 2010, **42**, 181–185.
67. E. Martinez-Garcia, R. Popovic, D. J. Min, S. M. Sweet, P. M. Thomas, L. Zamdborg, A. Heffner, C. Will, L. Lamy, L. M. Staudt, D. L. Levens, N. L. Kelleher and J. D. Licht, *Blood*, 2011, **117**, 211–220.
68. J. A. Oyer, X. Huang, Y. Zheng, J. Shim, T. Ezponda, Z. Carpenter, M. Allegretta, C. I. Okot-Kotber, J. P. Patel, A. Melnick, R. L. Levine, A. Ferrando, A. D. Mackerell Jr, N. L. Kelleher, J. D. Licht and R. Popovic, *Leukemia*, 2014, **28**, 198–201.
69. M. A. Dawson, T. Kouzarides and B. J. Huntly, *N. Engl. J. Med.*, 2012, **367**, 647–657.
70. K. I. J. K. Noel, S. Ooike, K. Sugahara, H. Nakamura and M. Daibata, *Mol. Cancer Ther.*, 2013, **12**(Suppl 11), C244.
71. A. Wyce, G. Ganji, K. N. Smitheman, C. W. Chung, S. Korenchuk, Y. Bai, O. Barbash, B. Le, P. D. Craggs, M. T. McCabe, K. M. Kennedy-Wilson, L. V. Sanchez, R. L. Gosmini, N. Parr, C. F. McHugh, D. Dhanak, R. K. Prinjha, K. R. Auger and P. J. Tummino, *PloS one*, 2013, **8**, e72967.
72. Z. Cheng, Y. Gong, Y. Ma, K. Lu, X. Lu, L. A. Pierce, R. C. Thompson, S. Muller, S. Knapp and J. Wang, *Clin. Cancer Res.*, 2013, **19**, 1748–1759.
73. W. W. Lockwood, K. Zejnullahu, J. E. Bradner and H. Varmus, *Proc. Natl. Acad. Sci. U. S. A.*, 2012, **109**, 19408–19413.

74. A. He, X. Shen, Q. Ma, J. Cao, A. von Gise, P. Zhou, G. Wang, V. E. Marquez, S. H. Orkin and W. T. Pu, *Genes Dev.*, 2012, **26**, 37–42.
75. K. Xu, Z. J. Wu, A. C. Groner, H. H. He, C. Cai, R. T. Lis, X. Wu, E. C. Stack, M. Loda, T. Liu, H. Xu, L. Cato, J. E. Thornton, R. I. Gregory, C. Morrissey, R. L. Vessella, R. Montironi, C. Magi-Galluzzi, P. W. Kantoff, S. P. Balk, X. S. Liu and M. Brown, *Science*, 2012, **338**, 1465–1469.
76. J. M. Lee, J. S. Lee, H. Kim, K. Kim, H. Park, J. Y. Kim, S. H. Lee, I. S. Kim, J. Kim, M. Lee, C. H. Chung, S. B. Seo, J. B. Yoon, E. Ko, D. Y. Noh, K. I. Kim, K. K. Kim and S. H. Baek, *Mol. Cell*, 2012, **48**, 572–586.
77. S. E. Ong, B. Blagoev, I. Kratchmarova, D. B. Kristensen, H. Steen, A. Pandey and M. Mann, *Mol. Cell. Proteomics*, 2002, **1**, 376–386.
78. A. C. Bishop, J. A. Ubersax, D. T. Petsch, D. P. Matheos, N. S. Gray, J. Blethrow, E. Shimizu, J. Z. Tsien, P. G. Schultz, M. D. Rose, J. L. Wood, D. O. Morgan and K. M. Shokat, *Nature*, 2000, **407**, 395–401.
79. W. Kim, G. H. Bird, T. Neff, G. Guo, M. A. Kerenyi, L. D. Walensky and S. H. Orkin, *Nat. Chem. Biol.*, 2013, **9**, 643–650.

CHAPTER 10

Dosing – When Less is More

KURINJI PANDIYAN[a,b], PETER A. JONES[b], AND STEPHEN B. BAYLIN*[a]

[a]The Sidney Kimmel Comprehensive Cancer Center, 1650 Orleans Street, Suite 5401, Baltimore, MD 21287, USA; [b]Norris Comprehensive Cancer Center, 1441 Eastlake Avenue, NOR 7341, Los Angeles, CA 90089, USA
*E-mail: sbaylin1@jhmi.edu

10.1 Introduction

Waddington was well ahead of his time when he recognised the importance of epigenetics, a term that he coined to describe "the causal interactions between genes and their products, which bring the phenotype into being".[1] The definition of epigenetics has been modified since then by Arthur Riggs and others as "the study of mitotically and/or meiotically heritable changes in gene function that cannot be explained by changes in DNA sequence". Most functionally, today, the term embodies not only the concept of stabilising cell phenotypes but doing so through providing a memory for, rather than rapid switches in, gene expression patterns that sub-serve these. In all of this, the basic concept posited by Waddington as to the importance of epigenetics in controlling gene expression has withstood the test of time.

In our current understanding, epigenetic mechanisms help control expression of both canonical coding genes and non-coding regulatory RNA species, and include DNA methylation changes at predominantly CpG sites in humans, modification of histone tails, packaging of DNA into nucleosomes, and control of nucleosome positioning or nucleosome

RSC Drug Discovery Series No. 48
Epigenetics for Drug Discovery
Edited by Nessa Carey

Published by the Royal Society of Chemistry, www.rsc.org

remodelling.[2–5] In concert, these dynamics define the epigenetic landscape of a cell, impacting gene expression and, hence, cell state definition.

Over the past decade or more, there has been much excitement over understanding how abnormalities of epigenetic mechanisms may contribute to disease states and especially cancer. It has long been believed that genetic changes are the primary causal events in tumorigenesis, and certainly these provide critical driving events. However, it is being increasingly established that a variety of epigenetic aberrations seen in cancers can complement genetic abnormalities in tumorigenesis and some are almost certainly driving events for malignant potential.[6–9] The first, and most studied, of these cancer epigenetic abnormalities are simultaneous losses and gains of DNA methylation.[8,10–12] While both may have their importance, more is currently understood about focal gains of this DNA modification that occur in CpG-rich regions, or CpG islands of gene promoters, 90% of which normally have no DNA methylation at these sites at any time during development.[8–12] Thus, since the identification of the first such cancer-specific, *de novo*, hypermethylated gene promoters, numerous papers have observed these epigenetic changes in genes that encode for proteins that are tumour suppressive in function, such as cell cycle checkpoint proteins,[13] DNA damage repair proteins[14] and adhesion proteins.[15] Global hypomethylation in tumours has also been implicated in destabilising the genome and activating proto-oncogenes.[16–18]

More recently, over-expression of the repressive Polycomb machinery that deposits the trimethylated mark on the 27th lysine of histone H3 (H3K27) has been associated with tumorigenesis.[19] Alterations in the expression state of other components of the epigenetic machinery, such as chromatin remodellers, histone methyltransferases, demethylases and deacetylases have all been observed in tumours.[6] Most excitingly, mutations in genes coding for proteins mediating many of the above regulatory steps for epigenetic control are being found frequently in many cancer types.[6,10,20] All of these findings indicate that the faithful inheritance of epigenetic processes is critical to the maintenance of the non-tumorigenic state.

A key aspect of dissecting all of the above steps in regulation of the cancer epigenome is the excitement revolving around the fact that, unlike mutations, epigenetic abnormalities are potentially reversible with respect to pharmacological intervention for cancer management. Even though epigenetic states are heritable, they are dynamic and do not change the primary DNA sequence, which makes them excellent targets for drug development.[21–24] This is well illustrated, especially for the mutations referred to above, in genes directly controlling the epigenome. Although the mutations themselves would not be easily reversible, their downstream consequences for epigenetic changes may be much more readily targeted. Pursuant to these possibilities, Figure 10.1 illustrates the anatomy and different components of the epigenome and some potentially druggable targets. In this chapter we will discuss the evolution of epigenetic therapy over the last four decades, our current understanding of the most effective way to use agents that are

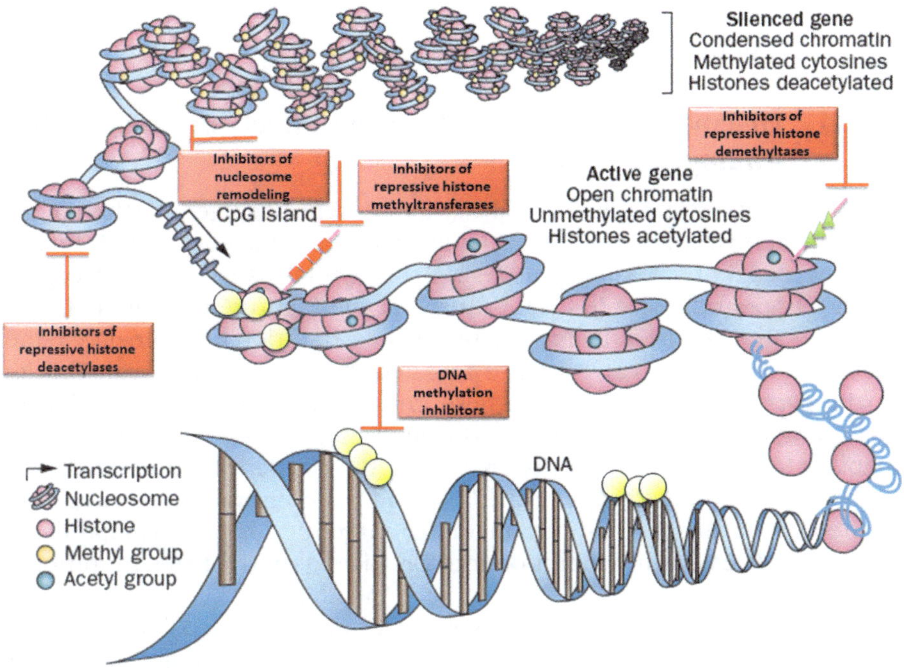

Figure 10.1 Anatomy of the epigenome, associated regulation of gene expression, and examples of druggable steps that could be potentially useful for cancer therapies. Gene expression is controlled in the promoter regions by a combination of DNA methylation and chromatin configuration. In normal cells, gene expression is silenced by condensing nucleosomes, methylating (at cytosine) DNA and deacetylating histones. By contrast, active genes, shown here for a CpG island-containing promoter, are those with open nucleosome spacing around the transcription start site, lack of DNA methylation in the region, and presence of acetylated histones. CpG islands that are rich in cytosine and guanine – and are typically unmethylated to promote gene expression – can be epigenetically silenced by hypermethylation in cancer. Steps in chromatin regulation that can be manipulated by small molecules which have been, or are being, developed are shown.

now clinically available and how these principles may help shape our future approaches towards newer drug development, regulatory studies and clinical efficacy assessment.

10.2 Shifting Paradigm in the Use of DNA Methylation Inhibitors

Although exploration, on a major scale, of targeting epigenetic abnormalities for cancer therapy is a relatively recent area of research, the field has seen a dramatic evolution in its brief history. The initial approach to epigenetic therapy involved treating with maximum tolerated doses of drugs available.

More recently the realisation has surfaced that lower doses of epigenetic drugs may confer the most therapeutic benefit.

10.2.1 Discovery of the Demethylating Effects of Nucleoside Analogues

One of the best-studied classes of drugs in the arena of epigenetic therapies is the DNA methyltransferase (DNMT) inhibitors, 5-azanucleoside (AZA, Vidaza) and the deoxyribose analogue, 5-aza-2′-deoxynucleoside (DAC, Dacogen). These compounds were first synthesised at the Institute of Organic Chemistry and Biochemistry in Prague in 1964 as traditional cytotoxic chemotherapeutic agents.[25] Later studies revealed that the C5 to N5 substitution in the pyrimidine ring renders these groups immune to methylation on incorporation of the drugs into DNA where they replace cytosines.[26,27] The inability to methylate these sites results in the covalent binding of DNMTs to these incorporated drugs in the DNA and the eventual cellular depletion of the DNMTs due to ubiquitin-mediated targeting for proteasomal degradation.[28–30] Because incorporation into DNA is required for the inhibition of DNMTs, these drugs are DNA replication-dependent.[31,32]

After initial synthesis, AZA was soon identified as a nucleoside antimetabolite with clinical specificity for acute myelogenous leukaemia.[33] It was only two decades later that the demethylating effects of these drugs were identified, a chance discovery when the drugs were found to induce muscle differentiation in mouse embryonic cells *via* DNA demethylation.[26,31] Subsequent studies revealed that these nucleoside analogues incorporate into DNA and covalently sequester DNA methyltransferases, targeting them for degradation.[27,34] The depletion of methyltransferases can thus reactivate the expression of genes that are aberrantly silenced in association with abnormal, promoter region DNA methylation.[35]

Since the discovery of the demethylating effects of AZA and DAC, these drugs have been approved by the Food and Drug Administration (FDA) for myeloid malignancies, in 2004 and 2006, respectively. The clinical efficacy leading to this approval emerged only after the realisation that doses needed to be lowered as outlined below. Because AZA and DAC are subject to intracellular deamination and are not soluble in the aqueous phase, more stable analogues with oral bioavailability such as Zebularine and SGI-110 have been developed. These have anti-tumour activity *in vitro* and are being tested in clinical trials.[36–38] Since the incorporation of nucleoside analogues into DNA is a concern, non-nucleoside inhibitors of DNMTs, such as SGI-1027 and RG108, have also been developed but have shown limited therapeutic potential.[39–41] Although the incorporation of drugs such as AZA and DAC into DNA is a concern from the perspective of DNA damage, this allows for the specific targeting of DNMT inhibition in dividing cells, such as tumour cells, which is a great advantage. Also, the degradation of the proteins *via* this process may result in loss of potential scaffolding functions of DNMTs, which have experimental transcription repression activities,[42–45] and this may also provide a therapeutic benefit.

10.2.2 Cytotoxicity of High Doses

Even prior to the identification of the demethylating effects of these nucleoside analogues, AZA was used in clinical trials as early as the 1960s. The instability of AZA in the aqueous phase made the drug very challenging to deliver.[46,47] Trials over the following decade used high doses of up to 500 mg m^{-2} per week, which produced disappointing results, principally due to severe adverse effects such as nausea, vomiting and hepatic toxicity.[48] Concurrently, reports revealing that these nucleoside analogues induced chromosome breakage[49] and that their by-products triggered cytotoxicity by interfering with *de novo* thymidylate synthesis were published.[50] Thus, documentation of unfavourable effects post-AZA treatment increased and the therapeutic benefit in solid tumours was found to be disappointing.[48] Some initial responses were seen in leukaemias such as AML, the pre-leukaemic disorder, myelodysplastic syndrome (MDS), and chronic myelogenous leukaemia (CML).[51] The dose regimens in these leukaemia trials involved 1- to 3 day high-dose treatments of the drug.

In the early days after the discovery of the hypomethylating effects of nucleoside analogues an interesting phenomenon was observed: AZA was found to be effective in inducing differentiation of myotubes only within a narrow dose range (Figure 10.2). Both higher and lower concentrations

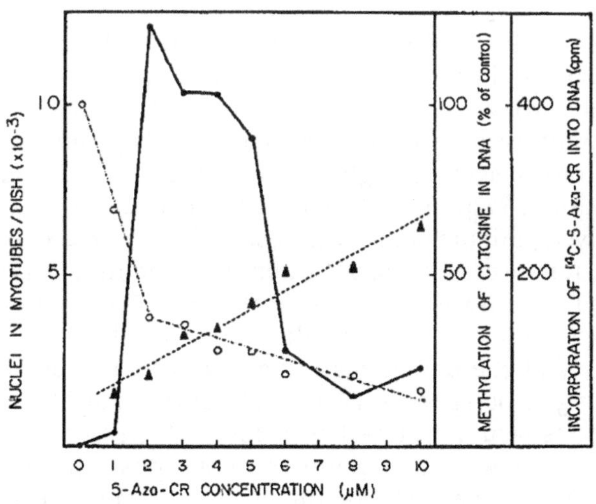

Figure 10.2 Induction of myotubes and production of DNA demethylation by 5-azacytidine (5-aza-CR). Ten T1/2 immortalised mouse fibroblast cells were exposed to the indicated concentrations of 5-aza-CR for 24 hours, maintained for 10–12 days and then fed and scored for the number of nuclei in multinucleated cells (closed circles, no./dish on left *y*-axis). The inhibition of incorporation of *H*-methyl groups into 5-methylcytosine in newly synthesised DNA (open circles) was measured for each concentration (proximal, right *y*-axis as % of control). The incorporation of C^{14}-5-aza-CR into DNA is also shown as a function of concentration (triangles, distal, right *y*-axis as cpm).

were ineffective at eliciting differentiation of these cells into multinucleated muscle cells, although the higher dosage corresponded to higher DNA incorporation.[26] This finding introduced the notion that higher doses of nucleoside analogues are less effective at inducing biological responses. Despite this observation, the epigenetics field, as noted above, moved towards escalating the dose of the drugs to the maximum tolerated dose (MTD) in clinical trials.

Studies in patients with osteogenic sarcoma and other solid tumours continued to reveal severe toxicity caused by the nucleoside analogues.[52,53] A study by Momparler and colleagues on patients with relapsed paediatric leukaemia demonstrated overall response rates of 37% with DAC treatment (37–81 mg kg^{-1}, 40–60 hours of treatment). Their results also suggested dose-dependent activity.[54] This triggered subsequent studies to continue the use of high doses of the agents in attempts to elicit therapeutic response. Wijermans and colleagues treated 21 elderly patients afflicted with MDS with 50 mg m^{-2} of DAC per day for 3 continuous days.[55] Although this produced a response rate of 54%, significant myelotoxicity resulted in the death of 17% of the patients. A subsequent study by the group using 45 mg m^{-2} of DAC in another cohort of MDS patients resulted in death due to toxicity in 7% of enrolled patients.[56] Similar high cytotoxicity and patient deaths were also observed in non-small cell lung cancer on treatment of patients with 67 mg m^{-2} of DAC along with 33 mg m^{-2} of cisplatin.[57]

After these decades of disappointing trials with high cytotoxicity and ambiguity as to whether AZA's efficacy as first line and salvage therapy in MDS, CML and AML was due to its cytotoxic or demethylation effects, the scientific community moved to the realisation that epigenetic drugs, unlike traditional cytotoxic agents, might require lower doses for administration of maximal patient responses.[51] Phase I/II sickle cell anaemia trials that followed involved patients being administered a low dose of 0.3 mg kg^{-1} for 5 days a week, for 2 weeks. Such treatment resulted in re-expression of foetal haemoglobin and no associated side effects.[58] This result reminded the field that early studies on myotube differentiation had already suggested that the efficacy of the demethylating agents was lost at higher doses.[26] With this changed perspective that maximal demethylation and, hence, anticancer effects would require lower doses of the demethylating agent, two landmark trials were done, aiming at the possibilities for reversing gene promoter DNA hypermethylation. Silverman and colleagues, followed by Issa and colleagues, showed that lower doses of DNA demethylating agents could result in high response rates in patients with various haematological malignancies.[59,60] The findings from this work significantly impacted the field and allowed for the widespread espousal of low-dose drug regimens in clinical trials. Figure 10.3 captures some key clinical and basic science developments that have shaped epigenetic therapeutics over the last few decades.

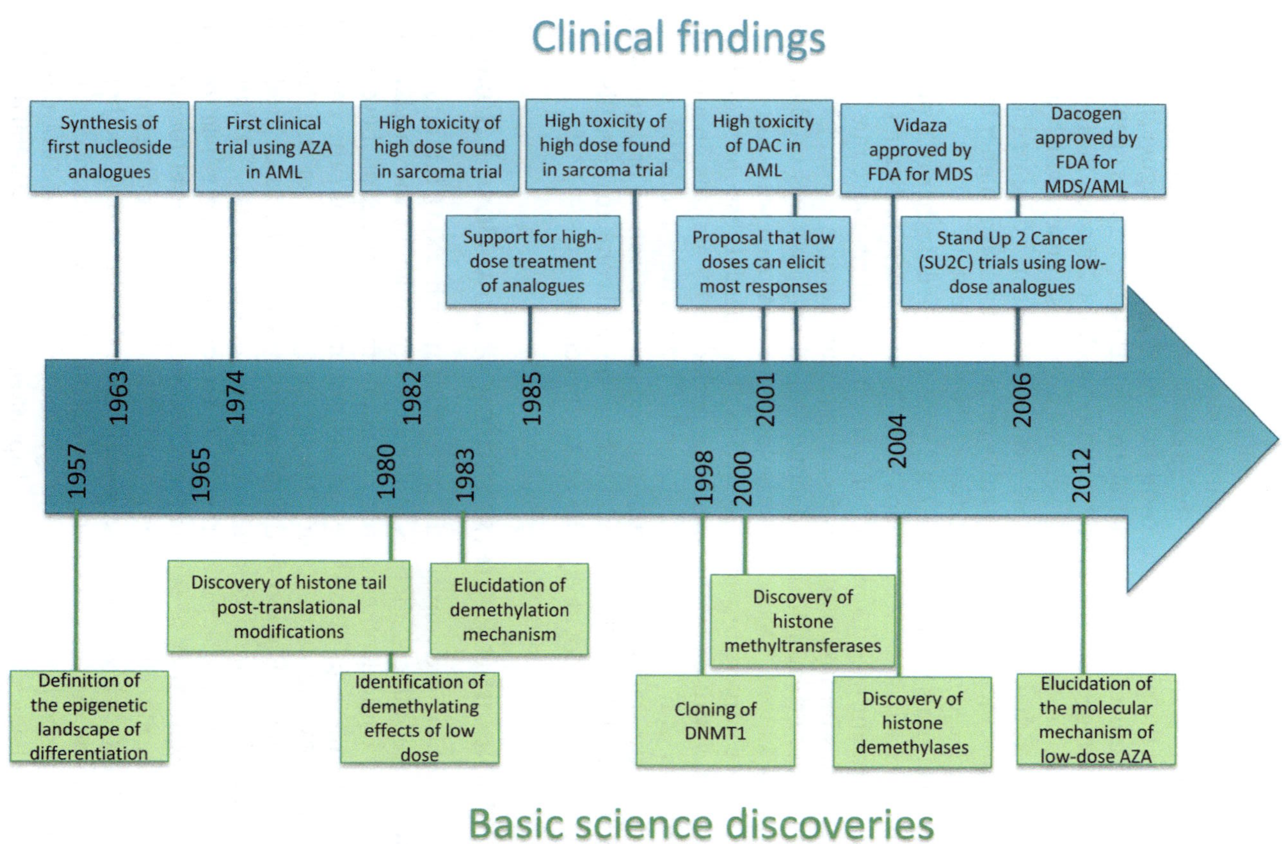

Figure 10.3 Schematic for the timing of some important basic discoveries in relationship to key observations for the concept of epigenetic therapy for cancer.

10.2.3 Therapeutic Benefits of Dose De-Escalation

Vidaza (AZA) was approved by the FDA for the treatment of MDS following the successful Cancer and Leukemia Group B (CALGB) trial 9221, wherein a complete response of 15.7% was demonstrated.[61–63] Concurrently, clinical trials confirmed that treatment with low-dose regimens of the nucleoside analogues were associated with the greatest responses and lowest toxicities.[59,60,62,64] In the phase I study conducted by Issa and colleagues, patients with relapsed and refractory leukaemias received DAC at 5, 10, 15 or 20 mg m^{-2} doses for prolonged durations of 5 days per week for 2 weeks. Besides myelosuppression, no adverse effects were seen at these doses, which are significantly lower than the MTD. Of interest, the 15 mg m^{-2} dose was found to induce the most therapeutic benefit (65% of patients), much higher than the response produced by the highest dose (11% of patients).[60]

The clinical trials with low-dose DAC were extended further in a phase II study that focused on patient groups with CML who were resistant to imatinib. On treatment with 15 mg m^{-2} of DAC, an overall response rate of 54% was observed, with myelosuppression as the only adverse effect.[59] The efficacy of low-dose treatment in inducing clinical response was validated by other groups as well.[65] More recently, low-dose AZA used with carboplatin resulted in a complete response as well as three partial responses and ten stable diseases with an average response period of 7.5 months, in a cohort of 29 patients with ovarian cancer.[66] Similar results were found by another group upon treatment of platinum-resistant ovarian cancer patients with carboplatin and DAC.[67] These results helped the field come to the realisation that higher dosing is not always the most efficacious.

10.3 Molecular Understanding of Low-Dose Efficacy

Although studies have confirmed that treatment with DNMT inhibitors can yield anticancer effects, there is still much debate about the mechanism behind these clinical observations. Multiple modes of action have been attributed to these inhibitors: cytotoxicity that results from the covalent binding to DNA and subsequent DNA damage;[68] depletion of the DNMTs that result in demethylation of the genome; or cytotoxicity associated with the analogue itself upon complete depletion of the DNMTs.[69] Doubt has remained for decades as to whether this cytotoxicity or demethylation followed by re-expression of tumour-suppressive genes is causal for tumour inhibition. The efficacy of low doses has led to the understanding that cytotoxicity induced by high doses does not allow for the effects of demethylation to reprogramme the cells to a non-tumorigenic state, which is necessary for a persistent response. Instead, off-target side effects such as DNA damage response result in cellular toxicity and apoptosis, as well as collateral damage to normal cells, hence preventing a durable anti-tumour effect.

Our group recently demonstrated in AML, breast, colorectal and lung cancer cell lines, as well as patient tissue, that transient, nanomolar doses of

DAC/AZA are sufficient to induce remarkable loss of tumorigenicity *in vivo* and *in vitro*, without accompanying cytotoxicity for normal haematopoietic cells.[70] Our results from serially enriching tumours for stem cell populations and treatments with DAC/AZA suggest that the drugs may be directly impacting the self-renewal capacity of a subpopulation of clonogenic and tumour-initiating cells.[70] This results in a 'memory' effect, wherein the progeny of the treated cells can, for weeks, maintain degrees of DNA methylation and gene expression change.[24] Furthermore, the genome-wide demethylation induced by the drugs correlated with increased expression of numerous pathways, including the immune response and cell differentiation that could aid in reshaping the epigenetic landscape and maintaining the treated cells in a non-tumorigenic state.[70] We believe that the re-expression of these different pathways is, at least in part, due to a reprogramming effect with potential for producing anti-tumour effects that may extend to a stem-like cell fraction.

The summary message from the multiple studies above is that DNA methylation inhibitors such as AZA and DAC require cell division for incorporation, hence any efficacy, with respect to specifically blocking DNMTs for anti-tumour responses, must be based on avoiding excessive doses which result in induction of very early apoptotic or cell cycle arrest responses. These latter most probably are off-target effects with respect to DNMT inhibition and many may be secondary to DNA damage.[68,69] The incorporation of the drugs into DNA, dependence on cell division and memory effects are likely to cause delayed therapeutic response in patients. It is critical to be cognizant of these factors while assessing the efficacy of the drug and determining the course of future treatment.[24] A recent study demonstrated that persistent treatment in MDS patients with nucleoside analogues, past the time of first response, showed significant responses in 48%, reiterating that a delayed response might be significant in effectively targeting the stem cell population of the tumour.[71]

10.4 Combination Epigenetic Therapy

Given that cancers are characterised by a multitude of epigenetic defects that are intimately related, DNA demethylation alone was not always found to be sufficient to re-express aberrantly repressed genes. Hence, the approach of combining DNMT inhibitors with other inhibitors has been adopted, the most frequent and synergistic combination being with histone deacetylase (HDAC) inhibitors.[64,72] These drugs, despite being tested extensively, have gained little therapeutic traction when used as single agents across a wide spectrum of human cancers.[73,74] However, two HDAC inhibitors, romidepsin and vorinostat, have received FDA approval for the treatment of cutaneous T-cell lymphoma, a setting where responses can be dramatic[75] for, as yet, not well understood reasons. Although, as outlined below, HDACs have shown some promise in combination clinical trials, they have shown little therapeutic benefit on their own outside of the above results for lymphomas.[24]

Against the above background, several clinical trials have attempted to uncover possible additive or synergistic effects of combining DNA demethylating agents with HDAC inhibitors.[39] These possibilities are driven by a cell-based laboratory paradigm. Essentially, giving a HDAC inhibitor subsequent to treatment with a low dose of a DNMT inhibitor results in additive effects for re-expressing DNA hypermethylated genes of the type under discussion.[72,76] The efficacy for the use of this drug combination is still being debated as results are inconsistent. In one clinical trial, overall response rates of 42% were observed in MDS and AML patients treated with AZA and valproic acid (VPA).[77] Another similar trial showed that increasing plasma concentrations of VPA corresponded to increased median survival rates.[78] However, a recent large trial in MDS/AML patients showed that AZA combined with the HDAC inhibitor entinostat may have actually had worse outcomes than AZA alone.[79]

More recently, patients with treatment-refractory, metastatic non-small cell lung cancer (NSCLC), who had failed standard of care treatment regimens, were treated with a combination of AZA and entinostat in a phase I/II trial. The study resulted in increased median survival rate of the cohort compared to historical data for the stage of NSCLC being treated and results included a complete remission and a robust partial response, both of which were extremely durable.[80] Perhaps, most importantly, the very exciting possibility emerged that the therapy sensitised patients to subsequent chemotherapy.[80] These observations come from trials with only a small number of patients and include the potential that there may be a dramatic sensitisation by the epigenetic therapy to an immunotherapy approach which breaks immune tolerance.[81] From associated laboratory studies, this latter paradigm could be due, in part, to a complex up-regulation of tumour immune recognition pathways in lung and other cancer cell types by low-dose AZA.[81,82]

A major focus in epigenetic therapeutics currently is rational design of inhibitors that target writers and erasers of histone methylation modifications, the histone lysine methyltransferases and histone demethylases. EZH2, which trimethylates histone H3 on lysine 27 as part of the Polycomb repressive complex 2, DOT1L, which trimethylates lysine 79 on H3, as well as G9A, which trimethylates lysine 9 on H3, have all been shown to contribute to chromatin repression.[83] Inhibitors against each of these above enzymes have been successfully developed and also for a reader for acetylated lysines, BRD4.[74,84–86] Those for DOT1L, EZH2 and BRD4 have now entered clinical trials and news of their clinical activities is eagerly awaited. In the laboratory, others show potential that they could have therapeutic efficacy when used as single agents and in combinatorial treatments with HDAC inhibitors.[87–90] Potent small-molecule inhibitors against a histone demethylase, LSD1[91,92] which targets principally HeK4me2, have been developed and are also nearing clinical trials. Current drug development efforts continue with a focus on inhibitors against other H3K4 demethylases, such as the JARID family. Our studies have suggested that the majority of DNA demethylation events do not lead to an open chromatin state.[93] If repressive histone methylation and acetylation are the reason for this repressed configuration, one can envision that combination treatment with the inhibitors described above could provide substantial synergistic benefit in chromatin opening. Although such combinatorial treatments hold great promise, numerous factors such as ideal dosage and schedule of the treatment regimen need to be carefully determined.

10.5 Perspectives for the Future: Need for Evolution in Drug Development, Clinical Efficacy Assessment and Regulatory Barriers

As greater efforts are invested into the development of novel drugs that target aberrations in the cancer epigenome, it is crucial to use lessons learned from recent studies to develop appropriate metrics to assess therapeutic efficacy. Often in preclinical work, immediate cell killing or apoptosis may not be the effect one should expect when using appropriately low doses of epigenetic therapy agents. Rather, gene expression, histone modification and nucleosome occupancy changes that result from altering of the cancer epigenome may take longer to provide a memory effect, which then induces anti-tumour effects.[70] Additionally, one hope raised by ongoing preclinical studies is that epigenetic therapies could target key subpopulations in tumours, such as stem-like tumour-initiating cells, and inhibition of these could take time to show up as reduction in bulk tumour.[70]

These above timing effects are critical to remember in both the use of existing and new epigenetic therapeutic agents when designing clinical trials and the metrics by which therapeutic efficacy is assessed. Work with the DNA demethylating agents amply illustrates these points. In the clinical trials that led to the FDA approval of AZA for the treatment of MDS, efficacy was measured by the overall response rate, complete response (CR) and partial response (PR), as defined by most haematological trials.[62,63] Other phenotypic changes such as time-to-event analyses of death, disease progression, relapse and transformation to AML were used to determine efficacy. The timing of these clinical analyses has proven to be critical, because epigenetic therapy using lower doses often does not produce results for months in patients with MDS/AML.[94] These data strongly hint that the Response Evaluation Criteria In Solid Tumors (RECIST) criteria that are classically used to assess clinical response within 6–8 weeks may therefore be inappropriate to monitor the ultimate efficacy of epigenetic therapy for patients with solid tumours. We have earlier provided data for early trials in patients with NSCLC that support this notion.[80] Hence, we may need to evolve new trial parameters to accommodate this new paradigm of treatment. Further, assessment of tumour size alone, as suggested by RECIST criteria, may not fully capture the extent of epigenetic therapy-induced responses and may result in false conclusions, premature termination of clinical studies and lack of regulatory approval.[24,95]

One helpful measure for early assessment of epigenetic therapy will be biomarkers that correlate with prediction of long-term efficacy. At present, there has been little success in developing these but also limited scenarios for developing the approaches. For patients with haematological tumours being treated with DNA demethylating agents where circulating tumour cells are accessible, measurements of global DNA demethylation or methylation changes in repeat sequences can at least help ensure that the tumour cells are demonstrating biological effects of the therapy agent. For example, the repetitive long interspersed nuclear element, LINE1, which is normally methylated in most cells, has been used to determine the level of DNA demethylation upon drug treatment.[96] This measure has been shown to be useful in solid and haematological tumours.[97]

Besides accounting for the changes in DNA methylation, much more needs to be understood about what constitutes a change induced by drug treatments that functionally impacts the cancer cells in reducing tumorigenicity. Some groups have found that many gene expression changes accompany treatment with AZA or DAC but not all of them are associated with DNA demethylation.[98] It is probable, as we have stressed in preclinical studies of low doses of DNA demethylating agents, that some of these responses occur in anti-tumour pathway genes lying downstream to those directly demethylated and re-expressed in response to AZA/DAC.[70] A major question requiring extensive exploration is which gene expression changes correlate with a therapeutic response and whether or not, and how, they are dependent on DNA demethylation events.

Other cellular and molecular responses to epigenetic agents like AZA/DAC are highly relevant to effects on tumour cells. It is important for both clinical and basic science studies, moving forward, to define systematically which demethylation events are associated with chromatin opening and the establishment of a nucleosome-depleted state. We have established the importance of this functional DNA demethylation during epigenetic therapy recently.[93] Identifying regions of chromatin opening will be crucial in understanding which epigenetic changes induced by drug treatment are functionally contributing to the inhibition of tumorigenesis. As the molecular definition of therapeutic response to epigenetic treatment evolves, we will probably learn to carefully discern, and predict early in treatment, patient populations that are benefitting from treatment. This ability would help us focus current therapeutic measures on patients who are more likely to respond to the treatment while developing new combinations of epigenetic therapies against the more resistant tumour types, leading to personalised approaches to epigenetic therapy.

Finally, as we have discussed, epigenetic therapy may serve as a priming approach for other therapy approaches. The early indications that such treatment may sensitise to conventional chemotherapy and immunotherapy in patients with advanced NSCLC, and to chemotherapy in patients with ovarian cancer, are examples.[67,80] It will also become important to additionally assess the potency of combinatorial epigenetic treatment of the epigenome for enhancing activities of not only conventional but targeted therapies.

10.6 Conclusions

The field of cancer management is moving towards personalising therapeutics with the discovery that each patient's tumour has taken a unique path of genetic abnormalities.[99] Similarly, the epigenetic changes that drive each patient's tumour may also be unique.[21] With more data surfacing that support the intricate link between genetics and epigenetics in cancer initiation and progression,[6,9,20] it is crucial to study these various facets holistically to determine the best treatment approach for each patient. Significant efforts have been invested by consortia such as the Roadmap Epigenome Project and The Cancer Genome Atlas (TCGA) to map epigenetic aberrations in a vast number and diversity of patients.[100–104] Because it will be years before it becomes feasible to study the genome and epigenome of every patient, finding from such consortia will help reveal overarching patterns in different tumour types, as well as guiding the tailoring of therapies to the specific cancer subtypes that are identified. With significant advances in the understanding of the mechanisms of action of epigenetic therapeutics and the rapid generation of vast amounts of 'omics' data on patient samples, epigenetic therapy will undoubtedly become a primary, or a combinatorial, addition to cancer therapy for many tumour types in the near future. As we move forward in developing epigenetic drugs and approving them for clinical use, we need to be fully cognizant of the mechanism of action of low-dose agents in reprogramming cells in what may often be, acutely, a non-cytotoxic manner.

References

1. C. H. Waddington, *Int. J. Epidemiol.*, 2012, **41**, 10–13.
2. B. E. Bernstein, A. Meissner and E. S. Lander, *Cell*, 2007, **128**, 669–681.
3. T. Kouzarides, *Cell*, 2007, **128**, 693–705.
4. M. M. Suzuki and A. Bird, *Nat. Rev. Genet.*, 2008, **9**, 465–476.
5. H. W. Hwang and J. T. Mendell, *Br. J. Cancer*, 2006, **94**, 776–780.
6. J. S. You and P. A. Jones, *Cancer Cell*, 2012, **22**, 9–20.
7. S. B. Baylin and P. A. Jones, in *Epigenetics*, ed. C.D. Allis, T. Jenuwein and D. Reinberg, Cold Spring Harbor Laboratory Press, Cold Spring Harbor, New York, 2006, pp. 457–476.
8. M. Esteller, *N. Engl. J. Med.*, 2008, **358**, 1148–1159.
9. H. Shen and P. W. Laird, *Cell*, 2013, **153**, 38–55.
10. S. B. Baylin and P. A. Jones, *Nat. Rev. Cancer*, 2011, **11**, 726–734.
11. P. A. Jones and S. B. Baylin, *Cell*, 2007, **128**, 683–692.
12. J. G. Herman and S. B. Baylin, *N. Engl. J. Med.*, 2003, **349**, 2042–2054.
13. V. Greger, E. Passarge, W. Hopping, E. Messmer and B. Horsthemke, *Hum. Genet.*, 1989, **83**, 155–158.
14. J. G. Herman, A. Umar, K. Polyak, J. R. Graff, N. Ahuja, J. P. Issa, S. Markowitz, J. K. Willson, S. R. Hamilton, K. W. Kinzler, M. F. Kane, R. D. Kolodner, B. Vogelstein, T. A. Kunkel and S. B. Baylin, *Proc. Natl. Acad. Sci. U. S. A.*, 1998, **95**, 6870–6875.

15. O. Galm, J. G. Herman and S. B. Baylin, *Blood Rev.*, 2006, **20**, 1–13.
16. A. Eden, F. Gaudet, A. Waghmare and R. Jaenisch, *Science*, 2003, **300**, 455.
17. G. Howard, R. Eiges, F. Gaudet, R. Jaenisch and A. Eden, *Oncogene*, 2008, **27**, 404–408.
18. S. Sharma, T. K. Kelly and P. A. Jones, *Carcinogenesis*, 2010, **31**, 27–36.
19. M. E. Valk-Lingbeek, S. W. Bruggeman and M. van Lohuizen, *Cell*, 2004, **118**, 409–418.
20. L. A. Garraway and E. S. Lander, *Cell*, 2013, **153**, 17–37.
21. H. Heyn and M. Esteller, *Nat. Rev. Genet.*, 2012, **13**, 679–692.
22. G. Egger, G. Liang, A. Aparicio and P. A. Jones, *Nature*, 2004, **429**, 457–463.
23. M. A. Dawson, T. Kouzarides and B. J. Huntly, *N. Engl. J. Med.*, 2012, **367**, 647–657.
24. N. Azad, C. A. Zahnow, C. M. Rudin and S. B. Baylin, *Nat. Rev. Clin. Oncol.*, 2013, **10**, 256–266.
25. F. Sorm, A. Piskala, A. Cihak and J. Vesely, *Experientia*, 1964, **20**, 202–203.
26. P. A. Jones and S. M. Taylor, *Cell*, 1980, **20**, 85–93.
27. S. M. Taylor and P. A. Jones, *J. Mol. Biol.*, 1982, **162**, 679–692.
28. S. Gabbara and A. S. Bhagwat, *Biochem. J.*, 1995, **307**(1), 87–92.
29. D. V. Santi, A. Norment and C. E. Garrett, *Proc. Natl. Acad. Sci. U. S. A.*, 1984, **81**, 6993–6997.
30. K. Ghoshal, J. Datta, S. Majumder, S. Bai, H. Kutay, T. Motiwala and S. T. Jacob, *Mol. Cell. Biol.*, 2005, **25**, 4727–4741.
31. P. G. Constantinides, P. A. Jones and W. Gevers, *Nature*, 1977, **267**, 364–366.
32. P. G. Constantinides, S. M. Taylor and P. A. Jones, *Dev. Biol.*, 1978, **66**, 57–71.
33. A. Cihak, *Oncology*, 1974, **30**, 405–422.
34. J. K. Christman, N. Mendelsohn, D. Herzog and N. Schneiderman, *Cancer Res.*, 1983, **43**, 763–769.
35. K. E. Schuebel, W. Chen, L. Cope, S. C. Glockner, H. Suzuki, J. M. Yi, T. A. Chan, L. Van Neste, W. Van Criekinge, S. van den Bosch, M. van Engeland, A. H. Ting, K. Jair, W. Yu, M. Toyota, K. Imai, N. Ahuja, J. G. Herman and S. B. Baylin, *PLoS Genet.*, 2007, **3**, 1709–1723.
36. C. B. Yoo, S. Jeong, G. Egger, G. Liang, P. Phiasivongsa, C. Tang, S. Redkar and P. A. Jones, *Cancer Res.*, 2007, **67**, 6400–6408.
37. C. B. Yoo, J. C. Chuang, H. M. Byun, G. Egger, A. S. Yang, L. Dubeau, T. Long, P. W. Laird, V. E. Marquez and P. A. Jones, *Cancer Prev. Res.*, 2008, **1**, 233–240.
38. J. C. Chuang, S. L. Warner, D. Vollmer, H. Vankayalapati, S. Redkar, D. J. Bearss, X. Qiu, C. B. Yoo and P. A. Jones, *Mol. Cancer Ther.*, 2010, **9**, 1443–1450.
39. X. Yang, F. Lay, H. Han and P. A. Jones, *Trends Pharmacol. Sci.*, 2010, **31**, 536–546.
40. J. Datta, K. Ghoshal, W. A. Denny, S. A. Gamage, D. G. Brooke, P. Phiasivongsa, S. Redkar and S. T. Jacob, *Cancer Res.*, 2009, **69**, 4277–4285.

41. B. Brueckner, R. Garcia Boy, P. Siedlecki, T. Musch, H. C. Kliem, P. Zielenkiewicz, S. Suhai, M. Wiessler and F. Lyko, *Cancer Res.*, 2005, **65**, 6305–6311.
42. M. R. Rountree, K. E. Bachman and S. B. Baylin, *Nat. Genet.*, 2000, **25**, 269–277.
43. K. D. Robertson, S. Ait-Si-Ali, T. Yokochi, P. A. Wade, P. L. Jones and A. P. Wolffe, *Nat. Genet.*, 2000, **25**, 338–342.
44. F. Fuks, W. A. Burgers, A. Brehm, L. Hughes-Davies and T. Kouzarides, *Nat. Genet.*, 2000, **24**, 88–91.
45. E. G. Clements, H. P. Mohammad, B. R. Leadem, H. Easwaran, Y. Cai, L. Van Neste and S. B. Baylin, *Nucleic Acids Res.*, 2012, **40**, 4334–4346.
46. R. J. Lin, L. Nagy, S. Inoue, W. Shao, W. H. Miller Jr and R. M. Evans, *Nature*, 1998, **391**, 811–814.
47. R. E. Notari and J. L. DeYoung, *J. Pharm. Sci.*, 1975, **64**, 1148–1157.
48. D. D. Von Hoff, M. Slavik and F. M. Muggia, *Ann. Intern. Med.*, 1976, **85**, 237–245.
49. M. Karon and W. F. Benedict, *Science*, 1972, **178**, 62.
50. J. K. Christman, *Oncogene*, 2002, **21**, 5483–5495.
51. V. Santini, H. M. Kantarjian and J. P. Issa, *Ann. Intern. Med.*, 2001, **134**, 573–586.
52. E. Velez-Garcia, W. R. Vogler, A. A. Bartolucci and S. N. Arkun, *Cancer Treat. Rep.*, 1977, **61**, 1675–1677.
53. U. Srinivasan, G. H. Reaman, D. G. Poplack, D. L. Glaubiger and A. S. LeVine, *Am. J. Clin. Oncol.*, 1982, **5**, 411–415.
54. R. L. Momparler, G. E. Rivard and M. Gyger, *Pharmacol. Ther.*, 1985, **30**, 277–286.
55. P. W. Wijermans, J. W. Krulder, P. C. Huijgens and P. Neve, *Leukemia*, 1997, **11**, 1–5.
56. P. Wijermans, M. Lubbert, G. Verhoef, A. Bosly, C. Ravoet, M. Andre and A. Ferrant, *J. Clin. Oncol.*, 2000, **18**, 956–962.
57. G. Schwartsmann, H. Schunemann, C. N. Gorini, A. F. Filho, C. Garbino, G. Sabini, I. Muse, L. DiLeone and D. R. Mans, *Invest. New Drugs*, 2000, **18**, 83–91.
58. Y. Saunthararajah, C. A. Hillery, D. Lavelle, R. Molokie, L. Dorn, L. Bressler, S. Gavazova, Y. H. Chen, R. Hoffman and J. DeSimone, *Blood*, 2003, **102**, 3865–3870.
59. J. P. Issa, V. Gharibyan, J. Cortes, J. Jelinek, G. Morris, S. Verstovsek, M. Talpaz, G. Garcia-Manero and H. M. Kantarjian, *J. Clin. Oncol.*, 2005, **23**, 3948–3956.
60. J. P. Issa, G. Garcia-Manero, F. J. Giles, R. Mannari, D. Thomas, S. Faderl, E. Bayar, J. Lyons, C. S. Rosenfeld, J. Cortes and H. M. Kantarjian, *Blood*, 2004, **103**, 1635–1640.
61. L. R. Silverman, E. P. Demakos, B. L. Peterson, A. B. Kornblith, J. C. Holland, R. Odchimar-Reissig, R. M. Stone, D. Nelson, B. L. Powell, C. M. DeCastro, J. Ellerton, R. A. Larson, C. A. Schiffer and J. F. Holland, *J. Clin. Oncol.*, 2002, **20**, 2429–2440.

62. E. Kaminskas, A. Farrell, S. Abraham, A. Baird, L. S. Hsieh, S. L. Lee, J. K. Leighton, H. Patel, A. Rahman, R. Sridhara, Y. C. Wang and R. Pazdur, *Clin. Cancer Res.*, 2005, **11**, 3604–3608.
63. E. Kaminskas, A. T. Farrell, Y. C. Wang, R. Sridhara and R. Pazdur, *Oncologist*, 2005, **10**, 176–182.
64. S. D. Gore, *Nat. Clin. Pract. Oncol.*, 2005, **2**(1), S30–S35.
65. A. F. Cashen, A. K. Shah, L. Todt, N. Fisher and J. DiPersio, *Cancer Chemother. Pharmacol.*, 2008, **61**, 759–766.
66. S. Fu, W. Hu, R. Iyer, J. J. Kavanagh, R. L. Coleman, C. F. Levenback, A. K. Sood, J. K. Wolf, D. M. Gershenson, M. Markman, B. T. Hennessy, R. Kurzrock and R. C. Bast Jr, *Cancer*, 2011, **117**, 1661–1669.
67. D. Matei, F. Fang, C. Shen, J. Schilder, A. Arnold, Y. Zeng, W. A. Berry, T. Huang and K. P. Nephew, *Cancer Res.*, 2012, **72**, 2197–2205.
68. T. E. Fandy, J. G. Herman, P. Kerns, A. Jiemjit, E. A. Sugar, S. H. Choi, A. S. Yang, T. Aucott, T. Dauses, R. Odchimar-Reissig, J. Licht, M. J. McConnell, C. Nasrallah, M. K. Kim, W. Zhang, Y. Sun, A. Murgo, I. Espinoza-Delgado, K. Oteiza, I. Owoeye, L. R. Silverman, S. D. Gore and H. E. Carraway, *Blood*, 2009, **114**, 2764–2773.
69. H. Shen and P. W. Laird, *Cell stem cell*, 2012, **10**, 353–354.
70. H. C. Tsai, H. Li, L. Van Neste, Y. Cai, C. Robert, F. V. Rassool, J. J. Shin, K. M. Harbom, R. Beaty, E. Pappou, J. Harris, R. W. Yen, N. Ahuja, M. V. Brock, V. Stearns, D. Feller-Kopman, L. B. Yarmus, Y. C. Lin, A. L. Welm, J. P. Issa, I. Minn, W. Matsui, Y. Y. Jang, S. J. Sharkis, S. B. Baylin and C. A. Zahnow, *Cancer Cell*, 2012, **21**, 430–446.
71. L. R. Silverman, P. Fenaux, G. J. Mufti, V. Santini, E. Hellstrom-Lindberg, N. Gattermann, G. Sanz, A. F. List, S. D. Gore and J. F. Seymour, *Cancer*, 2011, **117**, 2697–2702.
72. E. E. Cameron, K. E. Bachman, S. Myohanen, J. G. Herman and S. B. Baylin, *Nat. Genet.*, 1999, **21**, 103–107.
73. A. A. Lane and B. A. Chabner, *J. Clin. Oncol.*, 2009, **27**, 5459–5468.
74. M. A. Dawson and T. Kouzarides, *Cell*, 2012, **150**, 12–27.
75. R. L. Piekarz, R. Frye, M. Turner, J. J. Wright, S. L. Allen, M. H. Kirschbaum, J. Zain, H. M. Prince, J. P. Leonard, L. J. Geskin, C. Reeder, D. Joske, W. D. Figg, E. R. Gardner, S. M. Steinberg, E. S. Jaffe, M. Stetler-Stevenson, S. Lade, A. T. Fojo and S. E. Bates, *J. Clin. Oncol.*, 2009, **27**, 5410–5417.
76. H. Suzuki, E. Gabrielson, W. Chen, R. Anbazhagan, M. van Engeland, M. P. Weijenberg, J. G. Herman and S. B. Baylin, *Nat. Genet.*, 2002, **31**, 141–149.
77. A. O. Soriano, H. Yang, S. Faderl, Z. Estrov, F. Giles, F. Ravandi, J. Cortes, W. G. Wierda, S. Ouzounian, A. Quezada, S. Pierce, E. H. Estey, J. P. Issa, H. M. Kantarjian and G. Garcia-Manero, *Blood*, 2007, **110**, 2302–2308.
78. M. T. Voso, V. Santini, C. Finelli, P. Musto, E. Pogliani, E. Angelucci, G. Fioritoni, G. Alimena, L. Maurillo, A. Cortelezzi, F. Buccisano, M. Gobbi, L. Borin, A. Di Tucci, G. Zini, M. C. Petti, G. Martinelli, E. Fabiani, P. Fazi, M. Vignetti, A. Piciocchi, V. Liso, S. Amadori and G. Leone, *Clinical Cancer Res.*, 2009, **15**, 5002–5007.

79. T. Prebet, Z. Sun, M. E. Figueroa, R. Ketterling, A. Melnick, P. L. Greenberg, J. Herman, M. Juckett, M. R. Smith, L. Malick, E. Paietta, M. Czader, M. Litzow, J. Gabrilove, H. P. Erba, S. D. Gore and M. S. Tallman, *J. Clin. Oncol.*, 2014, DOI: 10.1200/jco.2013.50.3102.
80. R. A. Juergens, J. Wrangle, F. P. Vendetti, S. C. Murphy, M. Zhao, B. Coleman, R. Sebree, K. Rodgers, C. M. Hooker, N. Franco, B. Lee, S. Tsai, I. E. Delgado, M. A. Rudek, S. A. Belinsky, J. G. Herman, S. B. Baylin, M. V. Brock and C. M. Rudin, *Cancer Discovery*, 2011, **1**, 598–607.
81. J. Wrangle, W. Wang, A. Koch, H. Easwaran, H. P. Mohammad, F. Vendetti, W. Vancriekinge, T. Demeyer, Z. Du, P. Parsana, K. Rodgers, R. W. Yen, C. A. Zahnow, J. M. Taube, J. R. Brahmer, S. S. Tykodi, K. Easton, R. D. Carvajal, P. A. Jones, P. W. Laird, D. J. Weisenberger, S. Tsai, R. A. Juergens, S. L. Topalian, C. M. Rudin, M. V. Brock, D. Pardoll and S. B. Baylin, *Oncotarget*, 2013, **4**, 2067–2079.
82. H. Li, K. B. Chiappinelli, A. A. Guzzetta, H. Easwaran, R. W. Yen, R. Vatapalli, M. J. Topper, J. Luo, R. M. Connolly, N. S. Azad, V. Stearns, D. M. Pardoll, N. Davidson, P. A. Jones, D. J. Slamon, S. B. Baylin, C. A. Zahnow and N. Ahuja, *Oncotarget*, 2014, **5**, 587–598.
83. J. C. Black, C. Van Rechem and J. R. Whetstine, *Mol. Cell*, 2012, **48**, 491–507.
84. J. Bradner, *Clin. Adv. Hematol. Oncol.*, 2013, **11**, 375–376.
85. J. E. Delmore, G. C. Issa, M. E. Lemieux, P. B. Rahl, J. Shi, H. M. Jacobs, E. Kastritis, T. Gilpatrick, R. M. Paranal, J. Qi, M. Chesi, A. C. Schinzel, M. R. McKeown, T. P. Heffernan, C. R. Vakoc, P. L. Bergsagel, I. M. Ghobrial, P. G. Richardson, R. A. Young, W. C. Hahn, K. C. Anderson, A. L. Kung, J. E. Bradner and C. S. Mitsiades, *Cell*, 2011, **146**, 904–917.
86. P. Filippakopoulos, J. Qi, S. Picaud, Y. Shen, W. B. Smith, O. Fedorov, E. M. Morse, T. Keates, T. T. Hickman, I. Felletar, M. Philpott, S. Munro, M. R. McKeown, Y. Wang, A. L. Christie, N. West, M. J. Cameron, B. Schwartz, T. D. Heightman, N. La Thangue, C. A. French, O. Wiest, A. L. Kung, S. Knapp and J. E. Bradner, *Nature*, 2010, **468**, 1067–1073.
87. J. Tan, X. Yang, L. Zhuang, X. Jiang, W. Chen, P. L. Lee, R. K. Karuturi, P. B. Tan, E. T. Liu and Q. Yu, *Genes Dev.*, 2007, **21**, 1050–1063.
88. T. Yao, L. Yang, P. Q. Li, H. Wu, H. B. Xie, X. Shen and X. D. Xie, *Arch. Oral Biol.*, 2011, **56**, 73–78.
89. W. Fiskus, Y. Wang, A. Sreekumar, K. M. Buckley, H. Shi, A. Jillella, C. Ustun, R. Rao, P. Fernandez, J. Chen, R. Balusu, S. Koul, P. Atadja, V. E. Marquez and K. N. Bhalla, *Blood*, 2009, **114**, 2733–2743.
90. M. Vedadi, D. Barsyte-Lovejoy, F. Liu, S. Rival-Gervier, A. Allali-Hassani, V. Labrie, T. J. Wigle, P. A. Dimaggio, G. A. Wasney, A. Siarheyeva, A. Dong, W. Tempel, S. C. Wang, X. Chen, I. Chau, T. J. Mangano, X. P. Huang, C. D. Simpson, S. G. Pattenden, J. L. Norris, D. B. Kireev, A. Tripathy, A. Edwards, B. L. Roth, W. P. Janzen, B. A. Garcia, A. Petronis, J. Ellis, P. J. Brown, S. V. Frye, C. H. Arrowsmith and J. Jin, *Nat. Chem. Biol.*, 2011, **7**, 566–574.

91. Y. Huang, T. M. Stewart, Y. Wu, S. B. Baylin, L. J. Marton, B. Perkins, R. J. Jones, P. M. Woster and R. A. Casero Jr, *Clin. Cancer Res.*, 2009, **15**, 7217–7228.
92. J. T. Lynch, W. J. Harris and T. C. Somervaille, *Expert Opin. Ther. Targets*, 2012, **16**, 1239–1249.
93. K. Pandiyan, J. S. You, X. Yang, C. Dai, X. J. Zhou, S. B. Baylin, P. A. Jones and G. Liang, *Nucleic Acids Res.*, 2013, DOI: 10.1093/nar/gkt077.
94. J. P. Issa, *Nat. Clin. Pract. Oncol.*, 2005, **2**(1), S24–S29.
95. S. Diederich, *Cancer Imaging*, 2012, **12**, 347–350.
96. A. S. Yang, K. D. Doshi, S. W. Choi, J. B. Mason, R. K. Mannari, V. Gharybian, R. Luna, A. Rashid, L. Shen, M. R. Estecio, H. M. Kantarjian, G. Garcia-Manero and J. P. Issa, *Cancer Res.*, 2006, **66**, 5495–5503.
97. A. Aparicio, B. North, L. Barske, X. Wang, V. Bollati, D. Weisenberger, C. Yoo, N. Tannir, E. Horne, S. Groshen, P. Jones, A. Yang and J. P. Issa, *Epigenetics*, 2009, **4**, 176–184.
98. V. M. Komashko and P. J. Farnham, *Epigenetics*, 2010, **5**, 229–240.
99. J. E. Dancey, P. L. Bedard, N. Onetto and T. J. Hudson, *Cell*, 2012, **148**, 409–420.
100. P. S. Hammerman, D. N. Hayes, M. D. Wilkerson, N. Schultz, R. Bose, A. Chu, E. A. Collisson, L. Cope, C. J. Creighton, G. Getz, J. G. Herman, B. E. Johnson, R. Kucherlapati, M. Ladanyi, C. A. Maher, G. Robertson, C. Sander, R. Shen, R. Sinha, A. Sivachenko, R. K. Thomas, W. D. Travis, M. S. Tsao, J. N. Weinstein, D. A. Wigle, S. B. Baylin, R. Govindan and M. Meyerson, *Nature*, 2012, **489**, 519–525.
101. C. Kandoth, N. Schultz, A. D. Cherniack, R. Akbani, Y. Liu, H. Shen, A. G. Robertson, I. Pashtan, R. Shen, C. C. Benz, C. Yau, P. W. Laird, L. Ding, W. Zhang, G. B. Mills, R. Kucherlapati, E. R. Mardis and D. A. Levine, *Nature*, 2013, **497**, 67–73.
102. Cancer Genome Atlas Network, Comprehensive molecular characterization of human colon and rectal cancer, *Nature*, 2012, **487**, 330–337.
103. Cancer Genome Atlas Network, Comprehensive molecular portraits of human breast tumours, *Nature*, 2012, **490**, 61–70.
104. Cancer Genome Atlas Research Network, Genomic and epigenomic landscapes of adult de novo acute myeloid leukemia, *N. Engl. J. Med.*, 2013, **368**, 2059–2074.

CHAPTER 11

Histone Methyltransferase Activity Assays

HAICHING MA*[a], KONRAD T. HOWITZ[a], KURUMI Y. HORIUCHI[a], AND YUREN WANG[a]

[a]Reaction Biology Corporation, One Great Valley Parkway, Suite 2, Malvern, PA 19355, USA
*E-mail: haiching.ma@reactionbiology.com

11.1 Introduction

Epigenetic factors are enzymes or proteins that catalyse or recognise covalent modifications to chromatin DNA or proteins. Based on their functional activities, they can be divided into three broad groups, commonly referred to as the 'writers', 'erasers' and 'readers'. The writers are enzymes that add modifications to chromatin, such as methyl groups to DNA, or methyl or acetyl groups to histone proteins. These enzymes include DNA methyltransferases (DNMTs), histone methyltransferases (HMTs) and histone acetyltransferases (HATs). The erasers are enzymes that remove the moieties added to the DNA or histones by the writers. These enzymes include histone demethylases (HDMs), histone deacetylases (HDACs) and the newly discovered DNA demethylation enzymes, such as the ten-eleven translocation (TET) family of proteins. The readers are proteins that detect modifications such as methyl or acetyl groups, or the lack thereof, which are the consequences of writer and eraser activity. Bromodomains, for example, are acetyl lysine recognition modules found in a number of epigenetic factors. Domains recognising

RSC Drug Discovery Series No. 48
Epigenetics for Drug Discovery
Edited by Nessa Carey

Published by the Royal Society of Chemistry, www.rsc.org

methylated residues include PHD fingers and members of the 'Royal Family' (chromo-, MBT, Tudor and PWWP domains). Due to the complex activities of these epigenetic factors, assay development has posed different challenges for each class of these targets. This chapter focuses on assays that have been developed for studying HMT activity specifically, but many of the formats discussed here have been applied to other classes of epigenetic writers, erasers and readers.

There are over 70 HMTs encoded by the human genome.[1–4] By methylating either lysine residues (histone lysine methyltransferases or HKMTs) or arginine residues (protein arginine methyltransferases or PRMTs) in histones and other proteins, these enzymes play crucial roles in gene regulation. The activity of a number of HMTs has been implicated in the aetiology of diseases including cancer, inflammatory diseases, neurodegenerative diseases and other conditions (Figure 11.1).

All HMTs use *S*-adenosyl-L-methionine (SAM or AdoMet) as the methyl-donating co-substrate in a reaction that produces a methylated lysine or arginine protein product and *S*-adenosyl-L-homocysteine (SAH or AdoHcy) (Figure 11.2A). Each amine group of lysine can be modified by the addition of one (me1), two (me2) or three (me3) methyl groups (Figure 11.2B), while arginine methylation can produce monomethyl, symmetrical dimethyl or asymmetrical dimethyl states (Figure 11.2C). Symmetrical dimethylation of arginine refers to the addition of one methyl group to each ω-nitrogen of the guanidinium group (Type II PRMTs), whereas asymmetrical dimethylation refers to the addition of both methyl groups to one ω-nitrogen of the guanidinium group (type-I PRMTs) (Figure 11.2C). The enzymes PRMT1, PRMT2, PRMT3, PRMT4 (CARM1), PRMT6 and PRMT8 belong to the type-I group, while PRMT5, PRMT7 and PRMT9 are type-II enzymes.[5–9]

There are about a dozen HMT activity assays available (Figure 11.3) and methods of detection have generally focused on the methylated histone or SAH reaction products.[10] Since both the HMTs and protein kinases are transferases with protein substrates and adenosyl co-substrates, the considerations involved in choosing an assay and the pros and cons of various assay types are highly similar.[11]

11.2 Mass Spectrometric Assays

Mass spectrometry (MS) (Figure 11.3, Assay #1) is a commonly used technology in the study of histone post-translational modifications, including methylations and acetylations. MS analyses the mass-to-charge ratio and since analysing multiply charged large protein samples is problematic, the proteins are traditionally digested into smaller peptides before MS analysis ('bottom up' approach). To avoid losing any information during the digestion of the proteins, a full intact protein analysis is preferred ('top down' approach).[12] MS is a powerful tool in analysing and comparing protein isoforms, including histone isoforms.[13] It is not only able to indicate the presence of methylations/acetylations in proteins, a feature shared with radioactive and

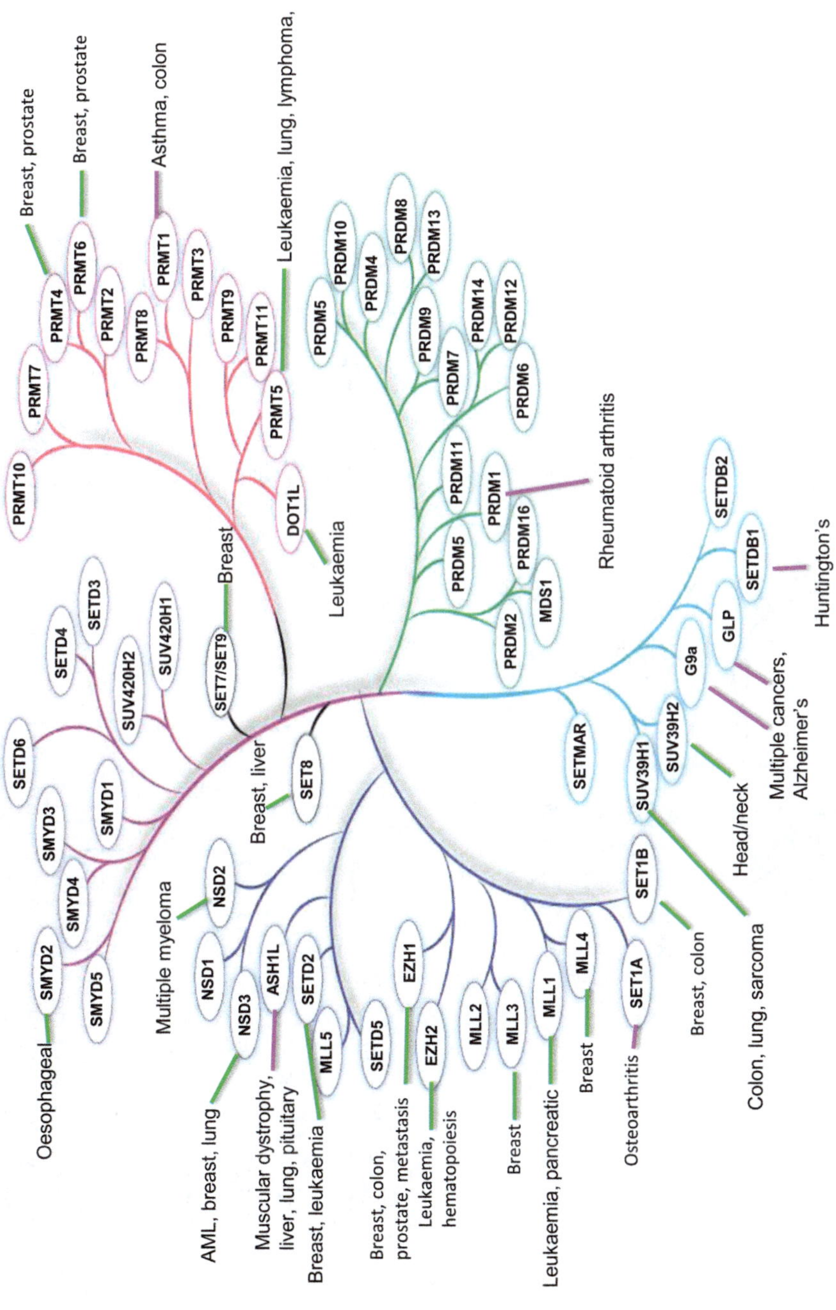

Figure 11.1 Human histone methyltransferases: evolutionary relationships and disease indications.

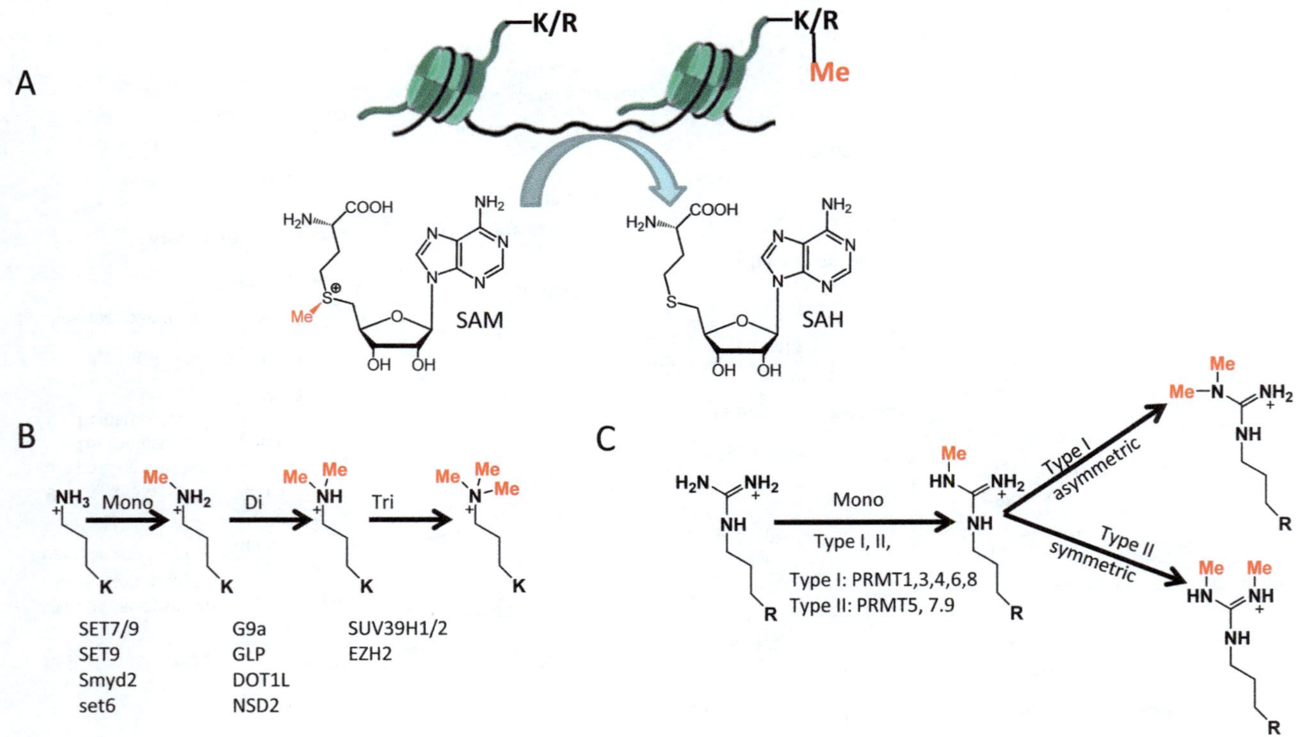

Figure 11.2 Histone methyltransferase reactions. (A) HMTs use *S*-adenosyl-L-methionine (SAM) as the methyl-donating co-substrate in a reaction that produces a methylated lysine or arginine protein product and *S*-adenosyl-L-homocysteine (SAH). (B) Lysine methylation can produce Mono, di and trimethylation states. (C) Arginine methylation can produce monomethyl, symmetrical dimethyl or asymmetrical dimethyl states.

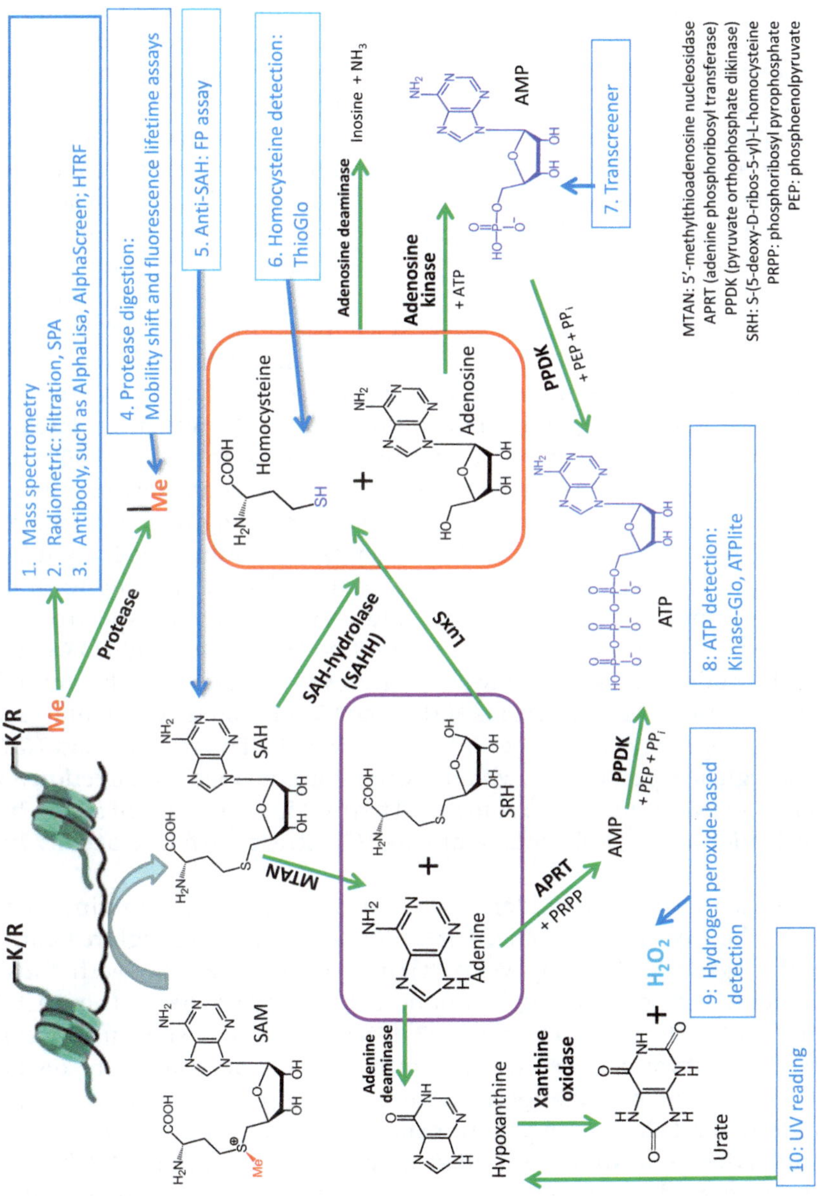

Figure 11.3 Histone methyltransferase assays.

immunodetection methods, but also able to confirm the identity of the protein and reveal the number of methylation/acetylation sites and states.

MS is the method of choice for precisely identifying the locations and types of methylated residues in proteins and for measurement of methylation stoichiometries.[14,15] MS can reliably distinguish between mono-, di- and trimethyl modifications to the same amino acid residue and has thus become essential for investigating the regulatory, physiological and pathological functions associated with protein methylation.[16,17]

For example, Ng *et al.* used matrix-assisted laser desorption/ionisation–time-of-flight (MALDI–TOF) MS to identify H3K79 methylation as a novel modification in the globular domain of mammalian H3 histones. They then showed that the same modification in yeast is due to the methylation activity of Dot1 (*D*isruptor *o*f *t*elomeric silencing *1*) and that this activity is required for telomeric silencing.[18] Jones and colleagues created mice deficient in DOT1L (mammalian Dot1 homologue, DOT1-Like), and analysed the H3K79 methylation pattern by using LTQ-FT MS (ThermoFisher Scientific, Waltham, MA).[19] DOT1L-deficient embryonic stem cells had almost no detectable methylated H3K79. In wild-type cells, about 11% of histone H3K79 was methylated, of which mono-, di- and trimethylated forms accounted for ~70%, ~30% and 0.1%, respectively. DOT1L-deficient stem cells were characterised by slow growth rate, increased apoptosis, aneuploidy and lengthened telomeres, while in the whole animal, DOT1L deficiency caused embryonic lethality.

MS has routinely been used to identify HMT substrates. Using recombinant nucleosomes, recombinant NSD2 and an Orbitrap mass spectrometer (ThermoFisher Scientific), Kuo *et al.*[20] identified H3K36me1 and H3K36me2 as the only products of NSD2 catalysis. NSD2-over-expression in the multiple myeloma cell line, KMS11, elevates H3K36me2 levels and is sufficient to activate four oncogenes (*TGFA*, *PAK1*, *MET* and *RRAS2*). The mammalian NSDs are homologues of yeast SET2, which, working in the opposite direction and using an LCQ DECA mass spectrometer (ThermoFinnigan), Strahl *et al.*[21] had previously identified as the source of H3K36 methyltransferase activity in a yeast cell extract.

MS has also been used to identify the sites of lysine and arginine methylation in the human proteome. For example, combining high-resolution MS technology (LTQ-Orbitrap Velos, ThermoFisher Scientific) with "heavy methyl stable isotope labeling by amino acids in cell culture" (hmSILAC), Bremang *et al.*[22] identified a total of 397 different sites of lysine and arginine methylation distributed over 139 separate proteins. Several recent reviews provide detailed technical approaches for the use of MS to analyse protein methylation[12] and chromatin post-translational modifications in general.[23,24]

Although the applications described above are more common, MS has also been used for studying small-molecule inhibition of HMTs and even as an HMT HTS assay. For example, Chang *et al.*[25] measured the inhibitory activity of the compound BIX-01294 against G9a and GLP by using MALDI-TOF. The RapidFire™ label-free mass spectrometer platform (Agilent Technologies, Santa Clara, CA) is one of the most popular HTS MS systems. With facilitated cycle times of <10 s per sample, it has been used for HTS analysis for a wide range of epigenetic factors. It can simultaneously detect multiple modifications on the same substrate, such as mono/di/trimethylation on a single lysine residue or acetylation at multiple sites within the same peptide.[26] Although most publications citing the use of RapidFire/MS have described inhibitor screening for protein demethylases, Cauchon *et al.*[27] reported a preliminary study using this platform to screen over 5000 compounds, including the LOPAC library, against phosphatidylethanolamine *N*-methyltransferase (PEMT), a SAM-dependent phosphatidylcholine synthetic enzyme and potential target for atherosclerosis.

Just as with HMTs, quantitative MS has been widely used for detection of demethylase activities because label-free detection eliminates many problems such as compound interference, associated with fluorescence detection and antibody-based assays.[28–30] MS is also a key technology in understanding the roles of acetylation by histone acetyltransferases (HATs) in chromatin remodelling, cell cycle, splicing, nuclear transport and actin nucleation.[31–33] Furthermore, although MS is used more for proteomic analysis rather than as a tool for screening, it has been playing important roles in the histone deacetylase (HDAC) inhibitor discovery process.[34–36]

11.3 Radioisotope-Based Assays

The two radioisotope-based assay formats, filter-binding and scintillation proximity (Figure 11.3, Assay #2), have been used for screening and the study of HMT activity. By using the co-substrate SAM with a radiolabelled methyl function, typically [methyl-^{3}H]-SAM, and capturing the labelled peptide, protein or nucleosomal product, both formats detect total methylation on the substrate, as opposed to MS which can distinguish mono-, di- and trimethylations. Each format has its pros and cons and can be readily interchanged based on the targets and purposes of the work.

11.3.1 Filter-Binding Assays

In filter-binding assays, the radioisotope labelled product is detected after the free radioisotope cofactor is washed away. The major advantage of this method is that it is a truly universal methyltransferase assay method, applicable to any methyltransferase and substrate without modification. It is therefore sometimes referred to as the gold standard assay for methyltransferases. This format has particular advantages for HMTs with a preference for nucleosomes, histone octamers or other proteins as substrate, because it is not simple to label or modify these substrates for other types of detection. Although there are exceptions,[37] most labs avoid this technology for HTS, due to the need for filtration and washing steps to remove the unreacted radiolabelled SAM before detection. However, in our laboratory, we have developed a miniaturised reaction system, the HotSpot^{SM} format, which reduces the labour-intensive washing process and can detect the labelled products from reaction samples in the nanolitre range.[10] These assays have been applied to >30 methyltransferases and have been used in profiling the selectivity of some of the first potent EZH2 inhibitors.[38] All told, due to the lack of compound interference with the assay and the assay's inherent sensitivity, filter-binding assays represent one of the more favourable assay methods for HMT profiling work. Filter-binding assays have also been successfully employed for HAT screening and profiling.[39–41]

11.3.2 Scintillation Proximity Assays

As a strategy to eliminate the washing and separation processes of the filter-binding assay format, GE and PerkinElmer have developed and commercialised products for use in two scintillation proximity assay (SPA) formats. The first format uses microtiter plates, such as ScintiPlates® or FlashPlates®, in which well interiors have a permanent coating of polystyrene-based scintillant. Most commonly, these will have a further coating of streptavidin in order to capture biotin-conjugated peptide substrates. Unless the HMT reaction is carried out in a separate plate and then transferred to the FlashPlate® for an end point determination, the reaction will be heterogeneous, *i.e.* the enzyme and ^{3}H-SAM will be in solution, but the peptide substrate will be bound to the surface. Radioactive methyl groups that have been transferred to the peptide substrate will be brought into close proximity to the scintillant-coated well surface by the binding of the biotin group to the streptavidin. These groups will efficiently excite the scintillant, whereas ^{3}H-SAM in solution does so far less efficiently and produces only a relatively minor background. Detection is carried out with a microplate-reading scintillation counter (*e.g.* Topcount NXT; PerkinElmer). The second format uses microscopic beads containing a scintillant, which, as with the FlashPlate®, can be stimulated by β-particles or auger electrons to emit light. Scintillation only occurs when radiolabelled molecules of interest are bound to the surface of the beads, which results in the emission of a light that can be detected using standard scintillation counters or with a CCD (charge coupled device) camera-based imaging instrument, such as the LEADseeker™ (GE) or ViewLux™ (PerkinElmer). An advantage over the FlashPlate® format is that the reaction can proceed homogeneously and, after addition of the beads, be counted in the same plate. Rathert *et al.*[42] were the first to report using a FlashPlate® methyltransferase assay, one in which the biotinylated peptide substrate was captured on streptavidin-coated FlashPlates. Because the substrate was pre-bound to the plate, the reaction was heterogeneous, but could be monitored continuously, and thus was well suited for kinetic studies. For SPA HTS studies of G9a, Ahrens *et al.*[43] chose an alternative approach, in which an end point FlashPlate® measurement was made after the G9a reactions were performed in a regular microtiter plate. Homogeneous reactions with full-length histone H3.1 were then stopped with an excess of unlabelled SAM and transferred to uncoated FlashPlates. These

FlashPlates were incubated for over 4 hours at 4 °C to allow binding of the H3.1 protein to the polystyrene surface, before being washed and read with a TopCount reader.

SPA formats with microscopic beads were utilised by GSK scientists for their recent EZH2 projects.[38,44] When biotinylated peptide was used as substrate, the peptide was preconjugated with streptavidin-coated PS imaging SPA beads (RPNQ0261), and the scintillation signal was acquired by ViewLux™. When HeLa nucleosomes were used as substrate, the methyltransferase reaction was performed in the absence of beads and then terminated by the addition of PS-PEI Imaging Beads (RPNQ0098). The positively-charged PEI (polyethyleneimine) coating captures the nucleosomes *via* binding to their DNA. For HTS, both assays were read as end points, although the peptide assay was also used with continuous reading for kinetic studies.

In general, SPAs are adaptable to HTS. Multiple substrate types can be used (peptide, protein or nucleosomes), although specific combinations of substrate modification and/or bead or plate coating may be required. Cost may be a drawback, given that the combination of FlashPlates or SPA beads, CCD-based instrumentation and the relatively large amounts of radiolabelled material required can be quite expensive. SPA formats are also used to study the activity of epigenetic factors such as HATs,[45] HDACs[46] and DNMTs.[47–50]

11.4 Anti-Methylation Antibody-Based Assays

With the rapid development of site-specific anti-methylation antibodies, antibody-based methods such as Western blotting, ELISA and ChIP are now widely used in publications on the molecular and cell biology of protein methylation (Figure 11.3, Assay #3).

PerkinElmer has developed many reagents capable of detecting the activity of various histone methyltransferases based on its AlphaLISA format.[51] Because the AlphaLISA format is homogeneous, highly sensitive, and requires no washing steps, it can be far more HTS-friendly than other antibody-based formats, such as an ordinary ELISA. A biotinylated histone H3 peptide is used as the substrate. After the HMT reaction, streptavidin-coated donor beads capture both the biotinylated substrate and its methylated reaction product. An anti-methylation antibody on the AlphaLISA acceptor beads recognises the methylation, thus forming a complex that brings the two types of beads into close proximity. Upon laser irradiation at 680 nm, the donor beads produce short-lived singlet oxygen molecules which generate an amplified 615 nm chemiluminescent signal from the acceptor beads, but only from nearby beads in the complex. Thus, the intensity of light emission is proportional to the level of substrate methylation.[51]

Anti-methylation antibodies, biotinylated or as acceptor bead conjugates, are available for H3K4me1-2, H3K9me2, H3K27me1-2, H3K27me3, H3K36me2 and H3K79me2. A DOT1L assay, with good signal to background at 2 nM enzyme, was recently developed using biotinylated anti-H3K79me2, streptavidin donor beads and anti-H3 (C-terminal) acceptor beads, with

HeLa oligonucleosomes as substrates (PerkinElmer Technical Note: AlphaLISA #25).[52] Customers can also use the various AlphaLISA 'epigenetic tool box' reagents to develop their own assays by pairing an optimised substrate with an appropriate antibody. For example, Quinn and Simeonov[53] have developed a G9a AlphaLISA assay in a 1536-well format and Simard *et al.*[54] have developed an AlphaLISA HTS assay for EZH2 using a biotinylated histone H3 peptide substrate.

As for any fluorescence- or luminescence-based method, certain compounds will interfere with the AlphaLISA detection system. In order to help identify false positives, PerkinElmer has developed a TruHits Kit (product no. AL900), which is used to detect compounds that may act as inner filters (compounds that either absorb light in the excitation or emission range of the assay[55]), light scatterers (insoluble compounds), singlet oxygen quenchers and biotin mimetics.

PerkinElmer and CisBio have also developed assay kits based on TR-FRET (time-resolved fluorescence resonance energy transfer). PerkinElmer's platform is referred to as LANCE® Ultra and CisBio's as HTRF®. The LANCE® assay uses a europium chelate-labelled anti-methylation antibody (Eu-Ab) as the donor (Ex. 320 or 340 nm/Em. 620 nm) and the red-shifted dye reagent U*Light*-streptavidin (U*Light*-SA) as the acceptor (Em. 665 nm).[51] In the HTRF® format, the anti-methylation antibody is labelled with Eu^{3+}-cryptate (Ex. 340 nm/Em. 590 nm), and the streptavidin is conjugated with the dye d2 (originally the modified phycobiliprotein) XL665, both with Em. 665 nm. The substrate is biotinylated in both assays. TR-FRET assays have two key advantages relative to simple fluorescence detection. The first advantage is decreased fluorescent interference from compounds or buffer because the emission is measured with a delay of 50–150 μs after excitation.[56] The second is that the signal data can be analysed ratiometrically (acceptor molecule emission divided by donor molecule emission), which will help to smooth errors arising from well-to-well differences in reagent addition or inner filter effects.[56,57]

As with any antibody-based assay system, an anti-methylation antibody's selectivity is a key factor in determining the assay's quality. Arguably it is more critical for protein methyltransferase assay development than for other types of targets such as kinases, because a methyltransferase reaction can produce more than one methylated product (Figure 11.2). If the antibody used can only detect one type of product (*e.g.* Kme2), the assay detection time window has to be optimised specifically for that product because progression towards the trimethyl product will decrease the dimethyl signal.

Commercial antibody-based assays are available for a variety of epigenetic factors other than HMTs. Antibody-based detection systems similar to those used for HMT assays have recently been applied to the development of HTS assays for demethylases, including the TR-FRET[58] and AlphaLISA[51,59] formats; the AlphaLISA format is also available for HAT targets.[51] Alpha (*a*mplified *l*uminescent *p*roximity *h*omogeneous *a*ssay) technology has also been used for screening compounds against epigenetic readers.[60–64]

For studies on HDAC targets, CisBio (Codolet, France) has developed a few HTS-amenable deacetylase assays by using HTRF (homogeneous time-resolved fluorescence).[56]

11.5 Protease-Coupled Assays

The blocking of particular peptidase activities at the sites of specific post-translational modifications has been exploited to develop various types of protease-coupled assays.[65,66] Common enzymes for these approaches include endopeptidases/endoproteinases (chymotrypsin, trypsin, Lys-C, Glu-C, *etc.*) and exopeptidases (carboxypeptidase, aminopeptidase, *etc.*). Well-known examples of protease-coupled screening assays include the trypsin-coupled HDAC assays and the Z′-LYTE™ Kinase Assay from Life Technologies.[67] Endoproteinase-LysC (Endo-LysC) is a serine protease that cleaves peptide bonds that are on the C-terminal side of lysine residues. Like trypsin, it cannot cleave at acetyl lysines and thus has also been used for HDAC/SIRT assays.[68] Of relevance to HMT assay development strategies (Figure 11.3, Assay #4) are the facts that Endo-LysC is unable to cleave at monomethyl lysine residues,[69] while trypsin, which can cut after monomethyl lysine,[70] is inactive at dimethyl lysine sites.[71]

Wigle and co-workers have described two assays based on the sensitivity of Endo-LysC activity to substrate methylation, one for the HKMT G9a and one for the demethylase LSD1.[69] The peptide substrates were labelled with fluorescent group, FAM (5/6-carboxyfluorescein), and the methylated and unmethylated (*i.e.* uncleaved and cleaved) forms were detected with a capillary electrophoresis system (Caliper LC3000) after Endo-LysC digestion. Almac Group (Craigavon, UK) uses a related approach for HMT assays (G9a, SET7/9, PRMT5) using their fluorescence lifetime technologies (FLT) based on the fluorescent dye 9-aminoacridine (9AA) and its derivatives.[72] The lysine residue is flanked by the long lifetime 9AA fluorophore on one side and an aromatic moiety, such as a tryptophan residue, on the other, which reduces the long lifetime fluorescence from 9AA. When the unmethylated substrate is cleaved by Endo-LysC (or Endo-ArgC for PRMT5), the fragment with 9AA will produce long lifetime fluorescence, whereas the short lifetime of the uncleaved methylated substrate is maintained.[43]

Major drawbacks of protease-coupled assays for the screening of HMTs are that peptide substrates must be used and that these need to be specially designed to be uncleavable after methylation. Quite a number of HMTs have poor activity even with peptides based on native target sequences (*e.g.* DOT1L, EZH2, NSD2, the MLLs) and others may not tolerate the sequence modifications (replacement of non-target lysines, fluorophore labelling) necessary for the coupled assay. In addition, it is still unclear what effect these artificial peptide substrates may have on HTS hit identification and characterisation. For example, Biofocus scientists have compared results obtained from screening G9a with a set of 2112 compounds using three different assays – FlashPlate with full-length histone H3.1, a protease-coupled fluorescence lifetime (FLT) assay and a protease-coupled electrophoretic mobility shift.[43,73] The FlashPlate/histone H3.1 screen found 23 hits, the FLT approach discovered 11 hits, and the mobility shift assay identified eight hits. Among these hits, six were identified by both the FLT approach and the mobility shift assay, perhaps to be expected because they employed peptide substrates that differed only in their fluorophore labelling. However, the FlashPlate/H3.1 and FLT assays shared only three hits, and the FlashPlate mobility shift assays had no hits in common. Therefore, there was no single common hit among the three assays.[73] Moreover, one particular compound's IC_{50} value determined with FlashPlate/H3.1 was 65-fold more potent than that determined by the FLT assay.[43] Although the screen was perhaps too small to draw any sweeping conclusions, it does highlight the possibility that the peptide-based protease-coupled assays may be missing many hits relative to radiolabelled assays employing full protein substrates. Of course, if the relevant difference is the use of a peptide rather than protein substrate, other assay types, such as AlphaLISA or FlashPlate assays with biotinylated peptide substrates, may be missing hits as well.

11.6 Detection of SAH

Thus far, the discussion of HMT assays has been limited to those that detect a methylated peptide, protein or nucleosome product. However, several types of assays have been based on the detection of SAH (Figure 11.3, Assay # 5–10), the product that is formed in 1 : 1 ratio with methylated lysine or arginine residues following transfer of the methyl function from SAM (*S*-adenosyl-L-methionine; the co-substrate in HMT reactions).

11.6.1 Anti-SAH Antibody-Based Assays

A simple and direct way to detect SAH produced in a HMT reaction is to develop antibodies or other macromolecules that can recognise SAH but not the structurally similar SAM (Figure 11.3, Assay #5). Abbott Laboratories developed a commercial diagnostic kit which determines plasma homocysteine by converting it enzymatically to SAH and then using an anti-SAH monoclonal antibody and fluorescein–SAH conjugate to measure SAH by competitive fluorescence polarisation.[74] Graves and colleagues optimised and validated an assay for catechol-*O*-methyltransferase (COMT) based on the anti-SAH and fluorescein–SAH from the Abbott kit.[75] The assay is homogeneous and HTS compatible and the antibody can detect as little as 5 nM SAH produced during the methyltransferase reaction. The monoclonal anti-SAH has a >150-fold preference for SAH over SAM, allowing the use of SAM at approximately ≤1 μM in the assay. Because the assay is universal for all SAM-utilising methyltransferases, it should be applicable to the HMTs. However, other than a reported screen for COMT inhibitors in which an *S*-adenosylcysteine-TAMRA conjugate was substituted for the fluorescein–SAH,[76] it does not seem to have found widespread use. The considerable length of time required to equilibrate the SAH and tracer binding to the antibody (1–3 hours),[75,76] together with the lack of a good non-SAH inhibitor stop for the HMT reaction, may make the assay more difficult to adapt for HMTs. (EDTA can be used to stop the COMT reaction.) However, given the typically high pH optima of the HMTs (frequently although not uniformly >8),[77,78] it may be possible to find conditions that sufficiently slow down or stop the HMT reaction without impeding the equilibration of the SAH antibody binding.

Cisbio Bioassays (Codolet, France) recently developed an assay based on displacement of a dye-labelled SAH (SAH-d2) from an anti-SAH antibody and now offers the assay reagents as a kit (EPIgeneous™ Methyltransferase Assay Kit), which has been validated with a number of HMTs and other methyltransferases.[79] In this case the assay signal is TR-FRET from the terbium cryptate-labelled anti-SAH antibody to the antibody-bound SAH-d2 tracer. Unlabelled SAH from the HMT reaction displaces the SAH-d2 from the antibody, so the signal is decreased in proportion to methyltransferase activity.

11.6.2 Detection of SAH with Coupled Enzyme Assays

There are two enzymes that can break SAH apart (Figure 11.3); one is *S*-adenosylhomocysteine hydrolase (SAHH) and the other is MTAN (5′-methylthioadenosine/S-adenosylhomocysteine nucleosidase). Each of these reactions, combined with one or more further enzymatic steps, has been used as the basis for various SAH-detecting HMT assays. While, as a group, these assays share certain advantages – lack of radioactivity, homogeneity, universality for methyltransferases, and flexibility with respect to the methyl recipient (peptide, protein, *etc.*) – the multiple coupling steps involved in signal generation may cause their own sets of difficulties. Coupling enzymes need to be counter-screened to eliminate false hits and, as Quinn and Simeonov have pointed out, coupled enzyme assays are inherently more difficult to optimise.[53] We'll discuss both the SAHH and MTAN assay 'pathways' in more detail below.

11.6.2.1 SAH-Hydrolase (SAHH) Pathway

SAHH catalyses the breakdown of SAH into homocysteine (HCy) and adenosine (Ado). Ado can be further metabolised to inosine plus ammonia by adenosine deaminase (Figure 11.3, Assay #6). Alternatively, Ado plus ATP can be converted by the Ado-kinase reaction to produce AMP and ADP (Figure 11.3, Assay #7).

The first approach in this pathway is to detect the HCy with thiol-reactive fluorescent dyes (Figure 11.3, Assay #6).[80] This assay utilises SAHH to hydrolyse the product AdoHcy to homocysteine (HCy) and Ado. Adenosine deaminase is added to the reaction to pull the SAHH reaction to completion.[80] The HCy concentration is then determined through conjugation of its free sulfhydryl moiety to a thiol-sensitive fluorophore, such as the ThioGlo® maleimide reagents (Covalent Associates, Corvallis, OR), or CPM (7-diethylamino-3-(4′-maleimidylphenyl)-4-methylcoumarin).[80] If colour detection is preferred, the reaction of the HCy with DTNB (5,5′-dithiobis(2-nitrobenzoic acid), colourless) to produce the yellow product TNB (5-mercapto-2-nitrobenzoic acid) could be used as an alternative.[81] A disadvantage of these assays is the slowness of SAHH catalysis, requiring the addition of large amounts of the coupling enzyme. In addition, the assay is sensitive to reducing agents (*e.g.* DTT) and, as with all fluorescence assays, subject to interference from fluorescent compounds or inner filter effects.

The second branch in this pathway is to employ Ado-kinase plus ATP to convert Ado to AMP and generate signal *via* AMP detection (Figure 11.3, Assay #7). Using an anti-AMP/GMP antibody and fluorescent AMP-conjugate tracer (Transcreener AMP/GMP Assay), BellBrook Labs has developed a generic competitive fluorescent polarisation assay for methyltransferases.[82,83] Currently the company has optimised assays for over 10 methyltransferases with this approach, including histone and DNA methyltransferases.

11.6.2.2 MTAN Pathway

SAH can also be cleaved by MTAN into adenine and *S*-(5-deoxy-D-ribos-5-yl)-L-homocysteine (SRH).[84] One route to deriving a signal from the adenine is to add phosphoribosyl pyrophosphate (PRPP) and convert these to AMP by the adenine phosphoribosyltransferase (APRT) reaction (Figure 11.3). Then, in the presence of added phosphoenolpyruvate and pyrophosphate, the AMP is further converted to ATP by PPDK (pyruvate phosphate dikinase).[85] The ATP can then be quantified with commercially available luciferase assay kits, such as ATPlite from PerkinElmer or Kinase-Glo from Promega.

An alternative path to measure the adenine generated in the MTAN/SAH reaction is to convert adenine to hypoxanthine with adenine deaminase. The associated decrease in absorbance at 265 nm can be monitored continuously with a spectrophotometer.[86] Hypoxanthine can be further converted to urate and hydrogen peroxide (H_2O_2) by xanthine oxidase. H_2O_2 can then be measured by fluorescent or colorimetric reagents that are available from many commercial sources, such as resorufin/HRP (Life Technologies), hydrogen peroxide detection kits (Enzo Life Sciences), and xanthine oxidase kit (Cayman Chemicals). A methyltransferase assay kit combining MTAN, adenine deaminase, xanthine oxidase and resorufin/HRP for H_2O_2 detection is offered by G-Biosciences (St. Louis, MO).

Finally, the MTAN pathway also provides an alternative route to generate HCy, which can then be detected with thiol-reactive dyes (Figure 11.3). The enzyme *S*-ribosylhomocysteinase (LuxS) cleaves the thioether bond in *S*-ribosylhomocysteine (SRH) to produce homocysteine and 4,5-dihydroxy-2,3-pentanedione.[87,88]

11.7 SAM-Binding Site Assays

Although the small-molecule interactions of the HMTs have been studied primarily through activity assays, active site binding-based assays are also beginning to emerge. For example, Bochar *et al.*[89] and Cayman Chemical have developed a fluorescent probe that will bind to the SAM-binding pocket of Set7/9, GLP, G9a and MLL with affinities ranging from 0.027 μM to 2 μM. Using the probe as the basis of a fluorescence polarisation assay for Set7/9 binding, the authors performed a proof of concept screening of a 14 400 compound library. A few compounds that displaced the probe from the binding pocket were identified, although both the hit rate and confirmation rate were low. Diaz *et al.*[44] have developed a ^{3}H-SAM SPA binding assay for EZH2, which uses PS-PEI-treated WGA (Type B) imaging beads (RPNQ0289; PerkinElmer) to bind the EZH2 and a ViewLux™ imager (PerkinElmer) to measure the scintillation signal from the EZH2-bound ^{3}H-SAM. Using this approach, the authors determined that sinefungin is a SAM-competitive inhibitor.

11.8 Cell-Based Assays

Cell-based assays are important tools in the drug discovery process and are undertaken with the general goal of characterising the complex changes occurring after targeted intervention. In comparison to the development of *in vitro* biochemical enzyme activity assays, the development of multiple cell-based assays to detect the activity of compounds on methylation and acetylation of endogenous targets, in addition to their impacts on cell physiology, can present a challenging task. It is well known that the epigenetic enzyme inhibitors significantly affect gene transcription and cellular function. For example, HDAC inhibitors generally reduce cell growth, increase differentiation, and induce apoptosis and cancer cell death, and these effects are associated with the ability of the inhibitors to increase histone acetylation.[90] Therefore, evaluation of the inhibitors of epigenetic enzymes in various multiplexed viability and cytotoxicity measurements, especially in cancer cell lines, can be an attractive approach. However, in this section, we will focus primarily on cell-based assays that are designed to measure the compound inhibitory effects on the activity of epigenetic enzymes in a cellular context. Because, in the cell, epigenetic enzymes are usually part of large multi-protein complexes, as opposed to the isolated form in which they are typically assayed *in vitro*, this represents an important way to assess the relevance of biochemical screening results to *in vivo* drug efficacy. Such assays may also be used as primary screens for epigenetic modulators.

Recently, two homogeneous cell-based assays using the AlphaLISA® and LanthaScreen® technologies to detect histone H3K27me3 have become commercially available, and a heterogeneous cell assay with a modified DELFIA® format has been reported as well.[91,92] The AlphaLISA cell assay detects epigenetic markers in cellular extracts by a biotinylated anti-histone H3 (C-terminus) antibody, and the AlphaLISA acceptor beads are conjugated to an antibody specific to the marker (see Section 11.3 for an explanation of the assay principle). This assay has a homogeneous format and is amenable to HTS. Besides H3K27me3, other assay kits, such as H3K4me2, H3K36me2, H3K79me2, H3K9me2 and H3K27me2, for different epigenetic markers, are available from PerkinElmer. The AlphaLISA format has also been adapted for cell-based screening of compounds against HATs and HDACs.

The LanthaScreen assay is another homogeneous cell assay suitable for HTS purposes. Invitrogen developed and introduced a BacMam histone H3K27me3 cellular assay kit, which utilises terbium-based TR-FRET technology in conjunction with the BacMam gene delivery system. Histone H3K27me3 protein is expressed as a fusion with green fluorescent protein (GFP), and anti-H3K27me3 antibody is labelled with terbium. The post-translational modification is detected upon cell lysis in the presence of a terbium-anti-histone H3K27me3 antibody. In this system, GFP acts as a FRET acceptor, and terbium as FRET donor. The advantage of this assay format is that the histone can be introduced to different cell backgrounds *via* transient transfection. A modified DELFIA assay detecting cellular H3K27me3 level was published by Xie *et al.*[92] Based on time-resolved fluorescence technology, DELFIA is a heterogeneous assay that can measure the modified histone and total histone simultaneously. This assay is, however, relatively low throughput and difficult to adapt to HTS. Qian *et al.*[93] recently performed a side-by-side comparison of AlphaLISA, LanthaScreen and DELFIA, in parallel with traditional Western blot detection of histone methylation inhibition by small-molecule HMA inhibitors. The data indicate that although all four formats produced acceptable results, the homogeneous AlphaLISA assay was best suited for high-throughput screening and compound profiling due to its wider window and ease of automation. The DELFIA and Western blot assays were useful as validation tools to confirm the cell activities and eliminate potential false-positive compounds.

Acknowledgements

We would like to thank Ms Tess Chadwick for her critical editing. The HotSpot technology and HMT protein production were partially supported by a US National Institutes of Health grant (R44CA139621 to H. M.).

References

1. C. H. Arrowsmith, C. Bountra, P. V. Fish, K. Lee and M. Schapira, *Nat. Rev. Drug Discovery*, 2012, **11**, 384.
2. S. Rea, F. Eisenhaber, D. O'Carroll, B. D. Strahl, Z. W. Sun, M. Schmid, S. Opravil, K. Mechtler, C. P. Ponting, C. D. Allis and T. Jenuwein, *Nature*, 2000, **406**, 593.
3. A. H. Lund and M. van Lohuizen, *Genes Dev.*, 2004, **18**, 2315.

4. R. A. Copeland, M. E. Solomon and V. M. Richon, *Nat. Rev. Drug Discovery*, 2009, **8**, 724.
5. M. T. Bedford and S. G. Clarke, *Mol. Cell*, 2009, **33**, 1.
6. C. D. Krause, Z. H. Yang, Y. S. Kim, J. H. Lee, J. R. Cook and S. Pestka, *Pharmacol. Ther.*, 2007, **113**, 50.
7. P. Trojer, M. Dangl, I. Bauer, S. Graessle, P. Loidl and G. Brosch, *Biochemistry*, 2004, **43**, 10834.
8. S. Pahlich, R. P. Zakaryan and H. Gehring, *Biochim. Biophys. Acta*, 2006, **1764**, 1890.
9. T. M. Lakowski and A. Frankel, *Biochem. J.*, 2009, **421**, 253.
10. K. Y. Horiuchi, M. M. Eason, J. J. Ferry, J. L. Planck, C. P. Walsh, R. F. Smith, K. T. Howitz and H. Ma, *Assay Drug Dev. Technol.*, 2013, **11**, 227.
11. H. Ma, S. Deacon and K. Horiuchi, *Expert Opin. Drug Discovery*, 2008, **3**, 607.
12. L. Afjehi-Sadat and B. A. Garcia, *Curr. Opin. Chem. Biol.*, 2013, **17**, 12.
13. D. Phanstiel, J. Brumbaugh, W. T. Berggren, K. Conard, X. Feng, M. E. Levenstein, G. C. McAlister, J. A. Thomson and J. J. Coon, *Proc. Natl. Acad. Sci. U. S. A.*, 2008, **105**, 4093.
14. T. Bonaldi, A. Imhof and J. T. Regula, *Proteomics*, 2004, **4**, 1382.
15. T. Bonaldi, J. T. Regula and A. Imhof, *Methods Enzymol.*, 2004, **377**, 111.
16. H. Tamaru, X. Zhang, D. McMillen, P. B. Singh, J. Nakayama, S. I. Grewal, C. D. Allis, X. Cheng and E. U. Selker, *Nat. Genet.*, 2003, **34**, 75.
17. X. Zhang, Z. Yang, S. I. Khan, J. R. Horton, H. Tamaru, E. U. Selker and X. Cheng, *Mol. Cell*, 2003, **12**, 177.
18. H. H. Ng, Q. Feng, H. Wang, H. Erdjument-Bromage, P. Tempst, Y. Zhang and K. Struhl, *Genes Dev.*, 2002, **16**, 1518.
19. B. Jones, H. Su, A. Bhat, H. Lei, J. Bajko, S. Hevi, G. A. Baltus, S. Kadam, H. Zhai, R. Valdez, S. Gonzalo, Y. Zhang, E. Li and T. Chen, *PLoS Genet.*, 2008, **4**, e1000190.
20. A. J. Kuo, P. Cheung, K. Chen, B. M. Zee, M. Kioi, J. Lauring, Y. Xi, B. H. Park, X. Shi, B. A. Garcia, W. Li and O. Gozani, *Mol. Cell*, 2011, **44**, 609.
21. B. D. Strahl, P. A. Grant, S. D. Briggs, Z. W. Sun, J. R. Bone, J. A. Caldwell, S. Mollah, R. G. Cook, J. Shabanowitz, D. F. Hunt and C. D. Allis, *Mol. Cell. Biol.*, 2002, **22**, 1298.
22. M. Bremang, A. Cuomo, A. M. Agresta, M. Stugiewicz, V. Spadotto and T. Bonaldi, *Mol. Biosyst.*, 2013, **9**, 2231.
23. L. M. Britton, M. Gonzales-Cope, B. M. Zee and B. A. Garcia, *Expert Rev. Proteomics*, 2011, **8**, 631.
24. M. Soldi, A. Cuomo, M. Bremang and T. Bonaldi, *Int. J. Mol. Sci.*, 2013, **14**, 5402.
25. Y. Chang, X. Zhang, J. R. Horton, A. K. Upadhyay, A. Spannhoff, J. Liu, J. P. Snyder, M. T. Bedford and X. Cheng, *Nat. Struct. Mol. Biol.*, 2009, **16**, 312.
26. P. T. Rye, L. E. Frick, C. C. Ozbal and W. A. Lamarr, *J. Biomol. Screening*, 2011, **16**, 1186.

27. Y. Zhao, B. Su, R. L. Jacobs, B. Kennedy, G. A. Francis, E. Waddington, J. T. Brosnan, J. E. Vance and D. E. Vance, *Arterioscler. Thromb. Vasc. Biol.*, 2009, **29**, 1349.
28. M. Plant, T. Dineen, A. Cheng, A. M. Long, H. Chen and K. A. Morgenstern, *Anal. Biochem.*, 2011, **419**, 217.
29. S. E. Hutchinson, M. V. Leveridge, M. L. Heathcote, P. Francis, L. Williams, M. Gee, J. Munoz-Muriedas, B. Leavens, A. Shillings, E. Jones, P. Homes, S. Baddeley, C. W. Chung, A. Bridges and A. Argyrou, *J. Biomol. Screening*, 2012, **17**, 39–48.
30. J. H. Zhang, T. D. Chung and K. R. Oldenburg, *J. Biomol. Screening*, 1999, **4**, 67.
31. C. Choudhary, C. Kumar, F. Gnad, M. L. Nielsen, M. Rehman, T. C. Walther, J. V. Olsen and M. Mann, *Science*, 2009, **325**, 834.
32. A. Lundby, K. Lage, B. T. Weinert, D. B. Bekker-Jensen, A. Secher, T. Skovgaard, C. D. Kelstrup, A. Dmytriyev, C. Choudhary, C. Lundby and J. V. Olsen, *Cell Rep.*, 2012, **2**, 419.
33. P. G. Shaw, R. Chaerkady, Z. Zhang, N. E. Davidson and A. Pandey, *Anal. Chem.*, 2011, **83**, 3623.
34. M. Bantscheff, C. Hopf, M. M. Savitski, A. Dittmann, P. Grandi, A. M. Michon, J. Schlegl, Y. Abraham, I. Becher, G. Bergamini, M. Boesche, M. Delling, B. Dumpelfeld, D. Eberhard, C. Huthmacher, T. Mathieson, D. Poeckel, V. Reader, K. Strunk, G. Sweetman, U. Kruse, G. Neubauer, N. G. Ramsden and G. Drewes, *Nat. Biotechnol.*, 2011, **29**, 255.
35. Z. A. Gurard-Levin, M. D. Scholle, A. H. Eisenberg and M. Mrksich, *ACS Comb. Sci.*, 2011, **13**, 347.
36. http://static.enzolifesciences.com/fileadmin/files/image/BML-AK518_insert2012Jan3.pdf.
37. S. K. Knutson, T. J. Wigle, N. M. Warholic, C. J. Sneeringer, C. J. Allain, C. R. Klaus, J. D. Sacks, A. Raimondi, C. R. Majer, J. Song, M. P. Scott, L. Jin, J. J. Smith, E. J. Olhava, R. Chesworth, M. P. Moyer, V. M. Richon, R. A. Copeland, H. Keilhack, R. M. Pollock and K. W. Kuntz, *Nat. Chem. Biol.*, 2012, **8**, 890.
38. M. T. McCabe, H. M. Ott, G. Ganji, S. Korenchuk, C. Thompson, G. S. Van Aller, Y. Liu, A. P. Graves, A. Della Pietra, 3rd, E. Diaz, L. V. LaFrance, M. Mellinger, C. Duquenne, X. Tian, R. G. Kruger, C. F. McHugh, M. Brandt, W. H. Miller, D. Dhanak, S. K. Verma, P. J. Tummino and C. L. Creasy, *Nature*, 2012, **492**, 108.
39. H. Yang, C. E. Pinello, J. Luo, D. Li, Y. Wang, L. Y. Zhao, S. C. Jahn, S. A. Saldanha, J. Planck, K. R. Geary, H. Ma, B. K. Law, W. R. Roush, P. Hodder and D. Liao, *Mol. Cancer Ther.*, 2013, **12**, 610.
40. A. M. Makowski, R. N. Dutnall and A. T. Annunziato, *J. Biol. Chem.*, 2001, **276**, 43499.
41. S. Ait-Si-Ali, S. Ramirez, P. Robin, D. Trouche and A. Harel-Bellan, *Nucleic Acids Res.*, 1998, **26**, 3869.
42. P. Rathert, X. Cheng and A. Jeltsch, *Biotechniques*, 2007, **43**, 602.

43. T. Ahrens, A. Bergner, D. Sheppard and D. Hafenbradl, *J. Biomol. Screening*, 2012, **17**, 85.
44. E. Diaz, C. A. Machutta, S. Chen, Y. Jiang, C. Nixon, G. Hofmann, D. Key, S. Sweitzer, M. Patel, Z. Wu, C. L. Creasy, R. G. Kruger, L. LaFrance, S. K. Verma, M. B. Pappalardi, B. Le, G. S. Van Aller, M. T. McCabe, P. J. Tummino, A. J. Pope, S. H. Thrall, B. Schwartz and M. Brandt, *J. Biomol. Screening*, 2012, **17**, 1279.
45. F. Turlais, A. Hardcastle, M. Rowlands, Y. Newbatt, A. Bannister, T. Kouzarides, P. Workman and G. W. Aherne, *Anal. Biochem.*, 2001, **298**, 62.
46. B. Nare, J. J. Allocco, R. Kuningas, S. Galuska, R. W. Myers, M. A. Bednarek and D. M. Schmatz, *Anal. Biochem.*, 1999, **267**, 390.
47. N. O. Reich and N. Mashhoon, *J. Biol. Chem.*, 1990, **265**, 8966.
48. D. Kuck, T. Caulfield, F. Lyko and J. L. Medina-Franco, *Mol. Cancer Ther.*, 2010, **9**, 3015.
49. S. Castellano, D. Kuck, M. Viviano, J. Yoo, F. Lopez-Vallejo, P. Conti, L. Tamborini, A. Pinto, J. L. Medina-Franco and G. Sbardella, *J. Med. Chem.*, 2011, **54**, 7663.
50. T. Yokochi and K. D. Robertson, *J. Biol. Chem.*, 2002, **277**, 11735.
51. N. Gauthier, M. Caron, L. Pedro, M. Arcand, J. Blouin, A. Labonte, C. Normand, V. Paquet, A. Rodenbrock, M. Roy, N. Rouleau, L. Beaudet, J. Padros and R. Rodriguez-Suarez, *J. Biomol. Screening*, 2012, **17**, 49.
52. http://www.perkinelmer.com/CMSResources/Images/44-150039TCH_AlphaLISA_DOT1L_Histone_H3.pdf.
53. A. M. Quinn and A. Simeonov, *Curr. Chem. Genomics*, 2011, **5**, 95.
54. J. R. Simard, M. Plant, R. Emkey and V. Yu, *Assay Drug Dev. Technol.*, 2013, **11**, 152.
55. M. O. Palmier and S. R. Van Doren, *Anal. Biochem.*, 2007, **371**, 43.
56. F. Degorce, A. Card, S. Soh, E. Trinquet, G. P. Knapik and B. Xie, *Curr. Chem. Genomics*, 2009, **3**, 22.
57. J. F. Glickman, X. Wu, R. Mercuri, C. Illy, B. R. Bowen, Y. He and M. Sills, *J. Biomol. Screening*, 2002, **7**, 3.
58. V. Yu, T. Fisch, A. M. Long, J. Tang, J. H. Lee, M. Hierl, H. Chen, P. Yakowec, R. Schwandner and R. Emkey, *J. Biomol. Screening*, 2012, **17**, 27.
59. A. Kawamura, A. Tumber, N. R. Rose, O. N. King, M. Daniel, U. Oppermann, T. D. Heightman and C. Schofield, *Anal. Biochem.*, 2010, **404**, 86.
60. P. Filippakopoulos, S. Picaud, M. Mangos, T. Keates, J. P. Lambert, D. Barsyte-Lovejoy, I. Felletar, R. Volkmer, S. Muller, T. Pawson, A. C. Gingras, C. H. Arrowsmith and S. Knapp, *Cell*, 2012, **149**, 214.
61. P. Filippakopoulos, J. Qi, S. Picaud, Y. Shen, W. B. Smith, O. Fedorov, E. M. Morse, T. Keates, T. T. Hickman, I. Felletar, M. Philpott, S. Munro, M. R. McKeown, Y. Wang, A. L. Christie, N. West, M. J. Cameron, B. Schwartz, T. D. Heightman, N. La Thangue, C. A. French, O. Wiest, A. L. Kung, S. Knapp and J. E. Bradner, *Nature*, 2010, **468**, 1067.
62. J. M. Herold, T. J. Wigle, J. L. Norris, R. Lam, V. K. Korboukh, C. Gao, L. A. Ingerman, D. B. Kireev, G. Senisterra, M. Vedadi, A. Tripathy, P. J. Brown,

C. H. Arrowsmith, J. Jin, W. P. Janzen and S. V. Frye, *J. Med. Chem.*, 2011, **54**, 2504.
63. A. M. Quinn, M. T. Bedford, A. Espejo, A. Spannhoff, C. P. Austin, U. Oppermann and A. Simeonov, *Nucleic Acids Res.*, 2010, **38**, e11.
64. T. J. Wigle, J. M. Herold, G. A. Senisterra, M. Vedadi, D. B. Kireev, C. H. Arrowsmith, S. V. Frye and W. P. Janzen, *J. Biomol. Screening*, 2010, **15**, 62.
65. G. Fischer, H. Bang and C. Mech, *Biomed. Biochim. Acta*, 1984, **43**, 1101.
66. M. Nishikata, K. Suzuki, Y. Yoshimura, Y. Deyama and A. Matsumoto, *Biochem. J.*, 1999, **343**(2), 385.
67. S. M. Rodems, B. D. Hamman, C. Lin, J. Zhao, S. Shah, D. Heidary, L. Makings, J. H. Stack and B. A. Pollok, *Assay Drug Dev. Technol.*, 2002, **1**, 9.
68. P. A. Marcotte, P. R. Richardson, J. Guo, L. W. Barrett, N. Xu, A. Gunasekera and K. B. Glaser, *Anal. Biochem.*, 2004, **332**, 90.
69. T. J. Wigle, L. M. Provencher, J. L. Norris, J. Jin, P. J. Brown, S. V. Frye and W. P. Janzen, *Chem. Biol.*, 2010, **17**, 695.
70. L. Benoiton and J. Deneault, *Biochim. Biophys. Acta*, 1966, **113**, 613.
71. L. Poncz and D. G. Dearborn, *J. Biol. Chem.*, 1983, **258**, 1844.
72. B. A. Maltman, C. J. Dunsmore, S. C. Couturier, A. E. Tirnaveanu, Z. Delbederi, R. A. McMordie, G. Naredo, R. Ramage and G. Cotton, *Chem. Commun.*, 2010, **46**, 6929.
73. http://www.biofocus.com/_downloads/posters/2012/epigenetic-protein-targets.pdf.
74. M. T. Shipchandler and E. G. Moore, *Clin. Chem.*, 1995, **41**, 991.
75. T. L. Graves, Y. Zhang and J. E. Scott, *Anal. Biochem.*, 2008, **373**, 296.
76. R. G. Robinson, S. M. Smith, S. E. Wolkenberg, M. Kandebo, L. Yao, C. R. Gibson, S. T. Harrison, S. Polsky-Fisher, J. C. Barrow, P. J. Manley, J. J. Mulhearn, K. K. Nanda, J. W. Schubert, B. W. Trotter, Z. Zhao, J. M. Sanders, R. F. Smith, D. McLoughlin, S. Sharma, D. L. Hall, T. L. Walker, J. L. Kershner, N. Bhandari, P. H. Hutson and N. A. Sachs, *ACS Chem. Neurosci.*, 2012, **3**, 129.
77. X. Zhang and T. C. Bruice, *Proc. Natl. Acad. Sci. U. S. A.*, 2008, **105**, 5728.
78. W. Xie, G. Merz and R. B. Denman, *J. Biophys. Struct. Biol.*, 2011, **3**, 49.
79. http://www.htrf.com/sites/default/files/ressources/cisbio-po-2013-epigeneous-methyltransferase.pdf.
80. E. Collazo, J. F. Couture, S. Bulfer and R. C. Trievel, *Anal. Biochem.*, 2005, **342**, 86.
81. P. R. Thompson, D. Wang, L. Wang, M. Fulco, N. Pediconi, D. Zhang, W. An, Q. Ge, R. G. Roeder, J. Wong, M. Levrero, V. Sartorelli, R. J. Cotter and P. A. Cole, *Nat. Struct. Mol. Biol.*, 2004, **11**, 308.
82. M. Staeben, K. M. Kleman-Leyer, A. L. Kopp, T. A. Westermeyer and R. G. Lowery, *Assay Drug Dev. Technol.*, 2010, **8**, 344.
83. T. A. Klink, M. Staeben, K. Twesten, A. L. Kopp, M. Kumar, R. S. Dunn, C. A. Pinchard, K. M. Kleman-Leyer, M. Klumpp and R. G. Lowery, *J. Biomol. Screening*, 2012, **17**, 59.
84. N. Parveen and K. A. Cornell, *Mol. Microbiol.*, 2011, **79**, 7.

85. G. Ibanez, J. L. McBean, Y. M. Astudillo and M. Luo, *Anal. Biochem.*, 2010, **401**, 203.
86. K. M. Dorgan, W. L. Wooderchak, D. P. Wynn, E. L. Karschner, J. F. Alfaro, Y. Cui, Z. S. Zhou and J. M. Hevel, *Anal. Biochem.*, 2006, **350**, 249.
87. J. Zhu, R. Patel and D. Pei, *Biochemistry*, 2004, **43**, 10166.
88. C. L. Hendricks, J. R. Ross, E. Pichersky, J. P. Noel and Z. S. Zhou, *Anal. Biochem.*, 2004, **326**, 100.
89. http://www.biotechniques.com/multimedia/archive/00177/BTN_0311-March_Post_177966a.pdf.
90. O. Khan and N. B. La Thangue, *Immunol. Cell Biol.*, 2012, **90**, 85.
91. T. Machleidt, M. B. Robers, S. B. Hermanson, J. M. Dudek and K. Bi, *J. Biomol. Screening*, 2011, **16**, 1236.
92. W. Xie, R. S. Ames and H. Li, *J. Biomol. Screening*, 2012, **17**, 99.
93. J. Qian, L. Lu, J. Wu and H. Ma, *Assay Drug Dev. Technol.*, 2013, **11**, 449–456.

CHAPTER 12

Safety Considerations for Epigenetic Mechanisms as Drug Targets: Are Existing Toxicology Studies Fit for Purpose?

PAUL BUTLER*[a], MICHAEL LAWTON[b], ROBERT E. CHAPIN[b], BRETT D. HOLLINGSHEAD[c], KAREN LEACH[d], PRASHANT R. NAMBIAR[c], MATHEW PLETCHER[e], AND AMY H. YANG[a]

[a]Drug Safety Research and Development, Pfizer Inc., San Diego, California, USA; [b]Drug Safety Research and Development, Pfizer Inc., Groton, Connecticut, USA; [c]Drug Safety Research and Development, Pfizer Inc., Andover, Massachusetts, USA; [d]Centers for Therapeutic Innovation, Pfizer Inc, Boston, Massachusetts, USA; [e]Rare Diseases Research Unit, Pfizer Inc., Cambridge, Massachusetts, USA
*E-mail: paul.butler@pfizer.com

12.1 Introduction

Epigenetics refers to molecular mechanisms that initiate and maintain heritable changes in gene expression without affecting the DNA sequence. As such, epigenetic mechanisms play critical roles during embryonic development and normal physiology. Dysregulation of epigenetic processes is implicated in various diseases including cancer,[1,2] rheumatic disease,[3] pain,[4,5] and

RSC Drug Discovery Series No. 48
Epigenetics for Drug Discovery
Edited by Nessa Carey

Published by the Royal Society of Chemistry, www.rsc.org

other conditions, making epigenetic mechanisms attractive therapeutic targets.[6] Indeed targeting epigenetic proteins, such as DNA methyltransferases (DNMTs), histone acetyltransferases (HATs), histone deacetylases (HDACs), histone methyltransferases (HMTs), histone demethylases (HDMs), and bromodomain-containing proteins (BCPs) is a highly active area of research within the pharmaceutical industry. Some drugs (targeting DNMTs or HDACs) have been approved for the treatment of cancer. However, the science and development of epigenetic mechanisms as drug targets are in their nascent stages, and regulatory guidance regarding the development of epigenetic drugs for the treatment of disease has yet to be developed. It is noteworthy that a number of commonly prescribed drugs (*e.g.* valproic acid for the treatment of epilepsy) are now recognised as epigenetic modulators, which may offer hope for the long-term safety of epigenetic therapy. Moreover, several epigenetic modulators, such as those targeting HDACs or BCPs, are in clinical trials for central nervous system (CNS) indications, atherosclerosis and other disease indications, where the risk–benefit tolerance is lower than that for oncology or other life-threatening diseases. Given the pleiotropic role of epigenetics in normal physiology, the impact of targeting epigenetic mechanisms, either adverse or adaptive, in non-clinical models and human populations is an important area to further understand.

This chapter will provide a framework for assessing the non-clinical safety of small-molecule compounds targeting epigenetic proteins as they advance through the drug development process, and could be equally applied to other modalities that target epigenetic mechanisms. The chapter will also highlight known target organ toxicities as well as areas where there is limited knowledge about long-term epigenetic effects, including those that could affect future generations.

12.2 Mechanisms of Epigenetic Regulation Targeted by Drugs

In order to understand the safety concerns of drugs targeting epigenetic mechanisms, it is important to briefly outline the classical mechanisms that maintain epigenetic states: covalent modifications of DNA (*e.g.* methylation) and DNA-packaging histones (*e.g.* changes in the state of histone acetylation, methylation, phosphorylation and others). Post-transcriptional regulation by non-coding RNAs such as microRNAs can also be added to the list of modifications,[7] but the safety issues of modulation of these events are not considered in this chapter.

DNA methylation is associated with transcriptional regulation due to covalent attachment of a methyl group at carbon 5 of cytosine by DMNTs and cofactor *S*-adenosylmethionine. Such DNA methylation causes disruption to the binding of transcription factors and related gene transcription machinery. The disruptive process is mediated by a family of methylcytosine-binding proteins including MBD1–4 and MeCP2.[8] The binding predominantly occurs at sites where a cytosine (C) precedes a guanine (G) base to form what is referred to as a CpG site, and is a key epigenetic modification, with hypermethylation of gene promoters associated with transcriptional silencing. In contrast to the well-defined methylation process, the various pathways of DNA demethylation are just being elucidated. DNA demethylation can occur *via* either passive (during cell division) or active processes through a series of enzymatic reactions.[9] Recently two DNMT inhibitors, 5-azacytidine (Vidaza) and 5-aza-2-deoxycytidine (Dacogen) have been approved for the treatment of myelodysplastic syndromes (MDS) and low blast count acute myeloid leukemia.[10]

The second and more widely therapeutically targeted epigenetic mechanism includes modifications of the DNA-packaging histone proteins. Two copies each of histones H2A, H2B, H3 and H4, wrapping around which are 146 base pairs of DNA, form the so-called nucleosome, which is the basic structure of chromatin. Histone proteins are post-translationally modified at specific sites, predominantly on the N-terminal tails, including lysine acetylation and ubiquitination, methylation of lysine and arginine methylation, and others.[11] Enzymes that catalyse these chemical modifications are termed 'writers'; conversely, those that remove them are termed 'erasers'.

Acetylation and deacetylation are regulated by HATs and HDACs, respectively. Acetyl groups are added to conserved lysine amino acids on histone proteins by HATs. HDACs modify the histone tail by the removal of an acetyl group to alter chromatin accessibility, thereby regulating gene expression. As a general rule, HDACs function as transcriptional repressors. Increased activity of HDACs has been associated with cancer,[12] probably due in part to the negative regulatory effects on tumour suppressor genes. HDAC inhibitors for the treatment of cancer were developed, with successful registration of several inhibitors.[13] HDAC inhibitors are now being explored for indications outside of oncology. For example, there are currently ongoing clinical trials applying these epigenetic modifiers to sickle cell disease, amyotrophic lateral sclerosis, Huntington's disease, Friedreich's ataxia, frontotemporal dementia and spinal muscular atrophy.[13–16]

HMTs catalyse histone methylation in a histone- and residue-specific manner, using *S*-adenosylmethionine as the methyl donor. The best-studied histone methylation events occur on lysine and arginine residues on the N-terminal tails of histones. The removal of histone methyl groups is catalysed by HDMs. Histone methylation acts as a signal integration platform to recruit various protein factors to target genes, leading to transcriptional activation or repression. The effects of histone methylation on target gene transcription are context-specific, dependent on the location of the methylated lysine or arginine residue on the histone, as well as the degree of methylation. Histone methylation is known to influence diverse biological processes in the context of cell type, developmental stage and environmental factors. It is not surprising that HMTs and HDMs have been implicated in various disease states, most notably in cancer where mutations, translocations and over-expression of the genes encoding these enzymes have been reported.[2]

In addition to cancer, histone methylation has been shown to be involved in intellectual disability and ageing, among other conditions.[17,18]

Finally, BCPs represent another class of epigenetic, transcriptional regulators. Many of these proteins contain multiple functional domains, including HAT activity, but the unifying activity of the 42 known BCPs is that they regulate gene expression by recognising acetylated histones in relaxed chromatin, and they facilitate transcriptional activation. This class of protein is commonly referred to as 'reader proteins'. There has been a growing interest in targeting BCPs for the purpose of treating cancer and other indications such as atherosclerosis.[19–21] The most advanced effort from a non-clinical perspective has been the discovery of a compound known as JQ1 that targets the closely related proteins BRD2, 3, 4 and BRDT[22] and several BCPs are now in early clinical development.[20]

12.3 Adverse Drug Effects of First-Generation Epigenetic Compounds and Lessons from Common Non-Epigenetic Drugs as Epigenetic Modulators

There has been recent clinical success with the approval of the first generation of epigenetic drugs for use in cancer patients. These are the HDAC inhibitors romidepsin (Istodax) and vorinostat (Zolinza), and the DNMT inhibitors decitabine (Dacogen) and azacytidine (Vidaza). Although these drugs have two distinct pharmacological actions, the clinical and non-clinical safety profiles are surprisingly similar. These compounds cause clinical adverse effects that are primarily haematological in nature (*e.g.* thrombocytopenia, leukopenia, neutropenia, *etc.*), in addition to manageable effects such as headache, nausea and diarrhoea. There are also possible class-specific hazards such as the association of electrocardiographic alterations observed with HDAC inhibition (see Section 12.5.1). Non-clinical studies for these compounds were typically performed to support the intended clinical dosing paradigm and as such were given *via* IV administration (vorinostat, the exception, is an oral daily dosed compound). These studies typically showed drug-induced mortality, reversible haematological toxicities and occasional liver, kidney and gastrointestinal (GI) effects. Additionally, effects on fertility and embryofetal development were noted for all four of these drugs, leading to a pregnancy category D warning on the label. Aside from romidepsin, the other three approved drugs were also mutagenic in at least one genotoxicity assay. Table 12.1 provides an overview of the drug label warnings and non-clinical toxicology and genotoxicity characteristics for these four drugs.

Although a number of second-generation epigenetic therapies, such as BCP, HMT and HDM inhibitors and more selective HDAC inhibitors are currently in clinical development, such as BCP, HMT and HDM inhibitors and more selective HDAC inhibitors, the availability of data related to the clinical and

Table 12.1 Registered epigenetic drugs and their clinical safety and non-clinical toxicology findings.

	Drug label		Non-clinical toxicology		
Drug	Developmental	Clinical safety warnings	Rat	Dog	Genotoxicity and carcinogenicity
Romidepsin (Istodax)	*Pregnancy category D*: Embryocidal in animal studies, and resulted in adverse effects on the developing foetus at exposures below those in patients at the recommended dose	Thrombocytopenia, leukopenia (neutropenia and lymphopenia), and anaemia, infections, including pneumonia and sepsis. Electrocardiographic changes (T-wave and ST-segment changes)	Mortality, ↓ WBCs, ↑ AST, ALT, LDH, CK, lymph degeneration/necrosis, bone marrow hypocellularity, testis degeneration, kidney tubular dilatation. Atrophy in the ovary, uterus, vagina and mammary gland at subclinical exposures	↓ WBCs, erythroblasts, ALT, AST, ↑ LDH, BUN, QT interval, lymph degeneration/necrosis	Negative in Ames and mouse lymphoma assay. Negative in *in vivo* bone marrow micronucleus assay tested up to maximum tolerated dose (1 mg kg^{-1} M^{-1}, 3 mg kg^{-1} F^{-1})
Vorinostat (Zolinza)	*Pregnancy category D*: Decreased foetal weights, incomplete ossifications of bones in rats and rabbits. The no observable effects level for these findings were <0.1 times the human exposure based on AUC	Pulmonary embolism and deep vein thrombosis, thrombocytopenia and anaemia, GI disturbances, including nausea, vomiting and diarrhoea, hyperglycaemia	Mortality, ↓ red cell mass, WBCs, ↑ CK, myeloid hypoplasia, erythroid hyperplasia, GI erosions, decreased ovarian *corpora lutea*, lymphoid depletion in lymphoid organs	Mortality, ↓ red cell parameters, WBC, bone marrow depletion, GI erosions, GI villous blunting, inflammation, crypt necrosis	Positive Ames (± metabolic activation), positive Chinese hamster ovaries (CHO) and human peripheral lymphocytes clastogen (± metabolic activation)

Decitabine (Dacogen)	*Pregnancy category D*: Decitabine was teratogenic, fetotoxic, and embryotoxic in preclinical studies in mice and rats	Neutropenia, thrombocytopenia, febrile neutropenia, and leukopenia, fatigue, pyrexia, nausea, cough, petechiae, constipation, diarrhoea and hyperglycaemia	Anaemia, testicular degeneration	Mortality (bone marrow depletion and enteropathy), ↓ WBCs, bone marrow depletion, GI enteropathy, necrosis	Increased mutation frequency in L5178Y mouse lymphoma cells, *in vivo* lac-I transgenic mice. Caused chromosomal rearrangements in larvae of fruit flies
Azacitidine (Vidaza)	*Pregnancy category D*: Congenital malformations in animals	Anaemia, thrombocytopenia, leukopenia, neutropenia	Toxicity in bone marrow, liver, kidney and lymphoid tissues	Mortality (bone marrow depletion), ↓ WBC, ↑ ALT, AST, BUN, liver deg., liver decreased glycogen, bone marrow and lymph node hypoplasia, lung inflammation secondary to infection	Positive in at least one Ames, micronuclease assay, clastogenicity, and transformation assay. Rat carci studies – tumours. Most common tumours were adenocarcinoma, lymphoma, malignant histiocytoma; most common locations were skin, testis, lung, liver, spleen and kidney

non-clinical safety profiles for these is limited. However, the high-level report from the phase II ENGAGE 501 trial for the class I specific HDAC inhibitor entinostat documented treatment-related thrombocytopenia, neutropenia and anaemia consistent with known effects of marketed pan-HDAC inhibitors.[23]

Where we may be able to gain further insight on the long-term safety of epigenetic modulators is the recent realisation that a number of commonly prescribed drugs are interfering with epigenetic mechanisms.[24,25] Drugs such as opioids, cannabinoids, valproic acid or cytostatics (*e.g.* gemcitabine) can induce alterations in DNA methylation patterns or histone tail modifications and their associated adverse effects may be related to epigenetic consequences and more importantly may influence the patient's disease phenotype. For example, chronic opioid treatment is associated with increased DNA methylation and correlated with increased clinical pain.[4] The antihypertensive drug hydralazine, which causes adverse immunological reactions such as drug-induced lupus-like syndrome, was found to be associated with DNA hypomethylation.[26,27] This hypomethylation leads to increased transcription and deregulated expression of proteins including LFA-1 (CD11a/CD11), leading to autoreactive T-cells and a breakdown in peripheral tolerance.[28,29] Purrucker and Mahlknecht have reviewed the case of valproic acid (VPA), the antiepileptic drug that acts as a HDAC inhibitor as well as inhibiting sodium channels and increasing the synaptic availability of GABA in the brain.[30] Their review discusses the effects of epigenetic modification on genomic stability in healthy cells and the finding that HDAC inhibitors may cause genomic instability and increase the rate of malignant transformations. Given the 30 years of clinical use with VPA, compared to the recently approved HDAC inhibitors, its toxicological profile is of special interest with regards to long-term safety of such drugs. The most common side effects include weight gain, dyspepsia and hepatic disturbances. Nausea, vomiting and indigestion are also commonly observed as well as sedation, fatigue and blood dyscrasia, which may occur due to VPA's influence on HDAC activity, gene expression and genomic stability of rapidly proliferating cells.[31,32] VPA also causes teratogenic side effects that have been attributed to HDAC inhibition, but the exact underlying mechanism has not yet been elucidated.[32,33] Currently, no increases in the frequency of solid or haematological neoplasias have been reported in association with administration of VPA; this offers hope for the long-term safety of HDAC inhibitors in non-oncology indications.

12.4 What are the Key Points to Consider When Designing Non-Clinical Safety Studies for Epigenetic Modulators?

12.4.1 Study Design and Data Interpretation

Importantly, the kinetics of epigenetic modifications need to be considered when designing *in vitro* or *in vivo* safety studies. Due to the transcriptional regulatory nature of epigenetic proteins, efficacy or safety end points may be expected to take longer to manifest compared to a direct protein inhibitory effect. Moreover, certain methylation marks have relatively slower turnover rates (days) compared to other post-translational modifications such as acetylation (within hours).[34,35] Therefore, when planning *in vitro* or *in vivo* studies or assessing reversibility of any observed 'phenotypic' changes, it is important to ensure that the study duration is appropriate to capture potentially persistent effects. For small molecules that target epigenetic mechanisms, short term (2 weeks or less), exploratory, non-clinical safety studies may not be of sufficient duration to assess modulation of certain targets. Pharmacodynamic data from non-clinical efficacy studies showing the timing of target modulation may be helpful in setting appropriate study duration. With all studies it is essential to establish both dose–response, particularly at *low* exposures, and reversibility, to aid risk assessment and dose setting for other non-clinical or clinical studies. It is important to develop an understanding of which epigenetic changes are causally linked to an adverse toxicity outcome and the relationship between the dose–response curve for the epigenetic changes and the organ toxicity observed.[36] In addition, drug development strategies may require targeted drug delivery to avoid exposing normal proliferating cells, or innovative dosing regimens that incorporate drug holidays in order to achieve greater therapeutic windows. For example, dose-limiting toxicities related to currently approved HDAC and DNMT inhibitors involve effects on rapidly proliferating systems (*e.g.* bone marrow, intestinal cells, testes) and show little to no therapeutic index even in short-term studies. It would appear that proliferating cells are also sensitive to the targeting of BCPs and testicular toxicity has been a reported outcome with the compound JQ1; however, the observed effects are completely reversible.[37] In contrast, for tissue compartments that are post-mitotic, adverse findings may take many weeks or months to manifest due to the epigenetic mechanisms mentioned; it will be a challenge for toxicologists to properly model long-term epigenetic effects.

A further major task is to understand the mechanism for driving inter-individual variations in response to drug exposure and distinguish between adverse and non-adverse signals. For example, a change in DNA methylation may be advantageous or deleterious and should not simply be equated with toxicity. Thus, given the state of our current limited scientific understanding of the fundamental control of gene expression and the mechanisms that maintain and generate epigenetic states, an epigenetic change in itself should not be conceptualised as adverse in the absence of a phenotypic correlate. For this reason it will be essential to understand the normal variability associated with epigenetic modifications and the influence of experimental variability that might confound data interpretation.[38]

12.4.2 Preclinical to Clinical Translation

A potential knowledge gap that is probably applicable to efficacy as well as safety is an appreciation for species-specific epigenetic controls that lack translation to humans. An important phenomenon that needs to be considered is genomic imprinting. This is an epigenetic process by which the male and female germline confer different DNA methylation marks and histone modifications onto specific gene regions, so that one allele of an imprinted gene is active and the other is silent. Because the dosage of the imprinted gene is important for normal development, growth and behaviour, the loss of the active allele can cause disease.[39] Bioinformatic approaches have predicted that the human genome contains fewer imprinted genes than the mouse. The types of imprinted genes may be strongly species dependent. For example, the *Insulin-like Growth Factor 2 Receptor* (*IGF2R*) gene is imprinted in mice, and only one copy of the gene is functionally expressed. However, in humans both copies of *IGF2R* are expressed. Because *IGF2R* is known to possess tumour suppressor activity one could hypothesise that mice would be more sensitive to epigenetic influences on *IGF2R* expression than humans.[40] In addition to genomic imprinting, epigenetic readers, writers and erasers may have binding partners, substrate specificity and direct target genes that are species-specific. These types of considerations will be important for interpretation of data generated using the currently available *in vitro* and *in vivo* safety models that are expected to identify potential adverse effects of drug candidates regardless of mechanism of action, epigenetic or not. This poses a challenge when considering the translation of drug-induced epigenetic changes (either adverse or adaptive) in preclinical models to adverse health outcomes in patient populations, a familiar problem for toxicologists.

12.4.3 Evaluation of Target-Mediated Molecular Signatures

A key feature of epigenetic modulation of gene expression is the one-to-many relationship between an epigenetic protein and its downstream target genes. Epigenetic proteins often form cell- or tissue-specific protein complexes. Tissue distribution of epigenetic proteins and the target genes they regulate therefore have implications for target organ toxicity. Although direct target genes for various epigenetic proteins and histone modification states are beginning to be elucidated, there is no systematic characterisation of the target genes in normal adult tissues across species. Addressing this gap by assessment of genome-wide distribution of histone marks, gene and/or protein expression changes along with functional or phenotypic end points following target modulation (*e.g.* by siRNA or chemical inhibition), could provide a mechanistic basis for a potential safety risk and help develop follow-on de-risking strategies when specific target organs have been identified. Using relevant *in vitro* models for target modulation, specific downstream end points could include non-specific profiling methods such as global methylation profiling, or more focused techniques such as chromatin immunoprecipitation with sequencing (ChIP-seq) to identify downstream target genes (*via* modifications of histone patterns) that might be involved in organ injury and linking histone patterns with biological function. Additionally, bioinformatics tools (*e.g.* pathway analysis) could help analyse the downstream genes for additional biological insight. These techniques generate high-resolution datasets that will allow comparison of different epigenomes. This should help elucidate the temporal dynamics of tissue-specific epigenetic landscapes following target modulation and how such epigenetic changes may ultimately be related to the organ toxicity observed.

12.4.4 Chemotype Safety

Assessing chemotype safety and necessary compound de-risking during the early stages of epigenetic target drug discovery is no different for any other typical small-molecule programmes, and currently available (standard) *in vitro* and *in vivo* safety testing batteries are expected to identify potential adverse effects of small-molecule drug candidates regardless of epigenetic mechanism of action. This is beyond the scope of this chapter but interested readers are referred elsewhere for comprehensive discussions on additional biological assays to help identify potential safety liabilities for small-molecule compounds.[41–46]

One area that is important to mention is around the need for *in vitro* pharmacological profiling to identify undesirable off-target activity profiles that could halt or hinder the development of drug candidates,[47,48] and the development of an *in vitro* safety panel for profiling small-molecule epigenetic modulators for pharmacological promiscuity at other epigenetic targets. Indeed, platforms are now being developed commercially that offer drug selectivity assays for 47 epigenetic targets covering readers (six assays), writers (16 assays) and erasers (25 assays).[49] As a way of mining potential epigenetic-related *in vivo* adverse drug reactions, up to a 1000 compounds from the BioPrint® database were profiled at 16 epigenetic targets. Preliminary results show that epigenetic targets not only cluster together by phylogeny but also by pharmacology, and the generated dataset has allowed the identification of epigenetically active compounds associated with safety liabilities.[49] Much more effort and a deeper dive into the science of epigenetic mechanisms is required in this area to better understand which epigenetic targets are the 'bad actors' leading to adverse drug reactions. This knowledge will then allow the establishment of an appropriate epigenetic drug safety panel to assess the undesirable off-target activity of epigenetic compounds. A number of companies now offer epigenetic drug screening tools and commercial assay kits.[50]

12.4.5 Identifying Safety Issues Related to Modulating Epigenetic Molecules

Hints about potential safety concerns associated with targeting epigenetic molecules can be uncovered with a literature review of target expression across tissues, genetic associations with disease in humans, knockout phenotypes in animal models, and other reports using tool or competitor compounds, if available. It should be noted that some epigenetic proteins have scaffold-dependent functions that help recruit and stabilise components within multi-protein complexes that are independent of their enzymatic/catalytic activities. Therefore, for drug discovery programmes aiming to target a molecule's catalytic function, it is advisable to confirm gene knockout or knockdown data gleaned from the literature using chemical tool compounds that target the enzymatic activity.

The critical role of many epigenetic molecules in embryogenesis and early development is reflected in the fact that genetic deletion of these molecules often results in embryonic lethality in model organisms.[51–54] Conditional deletion in specific cell lineages offers a glimpse into organ-specific functions.[55–57] Many of these molecules have been implicated in cardiovascular[58–60] and neural development.[61,62] A high-level survey reveals additional functions, such as myogenesis,[62–64] lipid metabolism,[65] hematopoiesis[66–68] and spermatogenesis,[69,70] to name just a few. From a safety perspective, the pleiotropic effects of these proteins warrant a proactive approach for both target safety assessment and risk mitigation when targeting them with a compound.

Also worth noting is the fact that many of the histone modifiers have non-histone substrates, which may be acetylated or methylated, with marked effects on protein function. Many transcription factors are substrates for acetylation, and where it has been studied, acetylation frequently has been associated with changes in DNA binding and in activation of target gene expression.[71] SMYD2 is an HMT that, in addition to activities on histone H3, also methylates the heat-shock protein HSP90, which in turn regulates the stability of the muscle structural protein titin,[72] with implications for cardiac and muscular functions.[73,74] The cAMP-response element protein (CREB) binding protein (CREBBP) bromodomain is known to interact with over 400 other proteins, many of which are broad transcriptional regulators in their own right such as p53 and CREB. Thus, target safety assessment should include an evaluation of both histone and non-histone substrates/partners.

In the next section we outline several known and anticipated safety concerns associated with targeting epigenetic mechanisms and provide general and specific recommendations for screening and de-risking approaches.

12.5 Known and Anticipated Organ Toxicities and Specific Risk Mitigation Strategies

12.5.1 Cardiovascular System

Emerging scientific data clearly suggest that epigenetic modifications play a key role in cardiac development and injury.[60,75] For example, several groups have examined the relationship between epigenetic modulation – DNA methylation and specific histone modification events – and cardiac hypertrophy and failure in both animal models and human subjects.[59,76–78] A study with Pax-interacting protein 1 (PTIP) knockout mice provides evidence that altering histone methyl marks in the adult heart is sufficient to alter critical cardiac parameters. Indeed, a cardiac-specific knockout of PTIP, an essential cofactor of the methylated histone H3 lysine 4 protein complex, impacts cardiac ion currents, action potential velocity and duration, cardiac contractility, and increases sensitivity to premature ventricular beats upon stimulation.[79] In mouse knockout studies, HDAC1 and HDAC2 have also been shown to redundantly regulate cardiac morphogenesis, growth and contractility; cardiac-specific deletion of both genes results in neonatal lethality, cardiac arrhythmias, dilated cardiomyopathy, and up-regulation of genes linked to hypertrophy.[55] HDAC inhibitors such as romidepsin have been associated clinically with cardiotoxicity such as T-wave flattening, ST segment depression and QT interval prolongation.[80] Although not completely understood, the mechanism of QT interval prolongation has been explained by aberrant cellular trafficking and/or functioning of the hERG K^+ channel.[81] For drugs known to modulate gene expression, such as HDAC inhibitors, altering the expression of hERG and its co-regulators may lead to QT prolongation even in the absence of any direct interaction with the hERG channel at therapeutic doses and above. Therefore, in addition to testing for any direct effects of an epigenetic modulator on hERG inhibition, if potential risks are identified for altering the expression of hERG and of its co-regulators, then additional studies to assess changes in the expression of such proteins are warranted.[82]

12.5.1.1 Screening and Risk Mitigation Strategies

When designing either *in vitro* or *in vivo* studies to assess cardiovascular safety, attention should be given to ensure that these are conducted over a long enough timeframe for expression levels of proteins to be altered and any subsequent cardiovascular changes to occur. Indeed, it may also be necessary to either overload (physical exercise) or stress the heart with, for example, β-adrenergic receptor activation to exacerbate the compromised cardiac function and elicit any pathological outcome(s). Assessment of target-mediated gene expression changes following treatment with an epigenetic inhibitor provides evidence of pharmacodynamic modulation. Furthermore, because modification of some targets may not cause the heart to exhibit any dysfunction or induce any histological findings, other end points should be considered. For example, the TUNEL assay can be used for assessing apoptotic cell death,[83] hypertrophy markers[84] or ubiquitin analysis, especially if modulation of the epigenetic target leads to any alterations of autophagy genes that could give rise to proteasomal dysfunction.[85] These additional end points should be conducted to evaluate any structural changes of the heart and certainly be considered when undertaking any *in vivo* repeat-dose study. Cardiovascular studies can be performed as either a stand-alone assessment (single- or repeat-dose studies using telemetry in, for example, rats or dogs)[86,87] or if required added to a large animal general toxicology study using jacketed external telemetry (JET).[88,89]

12.5.2 CNS Toxicity

DNA methylation and histone modifications (including acetylation and methylation) have been linked to a number of neurological and psychiatric disorders, suggesting their importance in normal brain development and function.[61,90–92] There is evidence that the epigenomic profiles within the brain, which has a large post-mitotic constituent, remain 'plastic' and that dynamic regulation occurs throughout all periods of brain development and ageing.[61] Epigenetic dysregulation has been implicated in a number of developmental and adult-onset neurological diseases, with manifestations of intellectual disability, cognitive impairment, emotional disturbance and neurodegeneration. For instance, GLP/EHMT1 and G9a/EHMT2 are histone methyltransferases that exist in a complex. Loss-of-function mutations in GLP cause the 9q34 subtelomeric deletion syndrome, characterised by intellectual disability, schizophrenia, psychiatric phenotypes and neurodegeneration.[93] Neuronal-specific deficiency of GLP and G9a in a mouse model leads to de-repression of neuronal progenitor and non-neuronal genes, as well as complex behavioural abnormalities, including defects in learning, motivation and environmental adaptation.[57] Interestingly, some psychotherapeutic drugs including clozapine (atypical antipsychotic), valproic acid (mood stabiliser), phenylzine (antidepressant) and cocaine have been shown to modulate histone methylation in the brain.[91] Certain HDAC inhibitors exhibit neuroprotective effects in animal models of neuronal injury and neurodegeneration, underlying the role of histone acetylation in various brain functions.[61,94] The role of epigenetics in adult neurogenesis also raises concern for altered regenerative capability following neuronal injury.[90]

12.5.2.1 Screening and Risk Mitigation Strategies

A key to limiting such CNS risks is to reduce the brain penetration of compounds but this may not be possible for indications in which the blood–brain barrier (BBB) in humans may be compromised (*e.g.* diabetes, oncology, CNS and peripheral inflammatory disorders) or if the primary indication requires CNS access, *e.g.* HDAC inhibitors are now being used in treatment of Huntington's and amyotrophic lateral sclerosis among other CNS diseases.[13,95] For epigenetic targets that are expressed in the CNS and may have the potential for unwanted adverse effects, it is recommended that brain and/or cerebral spinal fluid (CSF) exposures (if deemed appropriate)[96] be measured in relevant animal studies over multiple time-points to assess the rate of brain penetration over time to determine whether CNS issues will be a risk for modulating the epigenetic target. If a compound is shown to alter CNS function or activity, then evaluation of the risk for abuse potential[97,98] and possible suicidal ideation and behaviour, may need to be considered.[99,100]

For an initial early de-risking of any identified CNS issues, it is recommended that an functional observation battery (FOB) or neurofunctional assessment (FOB and quantitative assessment of spontaneous locomotor activity) should be added to a rodent toxicology study as a general screen for overt CNS (and peripheral nervous system) dysfunction.[101] A lack of findings in a FOB assessment with a CNS penetrant compound does not mean that CNS liabilities have been fully evaluated because effects, for example on memory and learning, are not assessed in the FOB. If any concerns are identified from such studies, detailed de-risking strategies should be constructed for those safety issues that are of concern and that may hinder a compound moving forward into development (*e.g.* neuropsychiatric adverse effects).

12.5.3 Stem Cell Homeostasis

Epigenetic targets expressed in embryonic and adult stem cells are involved in regulating key events during stem cell self-renewal, differentiation and tissue senescence.[102–104] Modulation of epigenetic targets has been known to be associated with embryonic and developmental toxicity.[51–54] Adult stem cells are distributed in most adult mammalian tissues as partially differentiated progenitor cells, including in the bone marrow, adipose tissue, heart, brain, liver, vasculature, and epithelial cells; they are involved in maintaining tissue homeostasis as well as aiding regeneration and repair during tissue damage. Because stem cells in each tissue rely on a different suite of transcription and epigenetic factors to maintain their own particular state of differentiation, modulation of the epigenetic target could lead to stem cell exhaustion, bone marrow suppression, immune suppression, GI toxicity, and impaired tissue regeneration and repair. For targets that have been shown to play a role in lineage-specific embryonic development, one may pose the question of whether they continue to function in the corresponding adult stems cells. For example, MMSET is an HMT that has a stem cell expression profile. Targeted deletion in mice leads to cardiac development defects,[105] and molecular investigation reveals that MMSET interacts with the cardiac-specific transcription factor Nkx2–5. Additionally, patients with Wolf–Hirschhorn syndrome harbour deletions of human chromosome 4p16.3, which include the MMSET gene, and also present with cardiac phenotypes.[106] Together, these data suggest a role for MMSET in adult cardiac stem cell function. Therapeutic modulation of MMSET in the adult population may raise concerns with repair capacity of cardiac tissue following cardiac ischemia reperfusion injury.[107]

12.5.3.1 Screening and Risk Mitigation Strategies

To understand such risks, a number of adult stem cell-based *in vitro* assays could be used to assess the effects of modulation of epigenetic targets on adult stem cell function. Commercially available stem cell-derived cardiomyocytes and neurons could be used to facilitate the evaluation of cardiovascular and

CNS toxicity, although studies to determine the value and translation of data generated using such screens are not currently available. There are also concerns about the ability to acquire sufficient quality of cardiac progenitor cells to feasibly develop an assay. Human primary neural progenitors would most likely be derived from foetal tissue, and thus may be difficult to undertake given ethical considerations, or from the olfactory bulb (but may not be commercially available). Additionally, human embryonic or induced pluripotent stem cells (hESC/hiPSC) may serve as a surrogate for adult progenitor/stem cells. While hESC/iPSC approximate an early stem cell, developmentally upstream of the adult progenitor/stem cells of typical concern, these cells have a defined epigenetic profile that is sensitive to perturbation[108,109] with a direct impact on self-renewal and differentiation capacity, the two key characteristics of all stem cells.

Whether assessing the impact on embryonic or adult stem cells, cells could be dosed with compounds and evaluated for a variety of end points including proliferation, cytotoxicity and differentiation, similar to the work described by Desbordes *et al.*[110] As with de-risking experiments proposed for other organ toxicities, it is important to carry out experiments to sufficient lengths to capture modulation of intended epigenetic mark(s) and any associated changes in monitored end points. Drug-induced changes in proliferation, cytotoxicity or differentiation in embryonic/pluripotent stem cells raise concerns for adverse developmental effects (see next section). However, *in vitro* to *in vivo* translation of similar changes in adult stem cells may be more difficult to predict. It may be necessary to employ appropriate *in vivo* injury/repair models to assess functional consequences of impacting adult stem cells following chemical inhibition of specific epigenetic molecules.

12.5.4 Developmental and Reproductive Effects

Many of the epigenetic proteins have critical roles during development. It is expected that modulation of these molecules would result in developmental defects, supported by the embryonic lethal phenotype of KO mice with targeted deletion or inactivation of these genes.[51–54] Functional relevance of such embryonic or perinatal phenotypes should be carefully considered for adult relevance, and not automatically assumed irrelevant. For one, epigenetic mechanisms important for embryonic stem cells may also play a role in maintaining the various tissue-specific adult stem cell populations. Additional information, such as conditional knockout or functional evaluation in the adult tissue of interest, will strengthen confidence in the safety concern. Reproductive toxicity (affecting the ability to make and deliver healthy gametes) is also a likely consequence for perturbing epigenetic mechanisms.

12.5.4.1 Screening and Risk Mitigation Strategies

Compounds should be tested in the latest low-bulk-requiring assays to test for (and probably confirm) developmental and reproductive toxicity. However, most current short-term *in vitro* prediction assays do a poor job of capturing longer-term or integrative changes, so drugs modulating epigenetic targets may or may not be positive in such assays. Prenatal exposure and postnatal evaluations, such as are common for environmental compounds, will probably be required to fully assess epigenetic modulators, and the assessments will surely go beyond structural changes into the evaluation of adult function and even the health trajectory during ageing.

There are two relatively short-term screening strategies for identifying developmental toxicity. These strategies for de-risking epigenetic targets in the conceptus are embryonic stem cells and zebrafish. Zebrafish are by far the more genomically tractable – there is strong conservation of gene and protein structure and function between zebrafish and humans, increasing our level of confidence in the applicability of this model.[111] For example, HDAC1 shares 91% sequence identity between the two species, while DNMT1 is 75% identical. Thus, qualitative estimates of 'target importance' can be easily gained from such studies. Zebrafish have the advantage of providing an intact organism that recapitulates the cell migration and pattern formation inherent in the mammalian conceptus; these are attributes lacking in stem cells. Stem cells, on the other hand, are most useful for evaluating small-molecule inhibitors, because they lack the exposure uncertainties that come with zebrafish, whose eggs can sometimes be difficult for compounds to penetrate. (Individual egg injection is a possible solution).

12.5.5 Transgenerational Transmission

Exposure to some environmental compounds or conditions has been shown to result in physiological effects in subsequent generations. For example, a low-protein diet,[112] dioxin,[113] vinclozolin,[114] bisphenol A, dibutyl phthalate and di-(2-ethylhexyl) phthalate, have all been reported to produce effects in the F3 and/or F4 generations after treatment of the pregnant F0 rodent dam. In some cases these 'transgenerational' effects are due to persistent changes in epigenetic control of gene expression.[115,116] Given this link between transgenerational effects and epigenetic mechanisms, concerns have been raised about the potential for epigenetic therapies to have potentially deleterious consequences in unexposed future generations.[13,117] Although these concerns would be minimised for drugs targeting life-threatening indications such as cancer, pharmaceutical companies are increasingly targeting epigenetic mechanisms for the treatment of other diseases where the tolerance for such risks is lower. Therefore, developing some assessment of the likelihood that these drug treatments might produce delayed effects on future generations, becomes increasingly important. Given the potential risks for transgenerational transmission, Tycko *et al.*[117] suggest that epigenetic therapy should be shown to not affect the germline, and that a patient's epigenome should be sequenced before and after treatment.

12.5.5.1 Screening and Risk Mitigation Strategies

Multigenerational rodent and/or rabbit studies, used primarily for developmental and teratogenicity assessments, might be appropriate for the identification of transgenerational effects. Given that the epigenome changes with age, it would be imperative to use concurrent age-matched controls and well-characterised strains. Some experts in the field have suggested that the mouse would be the preferred choice over the rat or rabbit as it is currently the best characterised of *any* species for evaluating epigenetic changes,[118] whilst others feel that the mouse has generally poor concordance to humans and prefer the rat for such studies.[117] Whatever species is used, such studies are long and therefore expensive, and fit poorly into any drug development paradigm. Zebrafish may be able to help truncate the duration of such studies, but this is not yet a widely used model for this application. As noted above in the Developmental and Reproductive Effects section (Section 12.5.4), zebrafish are genomically tractable, there is strong conservation of gene and protein structure and function between zebrafish and humans, and zebrafish have the additional advantage of providing an intact organism that recapitulates the cell migration and pattern formation. The use of zebrafish as a model for studying the potential transgenerational effects of pharmaceuticals should be explored further, the one drawback being that imprinted genes do not exist in this species.

Other model systems that might be employed to evaluate the ability of a compound to produce an epigenetic change affecting F1 and/or F3 generation was a topic of debate at the ILSI-HESI meeting on epigenetics.[118] The group discussed the use of the viable yellow agouti mouse model but concluded that these models may be oversensitive to compounds and not appropriate for screening purposes due to the presence of metastable epialleles which have not been identified in humans. Alternative models discussed were the zebrafish (see above), *Caenorhablitis elegans*, honeybees, *Drosophila* and embryonic stem cells, but as the epigenetic control mechanisms across species are not known, in addition to there being metabolism and tissue-specific differences it would be difficult to interpret directly the significance of these data to humans. *In vitro* models may have the best utility in proactive identification of a possible epigenetic liability but would not evaluate transgenerational effects, well-designed experimental approaches being needed in mammalian *in vivo* assessments to address the issue.

We need to recognise these possible transgenerational effects as a potential liability to differentiate the developmental or transgenerational toxic exposures from those that are not. It could be dealt with in one (or more) of the following ways: (i) watch this space for regulatory guidance, (ii) perform one of these studies to gain experience and learn about the likelihood of the persistence of the effect, or (iii) use gene knockdown or chemical tool compounds to undertake initial studies to modulate a single representative target from each class of epigenetic marks (*e.g.* DNMT, HAT, BCP) to generate sample data and gain experience and understand the relevance of such data and translation to patient populations. Although this approach assumes that all the targets from the same family would lead to the same effects on transgenerational transmission (some epigenetic marks may be more or less stable), undertaking such studies will help tease out what targets may be associated with transgenerational transmission and the types of studies needed to make such an assessment.

12.5.6 Carcinogenic Effects

Targeting epigenetic mechanisms is likely to have secondary effects, some of which are undesirable, including effects on cell development, growth and differentiation.[119] The modulation of cellular growth and differentiation is the reason why most of the drugs targeting epigenetic mechanisms are for oncology indications. Specifically, global alteration in DNA methylation at CpG islands and other promoter sites have accounted for marked changes in transcriptional patterns of several genes resulting in cancer. Some examples include hypermethylation of tumour suppressor genes such as hMLH1, BRCA1, p16INK4a, and a few microRNAs.[120,121] However, persistent modulation of such targets may affect cellular homeostasis by inhibiting growth or causing degeneration/necrosis that can lead to compensatory proliferative mechanisms that are risk factors for tumorigenesis. Another possibility of using compounds that alter the activity of epigenetic modifying enzymes is for the reactivation of endogenous retro-elements and latent viruses that could increase the risk of cancer. There are numerous reports that indicate the latency of viruses, such as Epstein–Barr virus, herpes simplex virus, Karposi's sarcoma virus and HIV, is regulated by histone modification and DNA methylation status.[122–125] Indeed, several groups have proposed the use of epigenetic modifying compounds as a means of reactivating latent viruses to make them more responsive to therapy and thereby reduce the virus reservoirs.[126,127] Some groups have also reported that retro-transpositional events have been seen in association with other human diseases, including some cancers.[128–130]

Because the risk–benefit profile of drugs in the development of chronic disease (*e.g.* diabetes) is completely different from drugs developed for oncology, carcinogenicity risk associated with epigenetic targets needs to be considered early in development. In other words, the default risk–benefit will be low unless proven otherwise with generation of *in vitro* (*e.g.* cell culture proliferation assays) and more importantly *in vivo* data.[131]

12.5.6.1 Screening and Risk Mitigation Strategies

Current *in vivo* assays for predicting tumorigenicity are the 2-year rodent bioassays and more recently the medium-term alternative mouse models (typically 6-month studies in rasH2 transgenic mice or p53 mutant mice).[132,133] While robust genotoxicity screening paradigms will screen out compounds with a genotoxic potential, it is a challenge to develop assays that can help assess tumour risk early in development. Currently, there are no such validated models or assays available, and a weight-of-evidence approach might be the only option. Risk assessment or mitigation strategies could be developed and may help position the value of the effort and prioritisation of epigenetic targets. Our recommendation is to prioritise a carcinogenicity risk assessment based on the stage of development and the tools or technologies that are readily available (*i.e.* if there is ready access to genetically engineered mouse models (GEMM), it might be value added to leverage the model to fully understand the risks of modulating the target protein). In contrast, it would not make sense for a project in the early stages of drug discovery to commission a GEMM or trigger a study in rasH2 mice. Finally, if tumours are found in a carcinogenicity study, then a component of the de-risking plan should include an assessment for activation of endogenous retro-elements and latent viruses as discussed above.

12.6 Regulatory Considerations

There has yet to be any formal regulatory guidance regarding the development of epigenetic drugs for the treatment of disease. While the nature and importance of at least some epigenetic changes are well established, many of the implications and mechanisms of epigenetics remain uncertain or speculative. Of particular interest is the potential transgenerational effects associated with epigenetic modulators. Discussions regarding the transgenerational epigenetic effects resulting from environmental exposure to substances such as endocrine disruptors (*e.g.* oestrogens, bisphenol A)[134] and consumption of food products derived from cloned animals (*e.g.* milk from cloned animals) continue to occur, and are significant topics of research by the chemical and agriculture technology industry. Some of the endocrine disruptor data are controversial, as initial reports have not yet been satisfactorily replicated, despite several attempts.[135–137] It is likely that regulators and companies developing compounds with epigenetic modulatory qualities will continue to have discussions regarding the short- and long-term risks associated with altering epigenetic signals. Indeed, the role of epigenetics in drug safety science was a topic of a workshop by ILSI-HESI entitled 'State of the Science: Evaluating Epigenetic Change', the proceedings of the workshop being summarised in a paper by Goodman *et al.*,[118] and further discussed by Priestley *et al.*[38] Although predominantly aimed at epigenetic screening in safety assessments, several common themes were highlighted that are relevant for compounds targeting epigenetic proteins: (i) incomplete understanding of the epigenetic process with key data and knowledge gaps; (ii) uncertainty of long-term effects on public health; (iii) insufficient validation of epigenetic tests for inclusion into the regulatory process, and specifically that no single test is ideal for all epigenetic effects. Overall, the workshop highlighted that the field of epigenetics is evolving at a rapid pace and that a great deal more needs to be learned prior to being able to rationally incorporate an epigenetic evaluation into safety assessments.

Thus, given our lack of knowledge of long-term epigenetic effects of epigenetic therapy, more basic research is needed on the elucidation of risk, especially for the treatment of chronic diseases. Tycko *et al.*[117] propose that as these drugs affect gene expression it is more appropriate to hold them to the standards of gene therapy, which entails an expectation to target only the genes, pathways and cells relevant for the disease as outlined in recent guidance from the US Food and Drug Administration (Guidance for Industry: Preclinical Assessment of Investigational Cellular and Gene Therapy Products, 2013). Taking such an approach may raise the bar impossibly high for the development of new epigenetic drugs given the ubiquity of epigenetic target distribution and current challenges to improve target specificity of compounds but it is a scientific debate that needs to happen.

12.7 Assessing the Risk–Benefit Considerations for Targeting Epigenetic Processes

Drugs developed for oncology and other life-threatening diseases have a much higher tolerance for risk. As already discussed, common liabilities associated with targeting epigenetic proteins, such as developmental, reproductive and carcinogenicity risks, are generally considered of acceptable impact for advanced cancer indications, or can be mitigated with appropriate exclusion criteria. Haematological and GI toxicities are also frequently observed. Although these are familiar to the physicians who care for oncology patients on standard non-epigenetic therapies, they could constitute dose-limiting toxicities that lead to dosing holidays or discontinuation of therapy. Effects on bone marrow, the GI tract and other proliferative tissues could be exacerbated when used in combination with other chemotherapeutic agents. Potential cardiovascular and CNS toxicities that are associated with the modulation of histone methylation levels will also need to be carefully assessed during the drug discovery process, as these can pose a significant impact on all drug development programmes.

However for other indications (*e.g.* chronic diseases, lifestyle drugs, *etc.*) the risk–benefit relationship should be carefully considered when evaluating any new target. The epigenetic drugs for which clinical safety and non-clinical toxicology data exist suggest, not surprisingly, that liabilities such as developmental, reproductive, haematological and carcinogenicity, are associated with compounds that regulate chromatin function. Current research related to rare and inflammatory diseases have the potential to alleviate these concerns through precision targeting of individual proteins compared to the pan-activity associated with currently marketed compounds (*e.g.* HDAC inhibitors). Still, for such targets, consideration should put early emphasis on the characterisation of the aforementioned risks in order to best understand the risk toleration for the intended indication.

As a case example, the double bromodomain-containing BET family of proteins is comprised of BRD2, BRD3, BRD4, and the testis-specific BRDT.[138] BRD4 is an epigenetic target of great interest for the treatment of inflammatory diseases,[138,139] with a small-molecule inhibitor having potential benefit for both COPD and asthma indications.[140–142] However, reviewing currently available literature raises concerns that BET inhibitors are likely to result in developmental and reproductive toxicities, in addition to other safety liabilities associated with altered nuclear hormone receptor function. Thus, such information would suggest an unacceptable risk–benefit relationship for the development of BET inhibitors for asthma; however in COPD patients, the patient demographics combined with the overall disease severity may pose an acceptable risk–benefit profile to continue exploring the mechanism for this indication.

12.8 Conclusions

This chapter discusses some of the key considerations for identifying and managing epigenetically mediated target safety concerns. The epigenetic targets explored to date for oncology indications have been those of a non-specific nature (*e.g.* pan-HDAC inhibitors and cytosine analogues). The non-specific targeting may lead to a broader toxicity profile than if drug action were targeted towards select protein family members. More selective next-generation targets currently being explored for oncology, rare diseases and inflammatory diseases have greater promise with regard to safety, but still have significant potential liabilities due to the mechanism of action. With the expansion of epigenetic targets into other indications that may require chronic treatment, experts in both safety pharmacology and toxicology should be engaged in early discussions to define and understand the level of risk toleration for the specific target and indication. As a result, early investment can be undertaken to understand programme-specific risks and develop target profiles both to maximise the chances of compound survival and to minimise the time to a go/no go decision.

Cardiovascular, CNS, stem cell homeostatic, developmental and reproductive, transgenerational, and carcinogenic effects are among the potential consequences of targeting epigenetic mechanisms. These are highlighted due to availability of scientific evidence supporting their occurrence, as well as the possibility of high impact on drug discovery programmes. A more focused review of individual epigenetic proteins of interest will undoubtedly reveal additional homeostatic functions and hence potential target-mediated toxicities. It is noteworthy that up until recent years, the prevailing view in the field has been that epigenetic modulation preferentially impacts proliferating cells or developmental processes. Emerging data in post-mitotic compartments, such as those in the heart and CNS, demonstrate that perturbation of epigenetic machinery in terminally differentiated cell types is capable of altering normal functions. In other words, the epigenome remains plastic throughout the lifespan of an organism.

From a drug safety perspective, current standard drug development paradigms, with particular regard to hazard identification, are appropriate for small-molecule epigenetic drugs and thus no fundamental changes to current processes is deemed necessary, but some customisation is needed. As drug candidates progress from concept to target validation and into exploratory and regulatory toxicity studies, target organ toxicities will be identified and will need to be further evaluated, particularly around mechanism(s) of epigenetic toxicity and causal links to any perturbation in tissue function. Epigenetic modulators perturb transcriptional events, which need to go through protein translation before effecting functional changes; they may also cause functional deficits without morphological abnormalities in the corresponding tissue or only in response to challenge. Therefore, when designing toxicity studies, it is important to anticipate delayed target-mediated toxicities, assess reversibility of these effects, and incorporate relevant functional end points. Epigenetic proteins form tissue-specific protein complexes and mediate tissue-specific gene expression patterns, which may shed light on mechanisms of organ toxicity and have utility as pharmacodynamics or toxicity biomarkers. Developing experience with these customised experimental approaches and specialised technologies for probing the epigenome will enable a fuller appreciation of target organ toxicities, as well as species-specific epigenetic mechanisms that may affect the preclinical to clinical translation of target-mediated effects.

Given the unique nature of epigenetic mechanisms, perturbation of these events, even in a selective manner, may have wide-reaching or long-term consequences not yet fully appreciated at present. Adapting an integrated discovery toxicology strategy not only paves the way for an increased confidence in drug safety, but also provides a basis for screening out undesirable off-target epigenetic activities in drug discovery programmes. Furthermore, the knowledge in basic epigenetic science and experience with emerging clinical epigenetic therapies will also help shape regulatory and policy considerations.[117,143] To move this whole area of science and policy forward, the pharmaceutical industry, academia and government agencies will need to work together to share knowledge and understanding around what comprises a comprehensive drug safety assessment so that drugs that modulate epigenetic proteins can be used with confidence beyond the target classes and indications of the currently approved epigenetic drugs.

Acknowledgements

We are indebted to our Pfizer colleagues Sandi Engle, Gina Yanochko, Wendy Hu and Mark Gosink, whose contributions to an internal white paper on this subject have been adopted and used in writing this book chapter. We thank both Lloyd Dethloff and Jon Cook for their valuable comments and helpful discussion when reviewing both the white paper and book chapter.

References

1. M. Brait and D. Sidransky, *Toxicol. Mech. Methods*, 2011, **21**, 275–288.
2. R. Kanwal and S. Gupta, *Clin. Genet.*, 2012, **81**, 303–311.
3. S. Viatte, D. Plant and S. Raychaudhuri, *Nat. Rev. Rheumatol.*, 2013, **9**, 141–153.

4. A. Doehring, G. Geisslinger and J. Lotsch, *Eur. J. Pain*, 2011, **15**, 11–16.
5. F. Denk and S. B. McMahon, *Neuron*, 2012, **73**, 435–444.
6. G. Egger, G. Liang, A. Aparicio and P. A. Jones, *Nature*, 2004, **429**, 457–463.
7. M. Esteller, *Genes Cancer*, 2011, **2**, 604–606.
8. A. Dhasarathy and P. A. Wade, *Mutat. Res.*, 2008, **647**, 39–43.
9. J. P. Thomson, J. G. Moggs, C. R. Wolf and R. R. Meehan, *Mutat. Res., Genet. Toxicol. Environ. Mutagen.*, 2014, **764**, 3–9.
10. J. M. Foulks, K. M. Parnell, R. N. Nix, S. Chau, K. Swierczek, M. Saunders, K. Wright, T. F. Hendrickson, K. K. Ho, M. V. McCullar and S. B. Kanner, *J. Biomol. Screening*, 2012, **17**, 2–17.
11. T. Kouzarides, *Cell*, 2007, **128**, 693–705.
12. M. Rodriguez-Paredes and M. Esteller, *Nat. Med.*, 2011, **17**, 330–339.
13. C. H. Arrowsmith, C. Bountra, P. V. Fish, K. Lee and M. Schapira, *Nat. Rev. Drug Discovery*, 2012, **11**, 384–400.
14. J. E. Bradner, R. Mak, S. K. Tanguturi, R. Mazitschek, S. J. Haggarty, K. Ross, C. Y. Chang, J. Bosco, N. West, E. Morse, K. Lin, J. P. Shen, N. P. Kwiatkowski, N. Gheldof, J. Dekker, D. J. DeAngelo, S. A. Carr, S. L. Schreiber, T. R. Golub and B. L. Ebert, *Proc. Natl. Acad. Sci. U. S. A.*, 2010, **107**, 12617–12622.
15. S. Lunke and A. El-Osta, *Neurotherapeutics*, 2013, **10**, 677–687.
16. S. Valente and A. Mai, *Expert Opin. Ther. Pat.*, 2014, **24**, 401–415.
17. E. L. Greer and Y. Shi, *Nat. Rev. Genet.*, 2012, **13**, 343–357.
18. H. Bokhoven, *Annu. Rev. Genet.*, 2011, **45**, 81–104.
19. I. Barbieri, E. Cannizzaro and M. A. Dawson, *Briefings Funct. Genomics*, 2013, **12**, 219–230.
20. J. M. Garnier, P. P. Sharp and C. J. Burns, *Expert Opin. Ther. Pat.*, 2014, **24**, 185–199.
21. K. G. McLure, E. M. Gesner, L. Tsujikawa, O. A. Kharenko, S. Attwell, E. Campeau, S. Wasiak, A. Stein, A. White, E. Fontano, R. K. Suto, N. C. Wong, G. S. Wagner, H. C. Hansen and P. R. Young, *PLoS One*, 2013, **8**, e83190.
22. P. Filippakopoulos, J. Qi, S. Picaud, Y. Shen, W. B. Smith, O. Fedorov, E. M. Morse, T. Keates, T. T. Hickman, I. Felletar, M. Philpott, S. Munro, M. R. McKeown, Y. Wang, A. L. Christie, N. West, M. J. Cameron, B. Schwartz, T. D. Heightman, N. La Thangue, C. A. French, O. Wiest, A. L. Kung, S. Knapp and J. E. Bradner, *Nature*, 2010, **468**, 1067–1073.
23. A. Younes, P. J. Rosen, F. J. Hernandez-llizaliturri, G. Bociek, Y. L. Kasomon, A. Copeland and G. T. Lee, *J. Clin. Oncol.*, 2010, **28**(15), TPS298.
24. A. B. Csoka and M. Szyf, *Med. Hypotheses*, 2009, **73**, 770–780.
25. J. Lotsch, G. Schneider, D. Reker, M. J. Parnham, P. Schneider, G. Geisslinger and A. Doehring, *Trends Mol. Med.*, 2013, **19**, 742–753.
26. C. Chang and M. E. Gershwin, *J. Autoimmun.*, 2010, **34**, J266–J275.
27. N. Singh, A. Duenas-Gonzalez, F. Lyko and J. L. Medina-Franco, *ChemMedChem*, 2009, **4**, 792–799.
28. C. Arce, B. Segura-Pacheco, E. Perez-Cardenas, L. Taja-Chayeb, M. Candelaria and A. Duennas-Gonzalez, *J. Transl. Med.*, 2006, **4**, 10.

29. E. Cornacchia, J. Golbus, J. Maybaum, J. Strahler, S. Hanash and B. Richardson, *J. Immunol.*, 1988, **140**, 2197–2200.
30. J. C. Purrucker and U. Mahlknecht, *Clin. Epigenet.*, 2010, **1**, 45–54.
31. L. Qiao, J. Schaack and J. Shao, *Endocrinology*, 2006, **147**, 865–874.
32. G. Rosenberg, *Cell. Mol. Life Sci.*, 2007, **64**, 2090–2103.
33. A. Ornoy, *Reprod. Toxicol.*, 2009, **28**, 1–10.
34. B. M. Zee, R. S. Levin, P. A. Dimaggio and B. A. Garcia, *Epigenet. Chromatin*, 2010, **3**, 22.
35. B. M. Zee, R. S. Levin, B. Xu, G. LeRoy, N. S. Wingreen and B. A. Garcia, *J. Biol. Chem.*, 2010, **285**, 3341–3350.
36. R. A. Alyea, N. P. Moore, M. J. LeBaron, B. B. Gollapudi and R. J. Rasoulpour, *J. Pharmacol. Toxicol. Methods*, 2012, **66**, 207–214.
37. M. M. Matzuk, M. R. McKeown, P. Filippakopoulos, Q. Li, L. Ma, J. E. Agno, M. E. Lemieux, S. Picaud, R. N. Yu, J. Qi, S. Knapp and J. E. Bradner, *Cell*, 2012, **150**, 673–684.
38. C. C. Priestley, M. Anderton, A. T. Doherty, P. Duffy, H. R. Mellor, H. Powell and R. Roberts, *Toxicol. Res.*, 2012, **1**, 23–31.
39. A. C. Ferguson-Smith, *Nat. Rev. Genet.*, 2011, **12**, 565–575.
40. R. L. Jirtle and M. K. Skinner, *Nat. Rev. Genet.*, 2007, **8**, 253–262.
41. A. S. Bass, M. E. Cartwright, C. Mahon, R. Morrison, R. Snyder, P. McNamara, P. Bradley, Y. Y. Zhou and J. Hunter, *J. Pharmacol. Toxicol. Methods*, 2009, **60**, 69–78.
42. J. Higgins, M. E. Cartwright and A. C. Templeton, *Drug Discovery Today*, 2012, **17**, 828–836.
43. G. J. Kaczorowski, M. L. Garcia, J. Bode, S. D. Hess and U. A. Patel, *Front. Pharmacol.*, 2011, **2**, 78.
44. J. Moggs, P. Moulin, F. Pognan, D. Brees, M. Leonard, S. Busch, A. Cordier, D. J. Heard, M. Kammuller, M. Merz, P. Bouchard and S. D. Chibout, *Expert Opin. Drug Metab. Toxicol.*, 2012, **8**, 1071–1082.
45. P. Y. Muller and M. N. Milton, *Nat. Rev. Drug Discovery*, 2012, **11**, 751–761.
46. C. E. Thomas and Y. Will, *Expert Opin. Drug Discovery*, 2012, **7**, 109–122.
47. J. Bowes, A. J. Brown, J. Hamon, W. Jarolimek, A. Sridhar, G. Waldron and S. Whitebread, *Nat. Rev. Drug Discovery*, 2012, **11**, 909–922.
48. J. U. Peters, J. Hert, C. Bissantz, A. Hillebrecht, G. Gerebtzoff, S. Bendels, F. Tillier, J. Migeon, H. Fischer, W. Guba and M. Kansy, *Drug Discovery Today*, 2012, **17**, 325–335.
49. A. Otto-Bruc, B. Mao, J. C. Migeon, F. Tillier and B. Fouchaq, *J. Pharmacol. Toxicol. Methods*, 2013, **68**, e18–e19.
50. J. Comley, *Drug Discovery World*, Spring, 2011, pp. 40–55.
51. M. Haberland, R. L. Montgomery and E. N. Olson, *Nat. Rev. Genet.*, 2009, **10**, 32–42.
52. D. Houzelstein, S. L. Bullock, D. E. Lynch, E. F. Grigorieva, V. A. Wilson and R. S. Beddington, *Mol. Cell. Biol.*, 2002, **22**, 3794–3802.
53. S. M. Kooistra and K. Helin, *Nat. Rev. Mol. Cell Biol.*, 2012, **13**, 297–311.
54. E. Li, *Nat. Rev. Genet.*, 2002, **3**, 662–673.

55. R. L. Montgomery, C. A. Davis, M. J. Potthoff, M. Haberland, J. Fielitz, X. Qi, J. A. Hill, J. A. Richardson and E. N. Olson, *Genes Dev.*, 2007, **21**, 1790–1802.
56. A. T. Nguyen, B. Xiao, R. L. Neppl, E. M. Kallin, J. Li, T. Chen, D. Z. Wang, X. Xiao and Y. Zhang, *Genes Dev.*, 2011, **25**, 263–274.
57. A. Schaefer, S. C. Sampath, A. Intrator, A. Min, T. S. Gertler, D. J. Surmeier, A. Tarakhovsky and P. Greengard, *Neuron*, 2009, **64**, 678–691.
58. C. P. Chang and B. G. Bruneau, *Annu. Rev. Physiol.*, 2012, **74**, 41–68.
59. S. A. Mahmoud and C. Poizat, *J. Pathol.*, 2013, **231**, 147–157.
60. K. Ohtani and S. Dimmeler, *Cardiovasc. Res.*, 2011, **90**, 404–412.
61. M. Jakovcevski and S. Akbarian, *Nat. Med.*, 2012, **18**, 1194–1204.
62. X. J. Sun, P. F. Xu, T. Zhou, M. Hu, C. T. Fu, Y. Zhang, Y. Jin, Y. Chen, S. J. Chen, Q. H. Huang, T. X. Liu and Z. Chen, *PLoS One*, 2008, **3**, e1499.
63. T. Fujii, S. Tsunesumi, K. Yamaguchi, S. Watanabe and Y. Furukawa, *PLoS One*, 2011, **6**, e23491.
64. V. Sartorelli and A. H. Juan, *Curr. Top. Dev. Biol.*, 2011, **96**, 57–83.
65. A. Ferrari, E. Fiorino, M. Giudici, F. Gilardi, A. Galmozzi, N. Mitro, G. Cermenati, C. Godio, D. Caruso, E. De Fabiani and M. Crestani, *Mol. Membr. Biol.*, 2012, **29**, 257–266.
66. S. M. Cullen, A. Mayle, L. Rossi and M. A. Goodell, *Curr. Top. Dev. Biol.*, 2014, **107**, 39–75.
67. K. Lund, P. D. Adams and M. Copland, *Leukemia*, 2014, **28**, 44–49.
68. G. Sashida and A. Iwama, *Int. J. Hematol.*, 2012, **96**, 405–412.
69. T. G. Jenkins and D. T. Carrell, *Reproduction*, 2012, **143**, 727–734.
70. E. Shang, H. D. Nickerson, D. Wen, X. Wang and D. J. Wolgemuth, *Development*, 2007, **134**, 3507–3515.
71. E. J. Chung, M. J. Lee, S. Lee and J. B. Trepel, *Expert Opin. Drug Metab. Toxicol.*, 2006, **2**, 213–230.
72. L. T. Donlin, C. Andresen, S. Just, E. Rudensky, C. T. Pappas, M. Kruger, E. Y. Jacobs, A. Unger, A. Zieseniss, M. W. Dobenecker, T. Voelkel, B. T. Chait, C. C. Gregorio, W. Rottbauer, A. Tarakhovsky and W. A. Linke, *Genes Dev.*, 2012, **26**, 114–119.
73. O. Cazorla, A. Freiburg, M. Helmes, T. Centner, M. McNabb, Y. Wu, K. Trombitas, S. Labeit and H. Granzier, *Circ. Res.*, 2000, **86**, 59–67.
74. H. L. Granzier and S. Labeit, *Circ. Res.*, 2004, **94**, 284–295.
75. J. M. Lorenzen, F. Martino and T. Thum, *Basic Res. Cardiol.*, 2012, **107**, 245.
76. R. Kaneda, S. Takada, Y. Yamashita, Y. L. Choi, M. Nonaka-Sarukawa, M. Soda, Y. Misawa, T. Isomura, K. Shimada and H. Mano, *Genes Cells*, 2009, **14**, 69–77.
77. M. Movassagh, M. K. Choy, D. A. Knowles, L. Cordeddu, S. Haider, T. Down, L. Siggens, A. Vujic, I. Simeoni, C. Penkett, M. Goddard, P. Lio, M. R. Bennett and R. S. Foo, *Circulation*, 2011, **124**, 2411–2422.
78. R. Papait and G. Condorelli, *Ann. N. Y. Acad. Sci.*, 2010, **1188**, 159–164.
79. A. B. Stein, T. A. Jones, T. J. Herron, S. R. Patel, S. M. Day, S. F. Noujaim, M. L. Milstein, M. Klos, P. B. Furspan, J. Jalife and G. R. Dressler, *J. Clin. Invest.*, 2011, **121**, 2641–2650.

80. R. L. Piekarz, R. Frye, M. Turner, J. J. Wright, S. L. Allen, M. H. Kirschbaum, J. Zain, H. M. Prince, J. P. Leonard, L. J. Geskin, C. Reeder, D. Joske, W. D. Figg, E. R. Gardner, S. M. Steinberg, E. S. Jaffe, M. Stetler-Stevenson, S. Lade, A. T. Fojo and S. E. Bates, *J. Clin. Oncol.*, 2009, **27**, 5410–5417.
81. M. L. Ponte, G. A. Keller and G. Di Girolamo, *Curr. Drug Saf.*, 2010, **5**, 44–53.
82. J. S. Cordes, Z. Sun, D. B. Lloyd, J. A. Bradley, A. C. Opsahl, M. W. Tengowski, X. Chen and J. Zhou, *Br. J. Pharmacol.*, 2005, **145**, 15–23.
83. O. Kepp, L. Galluzzi, M. Lipinski, J. Yuan and G. Kroemer, *Nat. Rev. Drug Discovery*, 2011, **10**, 221–237.
84. F. Cambronero, F. Marin, V. Roldan, D. Hernandez-Romero, M. Valdes and G. Y. Lip, *Eur. Heart J.*, 2009, **30**, 139–151.
85. Q. Zheng and X. Wang, *Panminerva Med.*, 2010, **52**, 9–25.
86. D. A. Huetteman and H. Bogie, *Methods Mol. Biol.*, 2009, **573**, 57–73.
87. A. Sivarajah, S. Collins, M. R. Sutton, N. Regan, H. West, M. Holbrook and N. Edmunds, *J. Pharmacol. Toxicol. Methods*, 2010, **62**, 12–19.
88. K. Derakhchan, R. W. Chui, D. Stevens, W. Gu and H. M. Vargas, *Br. J. Pharmacol.*, 2014, **171**, 509–522.
89. H. M. Vargas, H. R. Amouzadeh and M. J. Engwall, *Expert Opin. Drug Saf.*, 2013, **12**, 91–102.
90. D. K. Ma, M. C. Marchetto, J. U. Guo, G. L. Ming, F. H. Gage and H. Song, *Nat. Neurosci.*, 2010, **13**, 1338–1344.
91. C. J. Peter and S. Akbarian, *Trends Mol. Med.*, 2011, **17**, 372–379.
92. J. Xu and M. Andreassi, *Horm. Behav.*, 2011, **59**, 383–392.
93. T. Kleefstra, H. G. Brunner, J. Amiel, A. R. Oudakker, W. M. Nillesen, A. Magee, D. Genevieve, V. Cormier-Daire, H. van Esch, J. P. Fryns, B. C. Hamel, E. A. Sistermans, B. B. de Vries and H. van Bokhoven, *Am. J. Hum. Genet.*, 2006, **79**, 370–377.
94. A. Fischer, F. Sananbenesi, A. Mungenast and L. H. Tsai, *Trends Pharmacol. Sci.*, 2010, **31**, 605–617.
95. A. G. Kazantsev and L. M. Thompson, *Nat. Rev. Drug Discovery*, 2008, **7**, 854–868.
96. J. H. Lin, *Curr. Drug Metab.*, 2008, **9**, 46–59.
97. P. Moser, T. Wolinsky, M. Duxon and R. D. Porsolt, *J. Pharmacol. Exp. Ther.*, 2011, **336**, 588–595.
98. B. A. Rocha, *Behav. Pharmacol.*, 2013, **24**, 403–409.
99. C. Gassmann-Mayer, K. Jiang, P. McSorley, R. Arani, S. Dubrava, S. Suryawanshi, D. M. Webb and M. Nilsson, *Clin. Pharmacol. Ther.*, 2011, **90**, 554–560.
100. G. Turecki, C. Ernst, F. Jollant, B. Labonte and N. Mechawar, *Trends Neurosci.*, 2012, **35**, 14–23.
101. V. C. Moser, *Neurotoxicology*, 2000, **21**, 989–996.
102. L. Aloia, B. Di Stefano and L. Di Croce, *Development*, 2013, **140**, 2525–2534.
103. T. Chen and S. Y. Dent, *Nat. Rev. Genet.*, 2014, **15**, 93–106.
104. N. S. Christophersen and K. Helin, *J. Exp. Med.*, 2010, **207**, 2287–2295.

105. K. Nimura, K. Ura, H. Shiratori, M. Ikawa, M. Okabe, R. J. Schwartz and Y. Kaneda, *Nature*, 2009, **460**, 287–291.
106. A. D. Bergemann, F. Cole and K. Hirschhorn, *Trends Genet.*, 2005, **21**, 188–195.
107. L. M. Buja and D. Vela, *Cardiovasc. Pathol.*, 2008, **17**, 349–374.
108. G. Liang and Y. Zhang, *Cell Res.*, 2013, **23**, 49–69.
109. S. P. Medvedev, E. A. Pokushalov and S. M. Zakian, *Acta Nat.*, 2012, **4**, 28–46.
110. S. C. Desbordes, D. G. Placantonakis, A. Ciro, N. D. Socci, G. Lee, H. Djaballah and L. Studer, *Cell Stem Cell*, 2008, **2**, 602–612.
111. R. Mudbhary and K. C. Sadler, *Birth Defects Res., Part C*, 2011, **93**, 194–203.
112. E. D. Frantz, N. Peixoto-Silva and A. Pinheiro-Mulder, *Pancreas*, 2012, **41**, 1–9.
113. M. Manikkam, R. Tracey, C. Guerrero-Bosagna and M. K. Skinner, *PLoS One*, 2012, **7**, e46249.
114. M. D. Anway, A. S. Cupp, M. Uzumcu and M. K. Skinner, *Science*, 2005, **308**, 1466–1469.
115. D. H. Ho and W. W. Burggren, *J. Exp. Biol.*, 2010, **213**, 3–16.
116. N. A. Youngson and E. Whitelaw, *Annu. Rev. Genomics Hum. Genet.*, 2008, **9**, 233–257.
117. J. Tycko, D. Fields, D. Cabrera, M. Charawi and B. Kaptur, *Penn Bioethics J.*, 2013, **7**, 1–3.
118. J. I. Goodman, K. A. Augustine, M. L. Cunnningham, D. Dixon, Y. P. Dragan, J. G. Falls, R. J. Rasoulpour, R. C. Sills, R. D. Storer, D. C. Wolf and S. D. Pettit, *Toxicol. Sci.*, 2010, **116**, 375–381.
119. J. G. Moggs, J. I. Goodman, J. E. Trosko and R. A. Roberts, *Toxicol. Appl. Pharmacol.*, 2004, **196**, 422–430.
120. M. Berdasco and M. Esteller, *Dev. Cell*, 2010, **19**, 698–711.
121. S. Sharma, T. K. Kelly and P. A. Jones, *Carcinogenesis*, 2010, **31**, 27–36.
122. D. C. Bloom, N. V. Giordani and D. L. Kwiatkowski, *Biochim. Biophys. Acta*, 2010, **1799**, 246–256.
123. S. Hakre, L. Chavez, K. Shirakawa and E. Verdin, *Curr. Opin. HIV AIDS*, 2011, **6**, 19–24.
124. H. H. Niller, H. Wolf, E. Ay and J. Minarovits, *Adv. Exp. Med. Biol.*, 2011, **711**, 82–102.
125. S. N. Pantry and P. G. Medveczky, *Semin. Cancer Biol.*, 2009, **19**, 153–157.
126. S. Bouchat, J. S. Gatot, K. Kabeya, C. Cardona, L. Colin, G. Herbein, S. De Wit, N. Clumeck, O. Lambotte, C. Rouzioux, O. Rohr and C. Van Lint, *Aids*, 2012, **26**, 1473–1482.
127. D. M. Margolis, *Curr. Opin. HIV AIDS*, 2011, **6**, 25–29.
128. K. H. Burns and J. D. Boeke, *Cell*, 2012, **149**, 740–752.
129. D. C. Hancks and H. H. Kazazian Jr, *Curr. Opin. Genet. Dev.*, 2012, **22**, 191–203.
130. E. Lee, R. Iskow, L. Yang, O. Gokcumen, P. Haseley, L. J. Luquette 3rd, J. G. Lohr, C. C. Harris, L. Ding, R. K. Wilson, D. A. Wheeler, R. A. Gibbs, R.

Kucherlapati, C. Lee, P. V. Kharchenko, P. J. Park and N. Cancer, *Science*, 2012, **337**, 967–971.

131. Z. Herceg, M. P. Lambert, K. van Veldhoven, C. Demetriou, P. Vineis, M. T. Smith, K. Straif and C. P. Wild, *Carcinogenesis*, 2013, **34**, 1955–1967.
132. C. L. Alden, A. Lynn, A. Bourdeau, D. Morton, F. D. Sistare, V. J. Kadambi and L. Silverman, *Vet. Pathol.*, 2011, **48**, 772–784.
133. D. A. Eastmond, S. V. Vulimiri, J. E. French and B. Sonawane, *Crit. Rev. Toxicol.*, 2013, **43**, 611–631.
134. M. K. Skinner, M. Manikkam and C. Guerrero-Bosagna, *Reprod. Toxicol.*, 2011, **31**, 337–343.
135. K. Inawaka, M. Kawabe, S. Takahashi, Y. Doi, Y. Tomigahara, H. Tarui, J. Abe, S. Kawamura and T. Shirai, *Toxicol. Appl. Pharmacol.*, 2009, **237**, 178–187.
136. S. Schneider, W. Kaufmann, R. Buesen and B. van Ravenzwaay, *Reprod. Toxicol.*, 2008, **25**, 352–360.
137. S. Schneider, H. Marxfeld, S. Groters, R. Buesen and B. van Ravenzwaay, *Reprod. Toxicol.*, 2013, **37**, 6–14.
138. S. Muller and S. Knapp, *Med. Chem. Commun.*, 2014, **5**, 288–296.
139. R. K. Prinjha, J. Witherington and K. Lee, *Trends Pharmacol. Sci.*, 2012, **33**, 146–153.
140. M. Kabesch and I. M. Adcock, *Biochimie*, 2012, **94**, 2231–2241.
141. S. Rajendrasozhan, H. Yao and I. Rahman, *J. Chronic Obstruct. Pulm. Dis.*, 2009, **6**, 291–297.
142. M. M. Perry, A. L. Durham, I. M. Adcock and K. Chung, *Am. J. Respir. Crit. Care Med.*, 2013, **187**, A1257.
143. M. A. Rothstein, Y. Cai and G. E. Marchant, *Health Matrix*, 2009, **19**, 1–62.

Subject Index